186
Advances in Polymer Science

Advances in Polymer Science
Recently Published and Forthcoming Volumes

Polysaccharides I

Structure, Characterization and Use

Volume Editor: Thomas Heinze

With contributions by

H. Barsett · A. Ebringerová · S. E. Harding · T. Heinze
Z. Hromádková · C. Muzzarelli · R. A. A. Muzzarelli
B. S. Paulsen · O. A. El Seoud

 Springer

The series *Advances in Polymer Science* presents critical reviews of the present and future trends in polymer and biopolymer science including chemistry, physical chemistry, physics and material science. It is adressed to all scientists at universities and in industry who wish to keep abreast of advances in the topics covered.

As a rule, contributions are specially commissioned. The editors and publishers will, however, always be pleased to receive suggestions and supplementary information. Papers are accepted for *Advances in Polymer Science* in English.

In references *Advances in Polymer Science* is abbreviated *Adv Polym Sci* and is cited as a journal.

Springer WWW home page: http://www.springeronline.com
Visit the APS content at http://www.springerlink.com/

Library of Congress Control Number: 2005926094

ISSN 0065-3195
ISBN-10 3-540-26112-5 Springer Berlin Heidelberg New York
ISBN-13 978-3-540-26112-4 Springer Berlin Heidelberg New York
DOI 10.1007/b136812

Springer is a part of Springer Science+Business Media

springeronline.com

Cover design: *Design & Production* GmbH, Heidelberg
Typesetting and Production: LE-TEX Jelonek, Schmidt & Vöckler GbR, Leipzig

Printed on acid-free paper   02/3141 YL – 5 4 3 2 1 0

Volume Editor

Prof. Dr. Thomas Heinze
Kompetenzzentrum Polysaccharidforschung
Friedrich-Schiller-Universität Jena
Humboldtstraße 10
07743 Jena, Germany
Thomas.Heinze@uni-jena.de

Editorial Board

Prof. Akihiro Abe
Department of Industrial Chemistry
Tokyo Institute of Polytechnics
1583 Iiyama, Atsugi-shi 243-02, Japan
aabe@chem.t-kougei.ac.jp

Prof. A.-C. Albertsson
Department of Polymer Technology
The Royal Institute of Technology
10044 Stockholm, Sweden
aila@polymer.kth.se

Prof. Ruth Duncan
Welsh School of Pharmacy
Cardiff University
Redwood Building
King Edward VII Avenue
Cardiff CF 10 3XF
United Kingdom
duncan@cf.ac.uk

Prof. Karel Dušek
Institute of Macromolecular Chemistry,
Czech
Academy of Sciences of the Czech Republic
Heyrovský Sq. 2
16206 Prague 6, Czech Republic
dusek@imc.cas.cz

Prof. Dr. W. H. de Jeu
FOM-Institute AMOLF
Kruislaan 407
1098 SJ Amsterdam, The Netherlands
dejeu@amolf.nl
and Dutch Polymer Institute
Eindhoven University of Technology
PO Box 513
5600 MB Eindhoven, The Netherlands

Prof. Jean-François Joanny
Physicochimie Curie
Institut Curie section recherche
26 rue d'Ulm
75248 Paris cedex 05, France
jean-francois.joanny@curie.fr

Prof. Hans-Henning Kausch
EPFL SB ISIC GGEC
J2 492 Bâtiment CH
Station 6
1015 Lausanne, Switzerland
kausch.cully@bluewin.ch

Prof. S. Kobayashi
R & D Center for Bio-based Materials
Kyoto Institute of Technology
Matsugasaki, Sakyo-ku
Kyoto 606-8585, Japan
kobayashi@kit.ac.jp

Advances in Polymer Science
Also Available Electronically

For all customers who have a standing order to Advances in Polymer Science, we offer the electronic version via SpringerLink free of charge. Please contact your librarian who can receive a password or free access to the full articles by registering at:

springerlink.com

If you do not have a subscription, you can still view the tables of contents of the volumes and the abstract of each article by going to the SpringerLink Homepage, clicking on "Browse by Online Libraries", then "Chemical Sciences", and finally choose Advances in Polymer Science.

You will find information about the

– Editorial Board
– Aims and Scope
– Instructions for Authors
– Sample Contribution

at springeronline.com using the search function.

Preface

There is a wide range of naturally occurring polysaccharides derived from renewable resources possessing magnificent structural diversity and functional versatility. These molecules are amongst the key substances that make up the fundamental components of life. Some of these polysaccharides – in particular cellulose, starch and semi-synthetic derivatives thereof – are actively used in commercial products today, although many others still remain underutilized. With the recent rapid advancements in molecular and supramolecular characterization, in developing adequate isolation procedures for structurally uniform samples, in understanding structure property relationships, in designing synthesis pathways for the controlled derivatization, and in adapting and developing analytical tools for these biopolymers, new opportunities for the use of polysaccharides and their semi-synthetic derivatives are now being considered.

The structural and functional properties of polysaccharides are often superior to synthetic polymers. The cell wall architecture of plants, which involves a unique composite of cellulose and hemicellulose (and lignin) together with the mechanical properties of films and fibres of chitin are two impressive examples. The growing acceptance of the knowledge acquired over hundreds of years by people still using so-called *traditional medicine* has lead to the discovery of various new bioactive polysaccharides. From the polymer chemist's point of view, the unique structures of polysaccharides combined with such promising properties like hydrophilicity, biocompatibility, biodegradability (at least in the original state), stereoregularity, multichirality, and polyfunctionality – i.e. reactive functions (mainly OH-, NH-, and COOH moieties) that can be modified by various chemical reactions – provides an additional and important argument for the study of polysaccharides as a valuable and renewable resource for the future.

In recent years the socio-economical situation has changed to make natural polymer once again worth consideration for many applications. The increasing industrial interest in the field of polysaccharides has been well manifested by the establishment of, for example, the Centre of Excellence for Polysaccharide Research at Jena-Rudolstadt. Bayer AG – Wolff Cellulosics GmbH & Co. KG, Borregaard Chemcell, Dow Deutschland GmbH & Co. OHG, Hercules GmbH – Division Aqualon, and Rhodia Acetow GmbH are the industrial partners in this venture and have provided funding for a five-year period via project and programmatic support. The Centre is a joint project between the Friedrich Schiller University of Jena and the Thuringian Institute of Textile and Plastics Research located 40 km south of Jena in Rudolstadt. The aim of the Centre is to foster interdisciplinary fundamental research on polysaccharides and their application through active graduate student projects in the fields of carbohydrate chemistry, bioorganic chemistry, and structure analysis. The goal is also to encourage efficient international collaboration. As the director of the Centre, I would like to stress that the knowledge discussed in these review papers is not the final story. On the contrary, it is intended that the information summarized in this volume will stimulate scientists in academia and industry to continue with the search and development of new products and applications.

The collection of review papers in this volume addresses some of the current key concerns with regard the use of polysaccharides and their semi-synthetic derivatives. Thus the topics discussed focus on: hemicelluloses, bioactive pectic polysaccharides, cellulose esters synthesized via homogeneous conversions, chitin, and the modern analytical ultracentrifugation of polysaccharides. Although in terms of material design, cellulose and starch will always appear near the top of any list of polysaccharides, it is the editors opinion that consideration of the whole range of polysaccharides available is necessary to allow full advantage to be taken of this fascinating class of biopolymers. Starch will be considered in volume II of the special issue of *Advances in Polymer Science*. In this regard, as editor, I would like to take this opportunity to express my gratitude to the authors for their seminal and timely contributions.

I would also like, on behalf of the authors, to express my gratitude to Professor Nuyken, APS editorial board member, and to Springer, for agreeing to publish special issues with review papers on selected polysaccharide topics. In particular I would like to thank Dr. Marion Hertel (executive editor, Chemistry) and Ulrike Kreusel (desk editor, Chemistry) of Springer for their efficiency and their conscientious efforts to ensure timely completion of this book.

Jena, July 2005   *Thomas Heinze*

Contents

Adv Polym Sci (2005) 186: 1–67
DOI 10.1007/b136816
© Springer-Verlag Berlin Heidelberg 2005
Published online: 31 August 2005

Hemicellulose

Anna Ebringerová[1] (✉) · Zdenka Hromádková[1] · Thomas Heinze[2]

[1]Center of Excellence, Slovak Academy of Sciences, Institute of Chemistry,
Dúbravská cesta 9, 845 32 Bratislava, Slovakia
chemebri@savba.sk

[2]Center of Excellence for Polysaccharide Research, Friedrich Schiller University of Jena,
Humboldtstrasse 10, 07743 Jena, Germany
Thomas.Heinze@uni-jena.de

Abstract Hemicelluloses, comprising the non-cellulose cell-wall polysaccharides of vegetative and storage tissues of annual and perennial plants, represent an immense renewable resource of biopolymers. They occur in a large variety of structural types, divided into four general groups, i.e., xylans, mannans, mixed linkage β-glucans, and xyloglucans. The presented review summarized recent reports on hemicelluloses, including the arabinogalactan from larch wood, focused on new plant sources, isolation methods, and characterization of structural features, physicochemical and various functional properties. Attention was paid to derivatives prepared from these polysaccharides and to application possibilities of hemicelluloses or hemicellulosic materials for food and non-food applications, including the production of composite materials and other biomaterials.

Keywords Chemical functionalization · Hemicelluloses · Properties · Structures

Abbreviations

AFM	Atomic force microscopy
AG	Arabinogalactan
AGX	(L-Arabino)-4-O-methyl-D-glucurono-D-xylan
AX	L-Arabino-D-xylan
Cadoxen	Cadmium complex with ethylenediamine
CHX	Complex heteroxylan
CP/MAS	Cross-polarization magic-angle spinning ^{13}C-NMR spectroscopy
Cuoxam	Copper complex with ammonia
DMF	N,N-Dimethylformamide

DMSO	Dimethyl sulfoxide
DP	Degree of polymerization
DS	Degree of substitution
DSC	Differential scanning calorimetry
DTA	Differential thermal analysis
ESI-MS	Electrospray ion-trap mass spectrometry
FA	Ferulic acid
FeTNa	Ferric tartaric acid complex in alkaline aqueous solution
G'	Storage modul
GaM	D-Galacto-D-mannan
GAX	(4-O-Methyl-D-glucurono)-L-arabino-D-xylan
GFC/LLS	Gel filtration chromatography/laser light scattering
GGM	D-Galacto-D-gluco-D-mannan
GM	D-Gluco-D-mannan
GX	D-Glucurono-D-xylan
HMW-AX	High-molecular-weight arabinoxylan
HPAEC	High-performance anion-exchange chromatography
HPLC	High-pressure liquid chromatography
K'	Huggins constant
LMW-AX	Low-molecular-weight arabinoxylan
MALDI-PSD MS	Matrix-assisted laser desorption ionization post-source-decay mass spectrometry
MALDI-TOF MS	Matrix-assisted laser desorption ionization time-of-flight mass spectrometry
MALLS	Multi-angle laser light scattering
MGX	4-O-Methyl-D-glucurono-D-xylan
PAD	Pulsed amperometric detection
QCM-D	Quartz crystal microbalance with dissipation
R_g	Radius of gyration
SEC	Size-exclusion chromatography
Tg	Glass-transition temperature
TMP	Thermomechanical pulping
wis	Water-insoluble
ws	Water-soluble
X_3	β-(1→3)-D-Xylan
XG	D-Xylo-D-glucan
X_m	β-(1→3, 1→4)-D-Xylan
Sugars:	
Ara$f$	L-Arabinofuranose
Ara$p$	L-Arabinopyranose
Fuc$p$	L-Fucopyranose
Gal$p$	D-Galactopyranose
GlcA	D-Glucuronic acid
Glc$p$	D-Glucopyranose
Man$p$	D-Mannopyranose
MeGlcA	4-O-Methyl-D-glucuronic acid
Rha$p$	D-Rhamnopyranose
Xyl$p$	D-Xylopyranose

1
Introduction

Biomass has been pointed out as the most important source for a broad variety of advanced polymeric materials, which can be obtained by different routes employing:
(i) The native biopolymers (such as polysaccharides, proteins and lignin),
(ii) The synthesis of polysaccharides from carbohydrates released by hydrolytic treatments, and
(iii) The isolation of polymers from fermentation products.
Hemicelluloses, accounting for on average up to 50% of the biomass of annual and perennial plants, have emerged as an immense renewable resource of biopolymers. Their application potential, emphasized many times by leading polysaccharide scientists, has not yet been exploited on an industrial scale. During the last decade, an outstanding increased interest can be noticed in research of biopolymers from renewable sources among scientists from universities and research institutes as well as industrial companies. The future shortage of natural energy sources, replacement of petroleum-based products connected with the solution of worldwide environmental problems, and demands for healthy food and alternative medicines are the main driving forces for the immense activities in research of polysaccharides, including hemicelluloses.

In the past, research activities in the field of hemicellulose were aimed mainly at utilizing plant biomass by conversion into sugars, chemicals, fuel and as sources of heat energy. However, hemicelluloses, due to their structural varieties and diversity are also attractive as biopolymers, which can be utilized in their native or modified forms in various areas, including food and non-food applications.

The term hemicellulose was originally proposed by Schulze [1] to designate polysaccharides extractable, in comparison to cellulose, from higher plants by aqueous alkaline solutions. Earlier researches mistakenly regarded these polysaccharides as the precursors of cellulose. This is now known to be incorrect, but the term is still commonly used to refer to non-starch polysaccharides found in association with cellulose in the cell walls of higher plants. The term does not include pectic polysaccharides, as these are extractable by hot water, weak acids, or chelating agents. Less frequently the terms hemicellulose and pentosan are used interchangeable, mainly in food research and medicine.

Based on the current stage of knowledge, hemicelluloses can be divided into four general classes of structurally different cell-wall polysaccharide types, i.e.,
(a) xylans,
(b) mannans,

(c) β-glucans with mixed linkages, and

(d) xyloglucans.

They occur in structural variations differing in side-chain types, distribution, localization and/or types and distribution of glycoside linkages in the macromolecular backbone. Because polysaccharides might be differentiated from various points of view, such as function in plant tissues (reserve, supporting), primary structure, occurrence in plant tissues, etc., the same polysaccharide type has been ranked among hemicelluloses as well as gums. A typical example is the arabinogalactan from larch, an extracellular polysaccharide, classified as gum but also as hemicellulose. This name persisted from the early investigations of larch (Larix) wood species, when arabinogalactan was found to be their major non-cellulosic component.

Xyloglucans are classified as gum when they are extractable with hot water from seed endosperm cell walls, such as the tamarind seed xyloglucan, and as hemicelluloses because they are alkali-extractable from the cell walls of vegetative plant tissues where they are closely associated with cellulose [2]. Also β-glucans with mixed linkages appear under the name gum as well as hemicellulose in the literature.

Within the scope of this review, the contributions of the last decade concerning cell-wall polysaccharides isolated from woody and other plant tissues will be reviewed according to the above-proposed classification of hemicelluloses including larch arabinogalactans. The present review article updates and extends previous reviews [3–5] and will focus in particular on new investigated plant sources, isolation methods, structural features, physicochemical and various functional properties of hemicelluloses. Attention will also be paid to the modification of isolated hemicelluloses or hemicellulosic materials and the application possibilities of hemicelluloses and their derivatives, including their use for the production of composite materials and other biomaterials.

2
D-Xyloglycans

Xylan-type polysaccharides are the main hemicellulose components of secondary cell walls constituting about 20–30% of the biomass of dicotyl plants (hardwoods and herbaceous plants). In some tissues of monocotyl plants (grasses and cereals) xylans occur up to 50% [6]. Xylans are thus available in huge and replenishable amounts as by-products from forestry, the agriculture, wood, and pulp and paper industries. Nowadays, xylans of some seaweed represent a novel biopolymer resource [4]. The diversity and complexity of xylans suggest that many useful by-products can be potentially produced and, therefore, these polysaccharides are considered as possible biopolymer raw materials for various exploitations. As a renewable resource, xylans are being rediscovered not only in wood but also in all kinds of biomass. It is, therefore, desirable to confront the new findings with hitherto published information.

2.1
Occurrence and Structural Diversity

Xylans of the terrestrial plants are heteropolymers possessing a β-(1→4)-D-xylopyranose backbone, which is branched by short carbohydrate chains. They comprise D-glucuronic acid or its 4-O-methyl ether, L-arabinose and/or various oligosaccharides, composed of D-xylose, L-arabinose, D- or L-galactose and D-glucose. Based on the hitherto-reported review papers on the primary structure of xylans from various plant tissues, xylan-type polysaccharides can be divided into homoxylans and heteroxylans, which include glucuronoxylans, (arabino)glucuronoxylans, (glucurono)arabinoxylans, arabinoxylans, and complex heteroxylans. It should be mentioned that Figs. 1–5, 8, 9, 12, 14, and 16 illustrate only the typical structural features of the respective polysaccharide type and not a repeating unit of the polymer chains.

2.1.1
Homoxylans

Xylans as true homopolymers occur in seaweeds of the *Palmariales* and *Nemaliales*, however, their backbone consists of Xyl*p* residues linked by β-(1 → 3) (Type X_3, Fig. 1a) or mixed β-(1 → 3, 1 → 4)-glycosidic linkages (Type X_m, Fig. 1b). They are assumed mainly to have a structural function in the cell-wall architecture, but a reserve function cannot be ruled out [4]. From the microfibrils of green algae (*Siphonales*) such as *Caulerpa* and *Bryopsis* sp., X_3 was isolated and the structure confirmed by methylation analysis, ^{13}C-NMR spectroscopy [7], as well as by mass spectrometry of enzymically released linear oligosaccharides up to a degree of polymerization (DP) of

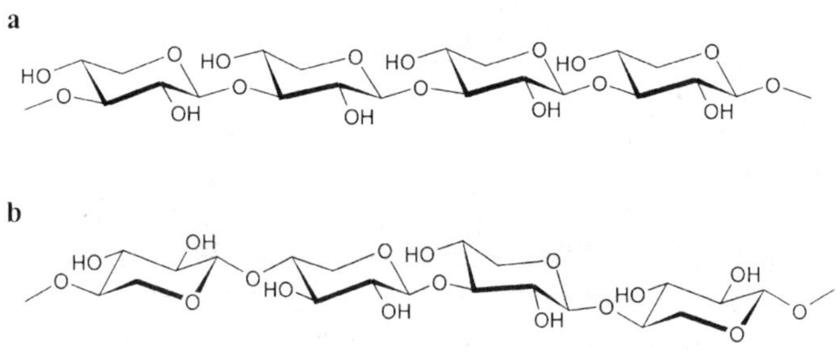

Fig. 1 Primary structure of **a** β-(1 → 3)-D-xylan type X_3 and **b** β-(1 → 3, 1 → 4)-D-xylan type X_m

25 [8]. Recently, from the edible red seaweed *Palmaria palmata* X_m was isolated and characterized [9, 10]. Analysis with the aid of HPAEC-PAD, ESI-MS and NMR spectroscopy of oligomeric fragments, produced by enzymic hydrolysis, indicated a regular distribution of 1,3-linkages idealized in a pentameric structure, which is the basis of the hydrogen bonds that principally link the xylans in the cell wall. However, minor amounts of phosphate, sulfate, and galactosylated and xylosylated peptides have been found in some xylan fractions as well [10, 11].

2.1.2
Glucuronoxylans

Most of the glucuronoxylans have single 4-*O*-methyl-α-D-glucopyranosyl uronic acid residues (MeGlcA) attached always at position 2 of the main chain Xyl*p* units (Fig. 2). This structural type is usually named as 4-*O*-methyl-D-glucurono-D-xylan (MGX). However, the glucuronic acid side chain may be present in both the 4-*O*-methylated and non-methylated forms (GlcA).

MGX represents the main hemicellulose component of hardwoods, showing Xyl : MeGlcA ratios from 4 : 1 to 16 : 1 depending on the extraction conditions used; on average, the ratio is about 10 : 1. Fractional precipitation of the bulk MGX or fractional extraction of the xylan component from the plant source usually yields xylan preparations with a broad range of Xyl : MeGlcA ratios [6]. An unusual MGX was isolated from the wood of *Eucalyptus globulus* [12, 13] as it contained, in addition to terminal MeGlcA residues, some of these residues substituted at *O*-2 with α-D-galactose.

In the native state the xylan is supposed to be *O*-acetylated. The content of acetyl groups of MGX isolated from hardwoods of temperate zones varies in the range 3–13% [4]. The acetyl groups are split during the necessary alkaline extraction conditions resulting in partial or full water-insolubility of the xylan preparations. But the acetyl groups may be, at least in part, preserved by treating with hot water or steam. Recently, water-soluble acetylated MGXs have been isolated from NaClO$_2$-delignified wood of birch and beech with DMSO [14]. Both from milled aspen wood [15] and the bast fibers of flax [16] water-soluble acetylated MGXs of low molecular weight were isolated by employing microwave treatment. The degrees of acetylation ranged

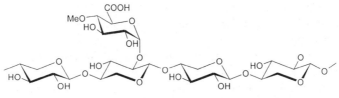

Fig. 2 Primary structure of 4-*O*-methyl-D-glucurono-D-xylan (MGX)

between 0.3 and 0.6 and the acetyl group occurred at various positions of the Xyl$p$ residues. MGX was isolated also from fruits and storage tissues such as the pericarp seed of the *Opuntia ficus-indica* pear [17], luffa (*Luffa cylindrica*) fruit fibres [18], date seed fibers (*Phoenix dactylifera*) [19, 20], sugar beet pulp [21], grape skin [22, 23], and hulls of Jojoba (*Simmondsia chinensis*) [24]. It was present in the complex of polysaccharides isolated from olive fruits at different ripening stages [25]. In the olive fruit cell walls [26] and leaf cell walls of the tree *Argania spinoza* [27], the MGXs were isolated in form of xylan-xyloglucan complexes.

Recently, the alkali-soluble hemicelluloses of hardwood dissolving pulps have been investigated [28]. Their composition and molecular properties depended on the pulp origin and steeping conditions. The MGX of the β-fraction from press lye had a low uronic acid content (ratio of MeGlcA to Xyl is about 1 : 20). The molecular weight of the hemicellulose fractions varied between 5000 and 10 000 g/mol.

As reported in the review [4], MGXs were found in various dicotyls such as in ground nut shells, jute baste and bark, sunflower hulls, flax fiber, medicinal plants *Althaea officinalis* and *Rudbeckia fulgida*, and kenaf. In MGX from kenaf additional terminal Rha$p$ and Ara$f$ side chains and $\sim 10\%$ of acetyl groups were shown to branch the xylan backbone. Acidic xylans with the glucuronic acid side chain both in the 4-O-methylated and non-methylated forms were isolated from olive pits, rape stem, red gram husk, and jute bast fiber [4]. The backbone of the xylan from the mucilage of quince seeds (*Cydonia cylindrica*) [18] is exceptionally heavily substituted with MeGlcA possessing a ratio of Xyl to MeGlcA of 2 : 1 and carries acetyl groups at various positions.

2.1.3
(Arabino)glucuronoxylan and (glucurono)arabinoxylan

Both (arabino)glucuronoxylan (AGX) and (glucurono)arabinoxylan (GAX) have single MeGlcA and α-L-Ara$f$ residues attached at position 2 and 3, respectively, to the β-(1 $\rightarrow$ 4)-D-xylopyranose backbone (Fig. 3), which might also be slightly acetylated. The AGX type occurs in appreciable quantity in coniferous species, but not as the dominant hemicellulosic component [6]. An exception is the hemicellulose of the tropic conifer *Podocarpus lambertii* [29], which contains an about equal proportion of mannans and xylans. Generally, AGX has the backbone more heavily substituted by MeGlcA than that of the hardwood MGX, with 5–6 Xyl residues per uronic acid group in the former and 10 on average in the latter.

AGX are also the dominant hemicelluloses in the cell walls of lignified supporting tissues of grasses and cereals. They were isolated from sisal, corncobs and the straw from various wheat species [4]. A more recent study on corncob xylans [30] showed the presence of a linear, water-insoluble polymer

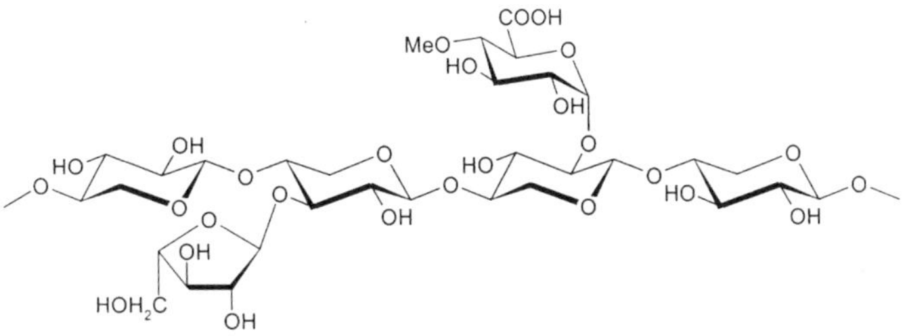

Fig. 3 Primary structure of (L-arabino)-4-O-methyl-D-glucurono-D-xylan (AGX)

(wis-AGX) with ~ 95% of the backbone unsubstituted, and a water-soluble xylan (ws-AGX) having more than 15% of the backbone substituted [31]. The uronic acid content was lower in the wis-AGX (~ 4%) than in ws-AGX (about 9%). A small proportion of the Xylp residues of the backbone are dis-ubstituted by α-L-Araf residues (Fig. 4). A peculiar structural feature of the ws-AGX is the presence of disaccharide side chains (1). This sugar moiety has been found usually esterified by ferulic acid (FA) at position 5 of the Araf unit and occur as a widespread component of grass cell walls [32]. Similar to acetyl groups, ester-linked FA is lost during the alkaline extraction of AGX. However, FA-containing ws-AGX preparations were isolated by ultrasonically assisted extraction of corncobs using hot water and very dilute alkali hydroxide solutions [33].

In contrast to AGX, the GAX consists of an arabinoxylan backbone, which contains about ten times fewer uronic acid side chains than α-L-Araf ones, and has some Xylp residues doubly substituted with these sugars. GAX are located in the non-endospermic tissues of cereal grains such as in wheat, corn, and rice bran. The degree and pattern of substitution of GAX appears to vary with the source from which they are extracted. These differences are reflected in the ratio of Ara to Xyl, the content of MeGlcA, and the presence

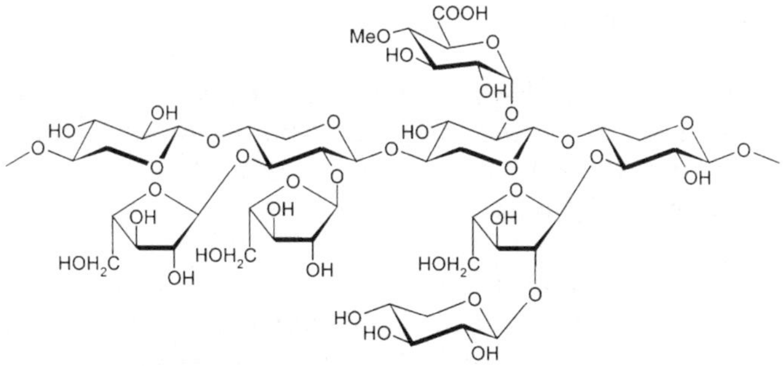

Fig. 4 Primary structure of water-soluble AGX (ws-AGX)

of disaccharide side chains (1) [4, 6] as well as the dimeric arabinosyl side chains (2) [32].

$$\beta\text{-D-Xyl}p\text{-}(1 \rightarrow 2)\text{-}\alpha\text{-L-Ara}f\text{-}(1 \rightarrow \qquad\qquad (1)$$

$$\alpha\text{-L-Ara}f\text{-}(1 \rightarrow 3)\text{-}\alpha\text{-L-Ara}f\text{-}(1 \rightarrow \qquad\qquad (2)$$

A recent study on wheat bran GAX [34] revealed the presence of lowly and highly substituted GAX fractions, which greatly differ in the amount and type of substitution by the Araf units, but not in the content of the glucuronic acid, half of which occurs as the 4-O-methyl ether. Feruloylated GAX fractions were isolated from wheat bran by cold water, steam and dilute alkali [35]. The yields of water and steam-extracted material were 1.5 and 20–31%, respectively, however, they contained 1.5% and 19–28%, respectively, of GAX. The alkali extraction yielded 10–30% material containing more than 90% of the GAX with moderate FA content.

GAX from maize-kernel cell walls [36] were reported to be highly substituted. Using a special enzymic preparation, Ultraflo-arabinoxylan degrader, in combination with digestion by endo-xylanase I, oligomers were prepared and analyzed by MALDI-TOF MS. The presence of arabinoxylan oligomers containing two glucuronic acid residues was indicated the first time. The uronic substituents were found to be evenly distributed over the xylan backbone.

2.1.4
Arabinoxylans

Arabinoxylan (AX) has been identified in a variety of the main commercial cereals: wheat, rye, barley, oat, rice, and corn, sorghum as well as in other plants such as rye grass, bamboo shoots, and pangola grass. They represent the major hemicellulose component of cell walls of the starchy endosperm (flour) and outer layers (bran) of the cereal grain. AX contents vary from 0.15% in rice endosperm to $\sim$ 13% in whole grain flour from barley, and rye, and up to 30% in wheat bran [3]. In the past decade, many detailed studies on the structural characteristics of AX from wheat, rye, millet, barley, oat spelt and other cereal grains appeared [37–46]. AXs occur as neutral as well as slightly acidic polymers, the latter usually being included in the GAX group. AX has a linear backbone that is, in part, substituted by α-L-Araf residues positioned either on O-3 or O-2 (monosubstitution) or on both O-2 and O-3 (disubstitution) of the Xylp monomer units (Fig. 5). In addition, phenolic acids such as ferulic and coumaric acid have been found to be esterified to O-5 of some Araf residues in AX [4, 47].

Due to the location and degree of integration in the different tissues of the grain, a part of AX is water-extractable, but the bulk can only be released by alkali solutions. Both the water- and alkali-extracted xylans consist of a family of related polymers differing in DS by α-Araf residues and substitution pat-

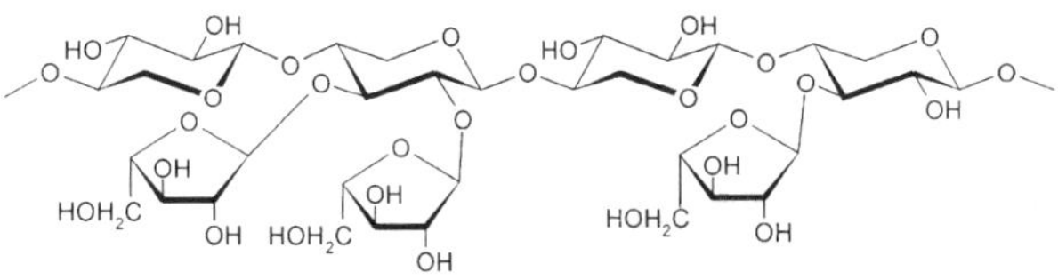

Fig. 5 Primary structure of water-soluble L-arabino-D-xylan (AX)

terns [48–50]. According to the DS, which affects solubility, AXs have been divided into two or three groups, depending on the isolation and fractionation procedures used as well as the distribution patterns [51]. In any case, the first group includes water-insoluble monosubstituted AXs (Ara : Xyl up to ~ 0.2–0.3) with α-L-Ara$f$ attached mainly at position 3 of the xylan backbone. The second and third groups comprise water-soluble xylans with a ratio of Ara to Xyl between 0.3 and 1.2 [48]. AXs of the second group with intermediate ratio of Ara to Xyl (0.5 – 0.9) have shorter sequences of disubstituted Xyl$p$ residues than that of the third group. The alkali-extractable AX samples [48] are of particular interest as they have similar but stronger bread-improving properties than the water-extractable AX, which were usually supposed to be the most responsible for the baking performance of cereal flours [41].

2.1.5
Complex Heteroxylans

The heteroxylans (CHX) present in cereals, seeds, gum exudates, and mucilages are structurally more complex [6]. They have a $(1 \rightarrow 4)$-β-D-xylopyranose backbone decorated, except of the single uronic acid and arabinosyl residues with various mono- and oligoglycosyl side chains. Reinvestigations of CHX isolated from corn bran [52] have confirmed that the xylan backbone is heavily substituted (at both positions 2 and 3) with β-D-Xyl$p$, β-L-Ara$f$, α-D-Glc$p$A residues and oligosaccharide side chains (1), (3) and (4).

$$\beta\text{-D-Gal}p\text{-}(1 \rightarrow 5)\text{-}\alpha\text{-L-Ara}f\text{-}(1 \rightarrow \qquad\qquad\qquad (3)$$

$$\text{L-Gal}p\text{-}(1 \rightarrow 4)\text{-}\beta\text{-D-Xyl}p\text{-}(1 \rightarrow 2)\text{-}\alpha\text{-L-Ara}f\text{-}(1 \rightarrow \qquad\qquad (4)$$

Several CHX samples were isolated from the leaves and barks of tropical dicots such as the *Litsea* species [3]. The mucilage-forming seeds of *Plantago* sp. contain very complex heteroxylans [53, 54]. For the CHX from *Plantago major* seeds [53], a $(1 \rightarrow 3, 1 \rightarrow 4)$-mixed-linkage xylan backbone has been suggested possessing short side chains attached to position 2 or 3 of some $(1 \rightarrow 4)$-linked D-Xyl$p$ residues. The side chains consist of β-D-Xyl$p$ and α-L-Ara$f$ residues, and disaccharide moieties (5) and (6). Further structural analysis of oligosaccharides generated by partial acid hydrolysis of the polysaccharide [55] revealed the presence of (1 → 4)-linked xylotrisaccharide and (1 → 3)-linked xylooligosaccharides with DP of 6 -- 11. These oligosaccharides were suggested as building blocks for the backbone of CHX. In addition, the presence of single β-D-Xyl$p$ O-2-linked to the backbone as well as of the acidic disaccharide (6) was confirmed.

$$\alpha\text{-L-Ara}f\text{-}(1 \to 3)\text{-}\beta\text{-D-Xyl}p\text{-}(1 \to \tag{5}$$

$$\alpha\text{-D-Glc}p\text{A-}(1 \to 3)\text{-}\alpha\text{-L-Ara}f\text{-}(1 \to \tag{6}$$

In contrast to the former Plantago CHX [55], the gel-forming CHX from psyllium (*Plantago ovata*) husks [54] was found to be a neutral, highly branched arabinoxylan with the (1 → 4)-β-D-xylopyranose backbone substituted at position 2 with single Xyl$p$ units and at position 3 with the trisaccharide moiety (7).

$$\alpha\text{-L-Ara}f\text{-}(1 \to 3)\text{-}\beta\text{-D-Xyl}p\text{-}(1 \to)\text{-}\alpha\text{-L-Ara}f\text{-}(1 \to \tag{7}$$

2.1.6
Fine Structure of Xylans

The fine structure, i.e., the distribution pattern of side chains in heteroxylans, was studied only in a few cases using the structure elucidation of oligosaccharides produced by acid or enzymatic hydrolysis as well as by the determination of the single-ion activity coefficient of calcium counter ions (γCa^{2+}) in solution of carboxyl group-containing polysaccharides [6]. The MeGlc$p$A side chains of the xylans from white willow bark and beech wood were found to be nonrandomly distributed, whereas a regular substitution pattern was suggested for the MGX from sunflower hulls. Recently, a uniform distribution was also established for the highly substituted MGX isolated from the leaves of marsh mallow (*Althaea officinalis*) [56] and the aerial parts of *Rudbeckia fulgida* [57], whereas it was found to be irregular for the MGX from mahony (*Mahonia aquifolium*) stems [58]. Large proportions of MeGlcA located at two contiguous Xyl$p$ units of the xylan backbone were reported for AGX from larch and spruce wood [6]. This indicates a rather blockwise distribution pattern similarly to the lowly substituted GAX from wheat bran [34] with regions of more than six contiguous unsubstituted Xyl$p$ units and regions containing two contiguous Xyl$p$ units substituted at position 3 by Ara$f$ units. For wood xylans, an irregular distribution of glucuronic acid in hardwood, and a regular distribution in softwood were suggested on the basis of MALDI mass spectrometry [59].

The substitution pattern of arabinosyl side chains in AX from cereal flours and bran, based on the structural analysis of oligomer fragments produced by xylan-degrading enzymes of known mode of action, was described by several authors [60–63], and various structural models were created [39, 60]. In a recent study [64] on the fine structure of wheat flour AX, a method was developed for fragmentation of the xylan chains into oligomeric products using a combination of periodate oxidation, Smith degradation and enzymatic hydrolysis. Blocks containing up to six substituted xylosyl residues were isolated from AX. The experimental results were compared to computed models. In accord with the conclusions in [60, 61], the results indicated that the distribution of the type of substitution (mono- or disubstitution) on the xylan backbone is not the result of a random process. The authors suggested that the biosynthetic mechanisms favor disubstitution and probably the arrangement of the type of substitution within contiguous substituted xylosyl residues. Considering only the distribution of substituted (irrespective of the type of substituents) and nonsubstituted residues, the distribution might be considered to be random.

Examination of $(1 \rightarrow 4)$-linked xylans have indicated a three-fold, left-handed helical structure [65]. In the case of the AX from rice endosperm flour [66], this structure was confirmed by both X-ray diffraction and conformational analysis using the PS79 computer program. Although it does not seems to be a desirable conformation for xylans to make a complex firmly associated with the two-fold ribbon-like structure of cellulose, the existence of such interactions was documented [67].

2.2
Resources and Isolation of Xylans

Potential resources of xylans are by-products produced in forestry and the pulp and paper industries (forest chips, wood meal and shavings), where GX and AGX comprise 25–35% of the biomass as well as annual crops (straw, stalks, husk, hulls, bran, etc.), which consist of 25–50% AX, AGX, GAX, and CHX [4]. New results were reported for xylans isolated from flax fiber [16, 68], abaca fiber [69], wheat straw [70, 71], sugar beet pulp [21, 72], sugarcane bagasse [73], rice straw [74], wheat bran [35, 75], and jute bast fiber [18]. Recently, about 39% hemicelluloses were extracted from vetiver grasses [76].

The extractability of xylan is restricted due to its physical and/or covalent interactions with other cell-wall constituents. In lignified tissues, xylan is ester-linked through its uronic acid side chains to lignin [3, 77]. Multiple forms of bonding between lignin and AX have been suggested to exist in gramineous plants [78, 79]. In the latter case, phenolic acid units provide some potential for covalent interactions of the xylan with other cell-wall polymers such as feruloylated, arabinose-containing pectic polysaccharides [32] or lignin [80]. Linkages to structural cell-wall proteins have been claimed to cause insolubility of maize bran heteroxylan [81] and AX–protein linkages were shown to produce the extremely high-molecular component of an AX fraction from rye bran [82]. Recently, the existence of covalent linkages between a highly branched AX and non-gluten protein has been reported [83].

A variety of extraction methods was elaborated to isolate the xylan component from plant cell walls in a polymeric form. Special extraction conditions were selected, depending on the plant source. For the isolation of xylans from hardwood, in particular, two-step procedures with a NaOH/H$_2$O$_2$ delignification step were shown to be more acceptable in practice than the hazardous delignification with sodium chlorite. These procedures were also successfully applied to lignified tissues of annual plants [4]. During recent years, the effects of conditions and alkali type on the yield and composition of xylan-rich hemicelluloses from wheat straw [70], sugar beet pulp [72], barley straw [84] have been extensively investigated. In a detailed study [85], the isolation procedure and material properties of the MGX from aspen wood were described. The application of ultrasound was shown to be very effective during the alkaline extraction of xylans from corncobs [33], the seed coat of buckwheat [86], corn hulls [33], wheat straw [87, 88], and sugarcane bagasse [73]. Due to the sonomechanical effect of ultrasound, the cell walls were disintegrated, achieving higher yields of xylans without substantial modification of their structural and molecular properties. Ultrasound was also applied during the alkaline organosolv extraction of wheat straw [89]. To improve the color of the isolated hemicelluloses, the products were posttreated with H$_2$O$_2$ activated with tetraacetylethylenediamine [87, 90].

From recent literature it is known that the disintegration of lignified cell walls can be achieved by steam explosion treatments resulting in solubilization of partially depolymerized hemicelluloses [91, 92]. The application of this method on wheat bran yielded feruloylated GAX with different ferulic acid content [93]. Partly depolymerized water-soluble, acetylated AGX was obtained from spruce wood by employing microwave treatment [94].

In spite of the absence or very low content of lignin, the isolation of the xylan component of seed endosperm (flours), seed coats (bran), hulls, husks, etc. might be complicated due to the presence of proteins or non-separated starch [4]. Nevertheless, large-scale isolation procedures have been developed for the isolation of xylans from wheat flour [95], rye whole-meal [96] and de-starched wheat bran [75]. Optimum extraction conditions were elaborated [97] for the isolation of the CHX (known as corn gum) from corn hulls (corn fiber), a by-product from the wet-milling isolation of cornstarch. This xylan represents a further valuable co-product next to fuel alcohol prepared from the corn fiber. Various conditions of non-pressurized and oxygen-aided alkaline extractions of oat spelt, containing from 22 to 39% xylan, have been studied and the results evaluated with regard to the yield, purity, and molecular weight of the isolated arabinoxylans [98]. Using a twin-screw extruder, xylans were isolated from poplar wood [99] and co-extracted from a mixture of wheat straw and wheat bran [100]. This procedure might be a technical approach to design a high-throughout process.

2.3
Physicochemical, Physical and Functional Properties

2.3.1
Molecular-Weight Average and Molecular-Weight Distribution

An essential problem in the molecular-weight determination of xylans is, as is usual for most polysaccharides, the solubility. It is known to be dependent on various factors described in the previous sections, such as the type and degree of substitution by the various glycosyl side chains and their distribution pattern, which is responsible for aggregation tendencies, the presence of acetyl groups, possible chemical linkages with lignin, and cross-links through phenolic acids. A recent study on alkali-extracted lignin-containing MGX from aspen wood [101] indicated that the lignin component, present in bound and unbound forms, is the main contributor to aggregate formation.

However, the solubility, and respectively the dissolvability, of xylan is also affected by the patterns of intra- and intermolecular hydrogen bonds that are native or created during the isolation and drying processes of the preparations as well as during storage [102]. For the splitting of the hydrogen-bond systems aqueous, aprotic, and complexing solvents, such as cuoxam, cadoxen, and FeTNa have been applied, and now also in combination with physical treatments such as ultrasonication and autoclave heating [103, 104].

The molecular-weight values reported show considerable variations and may vary depending on the estimation method even for the same sample [3, 4, 6]. For cereal AX and CHX, the M_w values were 64 000–380 000 g/mol. The M_w values for AGX and MGX were 30 000–370 000 g/mol and 5000–130 000 g/mol, respectively. However, even in water-soluble samples, fractions of extremely high M_w values (up to millions) were observed by light-scattering techniques, gel chromatography and coupled techniques such as SEC-MALLS. They were ascribed to chain aggregation and/or the existence of microgel particles. After their elimination by filtration, values of about 270 000–370 000 g/mol have been determined by static and dynamic light scattering for corn bran heteroxylans [105]. Comparable M_w values (150 000 and 162 000 g/mol) were obtained for the water-soluble corncob AGX by ultracentrifugation and viscometry coupled to static light scattering of fractions from gel chromatography, respectively [106, 107]. The authors suggested that the xylan adopted an extended worm-like conformation. Viscometric studies at different ionic strength on acidic corn bran CHX fractions [108] and the lower-branched wheat bran GAX fractions [109] gave values of the empirical stiffness parameter B ranging between 0.022–0.026 and 0.012–0.015, respectively. The values are indicative of semi-flexible random coils. Recently, a similar conclusion was presented [110] for a series of wheat flour arabinoxylans, based on the chain persistence length.

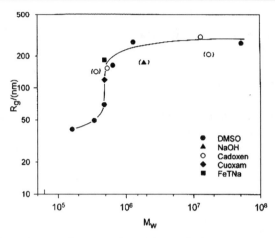

Fig. 6 Radius of gyration as a function of particle weight for AX-I in several solvents. The single, non-aggregated chains has $M_w = 62\,000$ g/mol, determined with the percarbanilated sample in dioxane [102]

The molecular-weight determination of very low-branched xylans is still an unsolved problem, because they are poorly soluble even in the complexing solvents commonly used for dissolution of cellulose. This has been demonstrated on the low-branched AX from rye bran [102] by viscometry and light-scattering techniques performed in various solvents as a function of time over a period of more than three years. The results suggested that the xylan had been isolated either as single strands with $M_w = 62\,600$ g/mol or at most dimerized strands. Such structures further aggregate and form clusters with time during storage, as indicated by the dependence of the radius of gyration (R_g) on the particle weights (Fig. 6). Complexing solvents used for cellulose dissolution were shown to dissolve the xylans down to a 6–7-strand bundle.

Most frequently, SEC with dextran-, pullulan-, or polystyrene calibration standards has been used to characterize the molecular properties of xylans. However, as for viscometric studies [108], a sufficient solvent ionic strength is a prerequisite for useful SEC measurements of charged polysaccharides, including glucuronoxylans [111–113]. An advantage of the SEC technique is that the presence of protein and phenolic components or oxidative changes can be detected by simultaneous ultraviolet (UV) detection.

2.3.2
Rheological Properties

The rheological behavior of xylans has rarely been investigated [4, 114, 115]. The water-insoluble hemicellulose from the viscose process (containing > 85% xylan) was reported to form thixotropic aqueous dispersions of high apparent viscosity, which decreased rapidly by the application of low shear rates. The MGXs from beech wood were shear-thinning and thixotropic, and behaved as a plastic material at higher concentrations (> 5 w/v %). This behavior mainly depends on the ionic state of the glucuronic acid carboxyl group and the content of the wis-MGX fraction. Strong thixotropy was also observed for low-branched corncob AGX and rye bran AX, whereas, the highly branched polysaccharide counterparts were shear-thinning without thixotropy. The deduced gel-like and liquid-like behavior, respectively, of both groups were confirmed by viscoelastic measurements [4].

Studies of the structure and molecular size of wheat AX [41] revealed that they are shear-thinning and exhibit two critical concentrations, which correspond to the onset of coil overlapping. The existence of three domains provided the evidence for the formerly suggested rigid, rod-like conformation of AX in solution. In a recent study [116], the previously reported conflicting suggestions on the conformation of AX were discussed.

The solution properties of two fractions of corncob xylan (fraction A and B) differing in the uronic acid content were investigated by viscometry using DMSO and 4% NaOH as solvents [117]. The polymers presented good stability in alkali without depolymerization during the measuring time. All solutions showed Newtonian behavior in the concentration and shear-rate ranges studied. Differences were found in the Huggins constant K', indicating flexible chains for A and rigid conformation for B, in accord with the low-branched and higher-branched backbone, respectively, of the fractions.

2.3.3
Interaction Properties with Other Polysaccharides

Starch/xylan interactions have been of particular interest as they affect the bread-making quality of flours [118–120]. Studies on the effect of ws-AXs on gelatinization and retrogradation of starch [121] revealed that addition of the xylan had no effect on gelation kinetics, but enhanced the storage modul G', and influenced starch retrogradation. The results suggest that AX increased the effective starch concentration in the continuous phase due to the different hydrodynamic volumes of the polymers. In a similar study [122], well-defined AX samples of low molecular weight (LMW-AX) and high molecular weight (HMW-AX) have been used as additives. Dough characteristics were more strongly affected by HMW-AX than by LMW-AX, as assessed by mixography. Small-scale bread-making tests showed that neither the water-holding capacity nor the oxidative gelation or viscosity was the main factor governing the LMW-AX functionality. The authors suggest that other mechanisms dictate the impact of both AX samples of bread making. More detailed studies are necessary to solve this problem.

The effect of X_m-rich ground husks form the seeds of *Plantago ovata* (is-abgol) on the bread-making properties of various flours was studied with the aim of substituting gluten in non-wheat bread [123]. Addition of isabgol alone or in combination with hydroxypropylmethylcellulose at 2% and 1% replacement level, respectively, significantly enhanced the loaf volumes of bread prepared from rice flour.

Interactions with xanthan were investigated for some GAX fractions of wheat bran [109]. Whereas, for lowly substituted GaMs a synergy in viscosity was observed at low total polymer concentrations, yielding a maximum of the relative viscosity at nearly equal proportions of both polysaccharides [124], the xanthan/xylan mixtures at the same experimental conditions showed no synergy. The observed decrease in the relative viscosity values upon addition of the xylan indicates that a certain interaction with xanthan takes place, but that it leads to a contraction in the hydrodynamic volume. The authors suggested that structural and conformational differences between GaM and GAX might be the reason for this observation.

The assembly characteristics of hemicelluloses, including xylans, and their interactions with cellulose have been intensively studied. Methods using spectral fitting of CP/MAS ^{13}C-NMR spectra [125] as well as AFM and QCM-D [126] were developed to diagnose and monitor interactions between cellulose I and xylan. Interactions are possible because the linear xylan backbone allows a partial alignment and formation of hydrogen bonds to cellulose microfibrils. In autoclave experiments of cellulose/GX mixtures under controlled conditions, the amount and localization of the globular-shaped xylan assemblies were visualized by immunolabelling and confocal laser microscopy [127 and references therein]. The xylans retained on the cellulose surfaces formed nano- and micro-sized particles. This is of practical importance as xylan-modified lignocellulosic fibres show improved wetting and liquid spreading properties.

2.3.4
Interfacial Properties

Although polysaccharides are not considered as surface-active agents, several AX preparations and beech wood MGX manifested such behavior [128]. They were reported to lower the surface tension (σ) of water from 72.0 up to 46.0 mN/m at the critical micelle concentration (c.m.c.) between 0.6 to 0.03% (w/v). Most of the xylans formed emulsions of the oil/water type with stability comparable to that of the commercial emulsifier Tween 20. The differences in the primary structure or viscosity of the xylans had no significant effect on the emulsifying activity. However, the presences of a low amount of proteins or lignin were supposed to contribute to these effects. The emulsion-stabilizing effect of several wheat bran GAX fractions [109] was compared to that of gum Arabic [129], a natural complex of an arabinogalactan type II-like polysaccharide with protein (about 2%). The expected effect of viscosity, which is suggested to be responsible primarily for stabilizing emulsions, was not confirmed. The GAX fractions showing the lowest emulsifying activity exhibited the highest intrinsic viscosity. Although the GAX fractions differed in their content of protein, uronic acids, and degree of substitution of the xylan chains, no relation was found between these analytical characteristics and the emulsifying activity.

The foamability of the xylans tested [128] was low in comparison to a commercial whipping protein D100. Only the highly viscous beech wood xylan and the rye bran AX–protein complex exhibited remarkable foaming activity, which was similar to that of gum arabic. As the MGX polymers contain considerable amounts of uronic acid side chains, this may play a role in their foaming activity together with the presence of low amounts of lignin.

The xylan samples mentioned [128] also showed a significant protective effect against thermal disruption of foams that are formed by the surface-active bovine serum albumin (BSA). The initially high foam volume of BSA decreased by heating of the samples tested. Addition of xylans also caused a decrease of the BSA foam volume. However, after heating the foam volume was restored by 70–100% with the xylans, except for the starch-containing highly branched corn hull heteroxylan (Fig. 7). The protective effect observed for a series of wheat endosperm AX fractions [41] was suggested to be due to their structural differences. Most important seems to be the association tendencies of xylans favored by the existence of low-branched chains or longer unsubstituted regions in the chains and/or by the presence of uronic acid substituents favoring electrostatic interactions. In a recent study [130] on

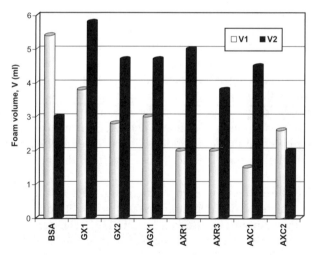

Fig. 7 Stabilization effect of various xylan types isolated from beech wood (GX1 and GX2), corn cobs (AGX1), rye bran (AXR1 and AXR3), and corn hulls (AXC1 and AXC2) on the protein (BSA) foam against thermal disruption; foam volume before (V1) and after (V2) heating at 95 °C for 3 min [128]

a mixture of BSA and the surfactant Tween 20, exhibiting high foaming stability at a very low mole ratio of both components, it was shown that addition of 0.2–0.3 mg/ml AX resulted in significant enhancement of the foam stability.

A great deal of effort has been made to investigate the role of xylans in bread making. Reviews on cereal xylans [39, 41, 118] have shown that the xylan component is primarily responsible for the effects on the mechanical properties of the dough as well as the texture and other end-product quality characteristics of baked products.

Film-forming properties of various pure xylans have been studied, but the results have not been promising [131, 132]. In a recent study [133], xylans from corncobs, grass and birch wood were added at levels up to 40% to wheat gluten to form biodegradable composite films. The results indicated that wheat/gluten composite films having different characteristics could be produced depending on the xylan type, composition, and process conditions without decreasing the film-forming quality and functionality such as the water-vapor transfer rate.

2.3.5
Thermal Behavior

The thermal properties of xylans have been studied in the past only in relation to problems in wood and pulp processing [4]. In some recent studies on xylans, the thermal behavior has been characterized [125, 134, 135]. Similarly to cellulose, xylans show no distinct thermal transitions. Several MGX fractions obtained by fractional extraction of steam-exploded poplar wood [136] as well as of AGX preparations isolated under various alkaline extraction conditions from wheat straw [134] have shown a broad onset of thermal degradation near 200 °C. In this study, the thermal behavior was supposed to be affected by the presence of lignin. For the lignin-rich hemicellulose fraction, the DSC curves showed a prominent effect at 250–540 °C with three maxima around 280, 435, and 500 °C corresponding to about 38, 20, and 13% of the total weight losses, respectively. The weight loss was interpreted as being due to decarboxylation in addition to dehydration and oxidation reactions of the carbohydrates and of less-condensed structures of the lignin component [68]. In the lignin-poor hemicellulose fraction, the peak maximum at 435 °C dominated.

The hemicellulose fraction, comprising about 75% of the hemicellulosic material, was isolated from delignified wheat straw. It consisted mainly of AGX and a minor amount of residual lignin [137]. The degradation of this fraction started at 220 °C and was completed at around 395 °C. The glass-transition temperature T_g of dry xylan was estimated to be in the range of 167–180 °C [138]. DTA of a flexible film prepared from birch O-acetyl-MGX after conditioning at room humidity, displayed endothermic behavior over a broad temperature range of 60–225 °C with a maximum at 150 °C [139], ascribed to water removal from the xylan hydrate noncrystalline state. Thermoplasticity was displayed also by AGX film from pinewood, the latter being far more extensible.

2.3.6
Biological Activity

The various physiological effects of xylan-rich dietary fibers as well as preparations of xylans isolated from various cereal bran and red algae are well known and have been reviewed [3, 4, 140 and references therein]. Heteroxylans from various plants, including edible tissues, showed a dose-dependent antimutagenic activity tested by the antibleaching effect of these polysaccharides against acridine orange- and ofloxacin-induced mutagenicity in the *Euglena* assay [141]. The feruloylated xylan fragments from corn bran [140] were reported to possess antioxidant activity that might contribute to some physiological functions of cereal dietary fiber.

Medicinal plants are known to be a potential source of biologically active polysaccharides [4, 30]. Anticomplementary activities have been reported for the complex heteroxylans of the seeds of *Plantago major* [53] as well as immunomodulating effects for the MGX from *Rudbeckia fulgida* [142]. Xylans from nonmedicinal plants also displayed immunostimulatory activity. The enzymically modified heteroxylan from rice bran (commercially known as MGN-3) possesses potent anti-HIV activity without any notable side effects [143, 144]. The ws-AGX from corncobs [145] exhibited significant immunostimulatory effects in the in vitro rat thymocyte tests. Its activity was comparable to those of the commercial immunomodulator Zymosan, a fungal

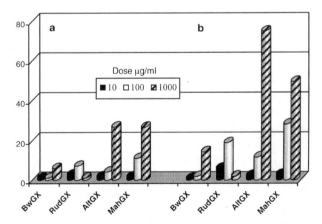

Fig. 8 Immunostimulatory activity of MGX from beech wood (BwGX), *Rudbeckia fulgida* (RudGX), *Altheae officinalis* (AltGX), and *Mahonia aquifolium* (MahGX) [146]; **a** Mitogenic activity, SI_{mit} and **b** comitogenic activity, SI_{comit}

β-glucan. In a recent paper [146], the responses in the comitogenic thymocyte test of the corn cob ws-AGX and MGXs prepared from beech wood and the herbs *Rudbeckia*, *Altheae*, and *Mahonia* were compared in terms of the molecular weight and the content and distribution of MeGlcA substituents (Fig. 8). However, no unequivocal relation either to the content of MeGlcA or to its distribution pattern was found for the MGX samples. The immunostimulatory effect of ws-AGX seems to be affected less by the molecular weight than by the presence of the disaccharide side chain (1) supposed to be important for the expression of the observed biological response [31].

Antitussive effects (tested on cats) have been reported for MGXs from *Rudbeckia* [147] and mahony [58]. Comparative tests performed under the same conditions with some drugs used in clinical praxis revealed a significantly higher activity than expressed by the non-narcotic synthetic drugs.

2.4
Application Potential of Xylans and Xylo-oligosaccharides

Some application possibilities have been suggested for various different xylan preparations based on their known functional properties [3]. The branan ferulate in its cross-linked form has already been marketed as a "super gel" for wound dressing [148]. Recently, the production of branan ferulate/alginate fibers by the wet-spinning technology has been reported [149]. Wheat-gluten composite films [133] are of potential use, particularly in food industry. The application of xylans as an additive in bread-making or as a drug carrier seems to be possible as well.

Xylans from beech wood, corncobs, and the alkaline steeping liquor of the viscose process have been shown to be applicable as pharmaceutical auxiliaries [3]. Micro- and nanoparticles were prepared by a coacervation method from xylan isolated from corncobs [150]. The process is based on neutralization of an alkaline solution in the presence of surfactant, which was shown to influence both the particle size and morphology. They are aimed at applications in drug delivery systems.

There are several reasons for the permanent research interest in oligosaccharides in general, including xylo-oligosaccharides. One reason is that oligosaccharides have been suggested as new functional food ingredients modifying food flavor and physicochemical characteristics [151]. Moreover, many of them possess properties beneficial to health. They show non-cariogenicity, a low caloric value, and the ability to stimulate the growth of beneficial bacteria in the colon, thus representing potential prebiotics [152]. Acidic xylo-oligosaccharides produced by family 10 and 11 endoxylanases were reported to exhibit antimicrobial activity [153].

Another reason is that xylo-oligosaccharides of defined structure are very important substrates that serve as model compounds for the optimization of hydrolytic processes and in enzymic assays. The enormous development in enzymic methodology for chemical modification and structural analysis [154–157] as well as in physical methods for the isolation, separation, and structural characterization of carbohydrates [158–164] offered the preparation of value-added oligomers of xylans and other hemicelluloses.

Classical acid hydrolysis [165], autohydrolysis at high temperatures based on acetic acid formed from the release of acetyl groups [166], endoxylanase treatment alone [153] or in combination with steaming and microwave heating [167] were used to prepare xylo-oligosaccharides from various xylan-rich materials. The hydrothermal treatment is of interest [168, 169] because it enables the fractionation of hemicellulose from cellulose and lignin, and separation of the formed oligosaccharides. This has been demonstrated, recently, on *Eucalyptus* wood and brewery spent grain [158, 170].

Nowadays, a strategic area of research is the development of polymers based on carbohydrates due to the worldwide focus on sustainable materials. Since the necessary multi-step synthesis of carbohydrate-based polymers is not economical for the production of commodity plastics, functionalization of synthetic polymers by carbohydrates has become a current subject of research. This aims to prepare new bioactive and biocompatible polymers capable of exerting a temporary therapeutic function. The large variety of methods of anchoring carbohydrates onto polymers as well as the current and potential applications of the functionalized polymers has been discussed recently in a critical review [171]. Of importance is that such modification renders not only functionality but also biodegradability to the synthetic polymers.

Serving as pre-polymers, promising carbohydrate candidates for functionalized synthetic polymers are oligomeric hemicelluloses produced from various renewable resources [172]. This was shown for hemicellulose oligomers obtained by steam explosion of spruce chips. The oligomers were modified with methacrylic functions and then subjected to radical polymerization with 2-hydroxyethyl methacrylate or poly(ethylene glycol) dimethacrylate in water using a redox initiator system [173, 174]. The hydrogels formed showed properties similar to those of pure poly(2-hydroxyethyl methacrylate)-based hydrogels.

3
D-Mannoglycans

The mannan-type polysaccharides of higher plants can be divided into
(i) galactomannans (Fig. 9) and
(ii) glucomannans (Fig. 10).
Whereas the backbone of the first type is made up exclusively of β-(1 → 4)-linked D-mannopyranose (D-Man$p$) residues in linear chains, the second type

has both β-(1 → 4)-linked D-Man$p$ and β-(1 → 4)-linked D-glucopyranose-(D-Glc$p$) residues in the main chain. As single side chains, D-galactopyranose (α-D-Gal$p$) residues tend to be 6-linked to the mannan backbone of both mannan-type polymers in different proportions. The resulting polymers are named galactomannans and galactoglucomannans. For such classification, a tolerance limit for the proportion of galactosyl groups (15%) has been suggested [6].

3.1
Sources, Structure and Isolation of Mannoglycans

3.1.1
Galactomannans

Mannans free of galactosyl side chains are rather rare. Slightly galactosylated mannans ($\sim$ 4% galactose), considered as linear β-(1 → 4)-D-mannans, have been isolated from the seed endosperm of vegetable ivory nut (*Phytelephas macrocarpa*) and date (*Phoenix dactylifera*) [175]. They are water-insoluble and resemble cellulose in conformation of the individual molecular chains, and occur in cellulose microfibrils of the plant tissues. High proportions of low-branched galactomannans (GaM), which have on average of one 23 Man$p$ residues galactosylated, were found in green arabica coffee beans [176].

GaM samples heavily substituted with Gal$p$ residues (30–96%) are water-soluble and abundant in the cell walls of storage tissues (endosperm, cotyledons, perisperm) of seeds. Notably, those from the endosperm of some leguminous seeds, such as guar (*Cyanopsis tetragonoloba*), locust bean or carob (*Ceratonia siliqua*), and tara gum (*Caesalpinia spinosa*), are widely used commercially [175]. During the last decade several studies have been devoted to GaM from the seeds of various plants used as traditional food or medicines, to find alternative sources of commercial gums, and to extend their application fields. The Man : Gal ratio is one of the main characteristics of GaM as it determines their physicochemical properties, such as solubility in water, density, and the viscosity of the solution. This ratio varies from 1 : 1 to 5.7 : 1 in leguminous plants [177]. GaMs have been isolated from the seeds of *Ipo-*

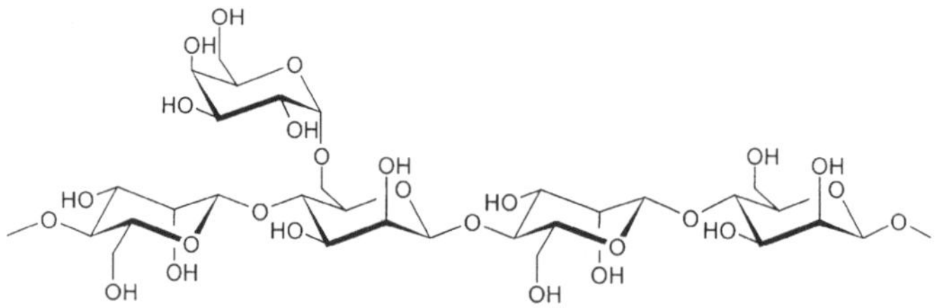

Fig. 9 Primary structure of D-galacto-D-mannan (GaM)

moea turpethum, a perennial plant [178], various *Cassia* species [179–181], date [182], *Mimosa scabrella* [183], and some *Strychnos* species [184, 185]. These studies aimed to elucidate the structural features of the polysaccharides that are a basic requirement for understanding their functional properties, particularly gelling and their interaction behavior.

The highly branched GaM of *Mimosa* seeds, isolated by a scaled-up process [186] at a yield of ~ 20% of seed weight, showed a Man : Gal ratio of 1.1 : 1 and structural features of the gum obtained on a laboratory scale [183]. Based on ^{13}C-NMR spectroscopy of GaM from *Cassia angustifolia* [179], with a Man : Gal ratio of 2.9 : 1, a blockwise pattern of the galactosyl branches was suggested.

The water-soluble GaM of green coffee bean (*Coffea arabica*) [187, 188] plays an important role in the retention of volatile substances. They contribute to the coffee brew viscosity and creamy sensation perceived in the mouth known as "body". The polysaccharides of roasted coffee beans [176] have been studied with the aim of providing information on the fate of these polysaccharides during roasting and on their conversion products. The results indicated that the mannan fractions isolated from hot-water extracts of roasted bean were very low-branched with Man : Gal ratios of 26 – 35 : 1, and contained some arabinogalactan. The mannan components retained, except for the branching degree, their original structural features as found before roasting. However, the question concerning the existence of true arabinogalactomannans or physical mixtures of both polysaccharides has not been answered yet.

The distribution of Gal*p* residues along the mannan backbone was shown to affect the molecular interactions of such chains (self-association) and with other polysaccharides (cooperative associations) in solution [175, 189]. Therefore, it is an important structural feature of GaM as a potential commercial gum. Recently, a simple method was developed that enables the enzymic determination of the Gal*p* distribution in GaM using endo-mannanase of *Aspergillus niger* for degradation and HPAEC of the fragments [190]. The method was applied on a large variety of commercial low- to high-substituted GaMs. In this way, the tara gum was found to have both random and blockwise distributions of Gal*p* residues, while the guar gum possess a blockwise, and cassia gum a very regular distribution. The greatest variations were observed among various batches of locust bean gum.

The slightly galactosylated mannans are essentially linear polymers. As a result of their cellulose-like (1 → 4)-β-D-mannan backbone, they tend towards self-association, insolubility, and crystallinity. Crystallographic study of *C. spectabilis* seed GaM [180] with a Man : Gal ratio 2.65 : 1 suggested an orthorhombic unit cell with lattice constants of $a = 9.12$, $b = 25.63$, and $c = 10.28$; the dimension b was shown to be sensitive to the degree of galactose substitution and the hydration conditions [180 and references therein, [191]].

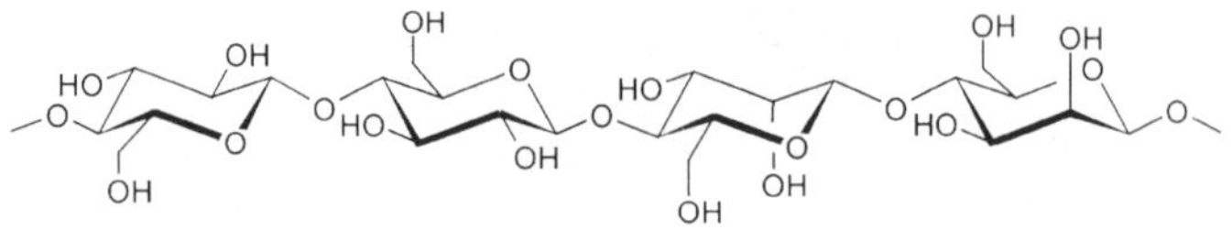

Fig. 10 Primary structure of D-gluco-D-mannan (GM)

3.1.2
Glucomannans and Galactoglucomannans

Glucomannans (GM) and galactoglucomannans (GGM), common constituents of plant cell walls, are the major hemicellulosic components of the secondary cell walls of softwoods, whereas in the secondary cell walls of hardwoods they occur in minor amounts. They are suggested to be present together with xylan and fucogalactoxyloglucan in the primary cell walls of higher plants [192]. These polysaccharides were extensively studied in the 1960s [6, 193].

GM and GGM show some variations in structural features depending on the plant species and stage of plant development. The extent of galactosylation governs their association tendency to the cellulose microfibrils and, hence, their extractability from the cell-wall matrix [194]. Alkali-extractable GMs from secondary cell walls are characterized by a low degree of substitution by galactosyl side chains and the Gal : Glc : Man ratio is about 0.1 : 1 : 3–4, whereas the water-extractable polymers have roughly equal proportions of galactose and glucose and the Glc : Man ratios vary between 1 : 1.4–3.1 [195, 196]. Mannoglycans isolated from the primary cell walls or from suspension cell cultures usually have equal proportions of all three sugar components [192]. The GM backbone is acetylated; however, acetyl groups are released by saponification during the frequently used alkali extraction.

The structural features of the GGM from wood have been investigated by glycosyl linkage analysis, ^{13}C-NMR spectroscopy, and partial acid hydrolysis due to the important biological activities of oligosaccharides derived from GGMs of poplar (*Populus monilifera*) [197] and spruce (*Picea abies*) [198 and references therein] in plant growth, development, and defense. The results indicate that the mannan backbone of GGM, extracted with strong alkali from the sodium chlorite-delignified spruce wood (holocellulose) [199], consists of galactose, glucose, and mannose in the mole ratio 1 : 8 : 33, and contains segments of more than four Man*p* residues, interrupted with the segment composed of both Man*p* and Glc*p* or two Glc*p* residues. Branches terminated by single Gal*p* residues were suggested to be attached at positions 2, 3, and 6 preferably of Man*p* and at positions 3 and 6 of Glc*p*. The GGM isolated by hot water from the spruce holocellulose [198] had a Gal : Glc : Man mole ratio of 1 : 3 : 17 and about 33% of acetyl groups attached at positions 2 and 3 of half of the Man*p* residues present.

The mannans of hardwoods, representing 3–5%, are usually co-extracted with the predominating AGX component. The GGM separated from the alkali extract of poplar holocellulose [197] consisted of galactose, glucose, and mannose in a molar ratio of 1 : 4.1 : 9.7. Acetylated mannan was found as a minor component of a fraction isolated from aspen wood by applying microwave heating [15, 94]. Low-molecular-weight GMs, acetylated at position 2 or 3 of some Manp, were isolated from aspen and birch wood by different procedures [200], including sequential extractions with DMSO and hot water, and fractional precipitation of GM from process water of mechanical pulping.

Soluble, often partially acetylated GMs also occur as constituents of bulbs, tubers, seeds, roots, and leaves of some non-gramineous monocotyl plants [6] such as the GM from Libyan dates [201] or the GMs from plants of the genus *Aloe* known for their medicinal value. The polysaccharide "acemannan" (a commercial product called aloe vera gel or Carrysin) is located in the parenchyma (filet) of the leaves as the main component of a mucilaginous jelly [202, 203]. The β-(1 → 4)-D-mannan backbone of acemannan is heavily substituted at positions 2 and 3 of the Manp residues by acetyl groups and has some hexosyl side chains attached at position 6. However, contradictory data concerning the hexosyl branches of acemannan have been published, suggesting either Galp or Glcp residues as candidates [204, 205 and references therein]. The compositional study of aloe vera leaves [206] revealed that the polysaccharides detected in the filet of the leaves and in the separated gel are similar and correspond to storage GaMs, whereas those isolated from the skin tissue correspond to GM located within the cell-wall matrix. This was confirmed in a recent study on the immunostimulatory activity of polysaccharides isolated from the filet and the skin of aloe vera leaves [207]. Differences in aloe cultivars, geographical location, and the season of collection as well as isolation and purification conditions of the mucilage polysaccharides may explain the different results reported by various authors. It should also be taken into consideration that the processing of aloe vera products, e.g., dehydration by convection drying at temperatures above 60 °C was shown to affect the sugar composition as well as the physicochemical properties of acemannan dramatically, which is explained by the loss of acetyl groups and galactose from the mannose backbone and changes of the molecular weight [208].

The GM known as konjac mannan in Japanese food applications was found in the tubers of the potato-like plant *Amorphophallus konjac*. This storage polysaccharide has sequences of Manp residues separated by Glcp units in the molecular backbone and is acetylated with one acetyl group per six Glcp residues groups [6]. Reinvestigations have been done concerning the structure, physicochemical characteristics, and properties of konjac mannan [209–211]. Chemical analysis of a purified konjac GM [212] revealed the ratio of Man to Glcp to be 1.6 : 1, and branching (about 8%) explicitly joined through the C-6 position of only the glucose residues. However, both Glcp and Manp residues (in the ratio of 1 : 2) were suggested to terminate the branches of the

Glc$p$ residues, what differs from previous investigations. Based on ^{13}C-NMR spectroscopy, the sequences of Glc$p$ and Man$p$ were estimated and a model structure for the GM was proposed.

In a recent paper [213], various GM-rich hemicellulose fractions containing 26–64% mannose and 1.4–2.2% galactose were isolated from ramie (*Boehmeria nivea*), which is a perennial grass. This is unexpected, because hemicelluloses of grasses contain predominantly xylans [4].

3.1.3
Isolation Processes

Since in structural plant tissues hemicelluloses are embedded in the cell walls, their isolation in polymeric form is difficult and requires multi-step extraction methods to separate them from lignin or proteins and other non-hemicellulosic polysaccharides. The different ways of hemicellulose isolation applied include physical methods such as steam treatment and microwave heating [214–219], and chemical methods such as the organosolv process [220]; but in most cases the polymer properties are lost due to degradation reactions. The water-soluble polysaccharides of Norway spruce wood and those accumulated into process water during thermomechanical pulping (TMP) have been reported in several publications [221 and references therein]. Water-soluble polysaccharides composed mainly of acetylated GGM have been isolated by multi-step sequential extraction procedures from native Norway spruce and spruce after TMP, and characterized. About 65% of the Man$p$ were acetylated at either positions 2 or 3 in a ratio of 2.2 : 1. The number of Gal$p$ residues linked mainly to Man$p$ was considerably lower for the GGM obtained by TMP than from the native wood. This was explained by the increased accessibility of low-branched mannans during the steaming process.

The microwave heat-fractionation process under various conditions has been investigated with the aim of finding conditions that allow the isolation of lignin-free acetylated GGM from spruce wood [216, 219]. The initial screening of conditions [216] was evaluated on the basis of the yield of dissolved mannan and the molecular weight of lignin-free hemicelluloses. At temperatures of 170–220 °C and with increasing the residence time, the yield of mannan increased, while SEC indicated depolymerization of the polymer. Lignin-free samples with M_w of 5000 and 1600 g/mol (Fig. 11) showed Gal : Glc : Man ratios of approximately 0.1 : 1 : 4 and acetyl groups distributed almost equally between C-2 and C-3 of one third of the Man$p$ residues. By varying the alkaline impregnation conditions of spruce chips and the temperature/time conditions of the microwave heat-fractionation process [219], various yields (3–78%, based on the amount in the wood) and M_w values (3800–14 000 g/mol) of acetylated GGMs were obtained. However,

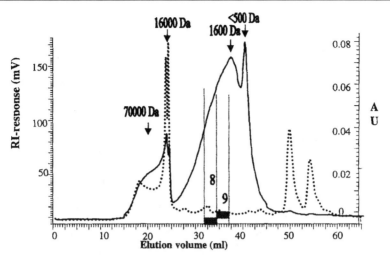

Fig. 11 Elution profile of the ws-material from microwave-heated spruce chips after SEC [218]. Detection by refractometry index (RI) (*dotted line*) and UV detection at 280 nm (*full line*). The arrows mark the elution volume of acetylated GGM fractions

by applying low-concentration (< 0.05%) NaOH only acetylated mannans were isolated. Recently, microwave and steam treatments were compared to optimize the extraction conditions of hemicellulosic oligosaccharides from spruce [167], which might serve as pre-polymers for the production of new biopolymers.

During the past decade, MALDI-TOF MS has proven to be an effective tool for the analysis of oligo- and polymeric mannoglucans (for extensive reviews see [222, 223]). SEC/MALDI mass spectrometry was employed in the analysis of hemicelluloses isolated by microwave heat-fractionation from spruce and aspen wood [94]. These methods allowed the separation and characterization of the oligo- and polysaccharide fractions derived from the xylan and mannan components of both woods [224].

3.2
Physicochemical and Functional Properties

3.2.1
Solution Properties

It is often difficult to compare the various mannan-type polysaccharides with regard to their molecular weight, because of the different analytical techniques used. Nevertheless, the M_w values reported for the GM and GGM isolated from woody tissues are rather low (1600 to 64 000 g/mol) [197–199, 216, 219, 221] in comparison to those of the water-extractable galactomannans from seeds with M_w 0.96×10^6 g/mol [179] and 1.26×10^6 g/mol [180]. Knowledge of the relation between the molecular structure of polysaccharides, including GaMs, and their behavior in sol and gel states have been broadened during the last years [180, 189, 225]. The viscosity of guar GaM was reported [177] to depend on many factors such as molecular weight, molecular-weight distribution (MWD), degree of branching, temperature, and pH [226]. Based on a comparative study of acid and enzymatic hydrolysis of guar gum [227], an acid hydrolytic procedure in ethanol slurry was developed which yielded GaM fractions with M_w in the range $5.7 \times 10^4 - 1.2 \times 10^6$ g/mol and limiting viscosity numbers $[\eta]$ in the range 272–1600 ml/g. The Huggins coefficient of the fractions, which is a measure of polymer–polymer interactions in solution, was found to be much lower (~ 0.4) than that of the native polymer (~ 0.79). The results indicated that the strong aggregation tendency between GaMs through unsubstituted regions on the backbone was weakened by the acid hydrolysis process. The problem with solubilization was solved using the cell solubilization method [228, 229].

Rheological studies of GaM from *Ipomoea* seeds [178] revealed that solutions of the polysaccharide were stable over a wide range of pH values. Reversible cohesive gels were formed with borate ions. Nonreversible gels were formed with transition metals depending on pH; thus, this gum resembles guar gum in behavior. The GaM from *Mimosa* seeds [186] showed a liquid-like behavior, in contrast to most other GaMs [129]. The difference was attributed to the higher galactose content of the *Mimosa* GaM (Gal : Man $\sim 1 : 1$).

In the linear konjac mannan, the degree of acetylation profoundly affects the solubility and flow properties of this hydrocolloid. The acetyl substitution prevents self-association of the mannan chains, but following deacetylation chain interactions become more energetically favorable [230].

3.2.2
Interaction Properties

The rheological behavior of GaM is of importance for the elucidation of their interactions with various polysaccharides in mixed hydrocolloid systems. The interactions of GaM such as guar gum, locust bean gum, and konjac GM with various hydrocolloids have been intensively studied and reviewed [175, 231, 232]. In the last decade, further reports on guar GaM and konjac GM interactions with starch have appeared. A large-deformation and dynamic viscoelastic study was performed on hydrocolloid systems containing konjac GM of different molecular weight, and the results were compared with those obtained for κ-carrageenan [233]. Konjac GM and guar or locust bean GaM were shown to affect the pasting behavior of starch (wheat, corn, waxy corn, tapioca and amaranth) using a rapid visco-analyzer (RVA) [234]. Addition of GaM to starch dispersions at sufficiently high concentrations leads to separation of the mixed gels into two phases and the formation of weak gels, as explained by thermodynamical incompatibility [235]. Later articles [236, 237] reported on the effects of konjac GM on the gelatinization and retrogradation rate of corn starch. From the rheological studies and DSC measurements, it was suggested that the GM promoted the retrogradation of starch during short-term storage and then retarded slightly after a certain storage time. In comparison to xanthan, the effects of GaM on the freeze/thaw properties of starch gels have been reported [238]. In a molecular modeling study acetan/GM interactions were described [239].

By analytical ultracentrifugation [240], interactions between locust bean GaM and the wheat protein gliadin in its natural and degraded forms have been investigated. The sedimentation velocity indicated weak non-covalent protein–polysaccharide interactions, which was supported by UV absorption spectrometry and gel filtration. Such interactions are of importance as they could protect sensitive persons from the harmful effects of this protein [241]. Small-deformation rotational oscillation was used to examine the effect of small additions (< 1%) of GaM on the glass transition of glucose syrup at a total level of solids of 83% [242]. The authors suggested that, due to the presence of the polysaccharide network, the rate of the relaxation process and diffusion mobility are lowered, accelerating the collapse of the free volume and thus inducing vitrification of the high-sugar/polysaccharide mixture at high temperatures.

3.2.3
Bioactive Properties

Significant physiological and therapeutic properties of mannoglucans and acemannan have been described in review articles; the guar GaM and konjac

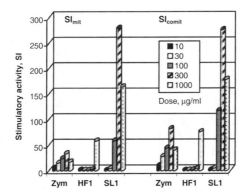

Fig. 12 Mitogenic (SI_{mit}) and comitogenic (SI_{comit}) activities of the acemannan isolated from the filet (HF1), the polysaccharide complex (SL1) of the skin of *aloe vera*, and the commercial fungal immunomodulator Zymosan (Zym) [207]

GM showed the ability to lower the level of plasma and liver cholesterol and the postprandial increase in plasma glucose [243, 244], the aloe vera acemannan exhibited immunomodulatory activities [203]. In a recent report [207], the GaM isolated from the filet of aloe vera (SF1), corresponding to acemannan, showed a comitogenic effect in the in vitro thymocyte test (Fig. 12). However, the immunostimulatory activity of the complex of polysaccharides isolated from the skin (HS1) was significantly higher. The GaM fractions prepared from fenugreek (*Trigonella foenum-graecum* L.) were reported to exhibit immunostimulatory activity as well [245].

3.3
Application Potential of Mannoglycans

The *Cassia angustifolia* GaM possesses the potential to become a new source of commercial gum due to its high content in the endosperm (about 50%) and its valuable rheological properties. It was suggested, in a similar way as for other GaMs, for usage as an additive in pharmaceutical formulations [188, 246]. Also the *Ipomoea* seed gum in its natural form, and after modification by grafting, has the potential to be used as a commercial gum [178].

The reported effect of konjac GaM on the glass transition of high-sugar/polysaccharide mixtures [242] can be utilized in sugar, hard-boiled and frozen confectionery products and might replace gelatin, which is refused by some consumers due to diet and health problems.

4
D-Xylo-D-glucans

4.1
Sources, Location and Isolation

Xyloglucan (XG) is a hemicellulosic polysaccharide found in all higher plants, where they represent a quantitatively major building material of the primary cell wall. The occurrence, isolation, structure, and functions of xyloglucans were reviewed in 1983 [6] and discussed in more recent papers [2, 247 and references therein]. XG makes up about 20–25% of the primary cell wall in dicotyledonous angiosperms such as *Sycamore* or *Arabidopsis thaliana*, somewhat less (2–5%) in grasses (monocotyledonous angiosperms), about 10% in the bulb cell walls of onion (a monocotyledonous angiosperm), and about 10% in the primary cell walls of fir trees (gymnosperms). In a selected subset of dicotyl plants, such as nasturtium and tamarind, the seeds have thick cell walls containing, instead of starch, XG as a storage polysaccharide that is mobilized during germination by the action of wall-bound enzymes.

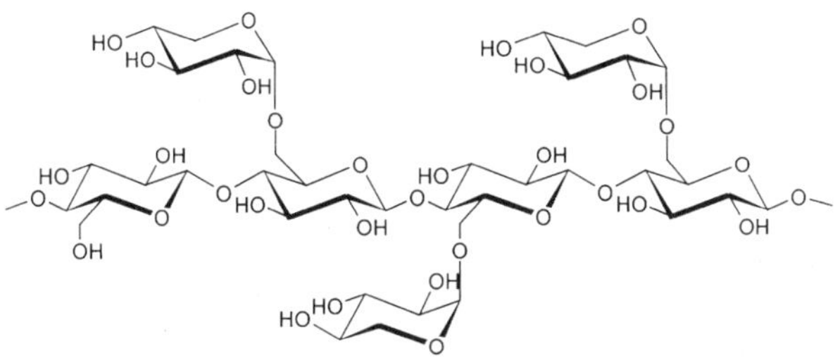

Fig. 13 Primary structure of D-xylo-D-glucan (XG)

XG have a cellulosic, i.e., $(1 \rightarrow 4)$-β-D-glucopyranan, backbone decorated with α-D-Xyl$p$ residues at position 6 (Fig. 13). Most XG are tightly hydrogen-bonded to the cellulose microfibrils of the cell wall [248]. In current models of the primary cell wall, XG and cellulose interact to form a key load-bearing network that prevents cells from rupturing under osmotic stress [249–251]. To elucidate cellulose–xyloglucan interactions, the adsorption of pea XG and tamarind XG to cellulose was studied [252]. The results indicated that XGs bind to cellulose as a monolayer and Fuc$p$ residues present in the pea XG contribute to the adsorption affinity [253]. The binding ability was demonstrated [254] on a series of XG oligosaccharides (reduced with tritiated sodium borohydride) and tested on amorphous microcrystalline cellulose, and filter paper; it increased with increasing DP of XG if the oligosaccharide contained more than four consecutive $(1 \rightarrow 4)$-β-D-Glc$p$ residues. Further studies [255, 256] revealed that storage xyloglucan–cellulose interactions are influenced by the galactosyl side chains and the molecular weight of the XG. In contrast to that of schizophyllan the adhesion of tamarind seed XG to cellulose was found to be largely independent of the molecular weight [253]. As the molecular weight of schizophyllan decreased, both its ability to form triple-helical structures and its adhesion to cellulose diminished and finally disappeared, indicating the need for high-molecular-weight domains.

The strong cellulose–xyloglucan bonding in the cell walls negatively affects the extractability of this hemicellulose component. XGs are extractable from vegetative plant cell walls only using concentrated alkali solutions, and more efficiently in combination with chaotropic agents [2]. In contrast, storage XGs are extractable by hot-water treatments from the seeds. Usually, stepwise alkaline extractions were applied to isolate non-seed XG. Extraction methods with optimized conditions were reported for the isolation of XG from depectinized apple pomace [257–259].

The difficulties in the extraction of XG from the cell walls as well as its separation from the other cell-wall polymers have been interpreted by various suggestions [260]. In addition to the existence of strong hydrogen bonds with cellulose and some hemicelluloses, various covalent bonds have been considered to fix the XG in the cell walls [261] such as esters with the COOH groups of pectin, benzyl sugar ethers of polysaccharide-bound dimers of ferulic acid or other phenolics, and alkali-stable (glycosidic) bonds with pectin or xylans and/or mixed-linkage β-glucans.

4.2
Structural Diversity

Many of the general structural features of XG are the same independent of the plant [154, 247]. In dicotyls about 60–75% of the Glc$p$ units are xylosylated, while the figure in grasses is about 30–40%. There is some regularity in the distribution of the xylose side chains, which enables the identification of two major types of xyloglucans in the plant cells walls, namely XXXG (Fig. 13) and XXGG. The XXXG type consists of repeating units of three consecutive xylosylated Glc$p$ residues separated by an unsubstituted Glc$p$. In the XXGG type, two unsubstituted Glc$p$ residues separate two xylosylated Glc$p$. The variety of side chains decorating the XG and their distribution on backbone are responsible for the large structural diversity of this polysaccharide.

The primary cell walls of most higher plant species contain XGs of the XXXG type, which bear trisaccharide side chains (8) on the backbone [247]. The seeds of many plants contain XXXG-type XGs, in which about 30% of the xylose units possess a β-D-Gal$p$ residue attached to position 2. Several plant species produce XGs that lack fucose and galactose, and have α-L-Ara$f$ attached to O-2 of some of the Xyl$p$ side-chains, such as XG isolated from olive fruit [262] and soybean (*Glycine maxima*) meal [263]. However, α-L-Ara$f$ residues occur also 2-linked directly to some of the Glc$p$ residues of the backbone [154].

$$\alpha\text{-L-Fuc}p\text{-}(1 \to 2)\text{-}\beta\text{-D-Gal}p\text{-}(1 \to)\text{-}\alpha\text{-D-Xyl}p(1 \to \qquad (8)$$

$$\alpha\text{-L-Gal}p\text{-}(1 \to 2)\text{-}\beta\text{-D-Gal}p\text{-}(1 \to \qquad (9)$$

A novel disaccharide moiety (9), attached to position 2 of the xylosyl side chains has been found in the jojoba seed XG [264]. Reinvestigation of tamarind seed XG [265] revealed the presence of the disaccharide moiety (3) linked directly at position 6 to the glucan backbone, in addition to α-L-Ara$f$ residues terminating the xylosyl side chains. In an earlier report [266], the arabinose component of this XG was ascribed to a contaminating arabinan. An unusual the dodecasaccharide subunit consisting of two glucose units bearing the fucose-containing trisaccharide side chain (8) and three unsubstituted glucose residues have been determined in apple fruit xyloglucan [267]. XG of the XXXG type from the leaves of *Argania spinosa* yielded fragments with a novel motif after treatment with specific endo-glucosidases [27]. The fragment contains xylobiosyl side chains (10) in the cellotetraose sequence. A new family of oligosaccharides has been isolated from the XG of the legume *Hymenaea courbaril* L. cotyledon [268, 269]. A combination of alternate treatments with various enzymes yielded a new oligosaccharide subunit XXXXG, carrying Gal*p* substituents in varying proportions, which is present in about the same amount as the typical XXXG subunit. Studies of the non-cellulosic cell-wall polysaccharides of various fruit tissues [270–274] and leaves [269] considerably broadened the knowledge on XG-type hemicelluloses.

$$\beta\text{-D-Xyl}p\text{-}(1 \rightarrow 2)\text{-}\alpha\text{-D-Xyl}p\text{-}(1 \rightarrow \qquad\qquad (10)$$

The cell walls of the *Solanaceae* plant species contain the XXGG type of XGs [247, 275]. Suspension-cultured tobacco cells secrete a XG, bearing disaccharide side chains (11) [276], as similarly reported for XGs of other solanaceous plants. The XG from suspension-cultured tomato cells is more complex, as it also contains the unusual $\beta\text{-L-Ara}f\text{-}(1{\rightarrow}3)\text{-}\alpha\text{-L-Ara}f\text{-}(1{\rightarrow}2)\text{-}\alpha\text{-D-Xyl}p$ trisaccharide side-chain (12) in addition to the prevailing dimeric moiety (13) [277].

$$\alpha\text{-L-Ara}f\text{-}(1 \rightarrow 2)\text{-}\alpha\text{-D-Xyl}p\text{-}(1 \rightarrow \qquad\qquad (11)$$

$$\beta\text{-L-Ara}f\text{-}(1 \rightarrow 3)\text{-}\alpha\text{-L-Ara}f\text{-}(1 \rightarrow 2)\text{-}\alpha\text{-D-Xyl}p\text{-}(1 \rightarrow \qquad\qquad (12)$$

$$\beta\text{-D-Gal}p\text{-}(1 \rightarrow 2)\text{-}\alpha\text{-D-Xyl}p\text{-}(1 \rightarrow \qquad\qquad (13)$$

Many XGs are acetylated [248, 278, 279]. XG isolated from the spent media of suspension-cultured sycamore cells Glc*p* contains acetyl groups at positions 3, 4, or 6 [2].

The distribution of various subunits in the XG backbone, representing the fine structure of the polymer, is responsible for the structural diversity in the case of relatively low-branched XGs, as documented by the variety of oligosaccharide subunits released with a $\beta\text{-}(1 \rightarrow 4)$-endoglucanase [280, 281]. During the last decade a variety of enzymes for the hydrolytic fragmentation of xyloglucan chains [154, 282, 283] in combination with chromatographic fractionation techniques of the released oligosaccharides was reported. Sophisticated analytical methods for their structural analysis, such as Dionex high-pH HPAEC-HPLC, HPAEC-PAD, MALDI-TOF MS, MALDI-PSD MS, and 1D- and 2D-NMR spectroscopy [265, 276, 277, 284–287], has supplied more detailed information on the structural differences of XG observed with a variety of plant tissues of different origin.

4.3
Physicochemical Properties

4.3.1
Hydrodynamic Properties

Most of the numerous applications of tamarind seed XG, e.g., as thickening, stabilizing and gelling agent in food, in textile sizing and weaving, and as an adhesive or binding agent in industry have been related to the solution properties [288]. Despite these facts, few studies on the molecular as well as the rheological properties of this polysaccharide have been published. Light scattering [266] indicated single-stranded molecules of XG in solution. The marked stiffness of the chain (C_∞ 110) was ascribed to restriction of the motion of the cellulosic backbone by its extensive ($\sim 80\%$) glycosylation. In a further light-scattering study, it was shown that the relation between the thermodynamic and hydrodynamic behavior of the XG is different from that of Gaussian coils and is rather that of polymers, which are known to have high chain stiffness [289]. Small-angle X-ray scattering and synchrotron-radiation scattering experiments [290] confirmed that tamarind XG in aqueous solution consists of multi-stranded aggregates with a high degree of particle stiffness.

No reproducibility of molecular weight of the tamarind seed XG was achieved. Various M_w values, 115 000 g/mol [129], 880 000 g/mol (light scattering) [266] or even 2.5×10^6 g/mol [290] were reported. These variations might be explained by the strong tendency to self-association of the polysaccharide, or by using non-molecular solutions in which the materials were not fully solubilized. However, various preparation conditions and/or tamarind seed sources might contribute to the observed differences in molecular weight as well.

The XG from *Detarium senegalense* Gmelin, used as a thickener in food, has a similar structure as the tamarind seed XG. The main differences are in the lower ratio of galactose to xylose (0.46) and the higher ratio of XXLG to XXXG subunits (5.6) compared to the respective data for tamarind seed XG, which are 0.51 and 2.07 [291]. Recently, the hydrodynamic properties of tamarind seed XG have been studied by capillary viscometry and light-scattering techniques using the pressure-cell solubilization method [103, 104]. Using a more complete range of pressure-cell heating treatments, XGs were recharacterized. The Mark–Houwink and Flory exponents (0.67 and 0.51, respectively) of XGs were consistent with linear random coil behavior. However, the previously reported branched chain profile for *Detarium* gum [104] was not confirmed, although it was suggested that this was affected by the different mathematical evaluation methods.

4.3.2
Rheological Properties

The rheological behavior of storage XGs was characterized by steady and dynamic shear rheometry [104, 266]. Tamarind seed XG [266] showed a marked dependence of zero-shear viscosity on concentration in the semi-dilute region, which was similar to that of other stiff neutral polysaccharides, and ascribed to hyper-entanglements. In a later paper [292], the flow properties of XGs from different plant species, namely, suspension-cultured tobacco cells, apple pomace, and tamarind seed, were compared. The three XGs differed in composition and structural features (as mentioned in the former section) and molecular properties, determined by the MALLS technique. R_g increased with increasing degree of branching of the XG according to the restricted mobility of the molecules. Whereas solutions of the tamarind and apple pomace XGs were shear thinning, solutions of *Nicotonia* XG were considerably less viscous and displayed near-Newtonian behavior.

The solution behavior of XGs resembles that of GaMs. Precipitation of GaM by self-association occurred at degrees of Gal*p* substitution below 11% [175], and gelation of XG occurred at a similar degree of Gal*p* substitution (achieved by treatment with β-galactosidase) of the xylose branches [293]. The results of the comparative study [292] indicated the molecular weight to be the major factor influencing the solution behavior of the tested XGs, whereas difference in the types of side chains played a minor role.

An important property of storage XGs is their gelling ability. Tamarind seed XG dissolves in water, yielding highly viscous solutions, but no gelation was observed unless other substances are added. Gelation occurred by removing $\sim$ 35% of the Gal*p* residues by fungal β-galactosidase [283]. The gel strength became greater with increasing removal of galactose from XG. This gel had the unique property of forming a gel on heating and reverting to a sol state on cooling. The phase transition between sol and gel is reversible and depends on the degree of Gal*p* removal. Gelation was believed to be induced by the association of main chains through hydrophobic binding. In the presence of ethanol, the XG gels were reported to gel at a lower temperature [294]. This thermoreversible gelation was studied by time-resolved small-angle X-ray scattering and the results evaluated on basis of a two-phase concept model, i.e., single chains (dilute phase) and domains with random aggregates (condensed phase). Complete dissolution of the aggregates corresponded to the gel–sol transition. The gelation mechanisms were assumed to be different. The cross-linking domains in the enzymically formed gels are composed of aligned, stripped XG chains in the shape of flat plates, while no ordered structure was found in the ethanol-formed gels. In the last case, the cross-linking domain seems to be formed by random aggregation of XG chains due to poor solubility in ethanol.

4.3.3
Interactions

Single-phase gels were prepared from XG/gellan mixtures [295]. Both polymers alone did not form a gel at a concentration below 0.75% and 0.5%, respectively, while a mixture of 0.05% gellan and 0.07% XG formed a single-phase gel demonstrating synergistic interactions. XG increased the effective concentration of gellan in the mixture, presumably via volume exclusion effects.

Tamarind seed XG has been used to affect the rheological properties of other hydrocolloids such as starch [296, 297]. Rheological experiments and

DSC revealed that the XG and starch did not interact synergistically and hence did not promote the formation of three-dimensional network structures. However, the hydrocolloid significantly decreased the retrogradation and syneresis of the starch paste, particularly in blends with a starch/XG ratio of 8.5/1.5. Mixing 1% or 2% tamarind XG with 9% cornstarch resulted in an increase in the paste viscosity from 385 to 460 and 560 BU (Brabender units), respectively [298]. The XG is associated with starch, as was evident from the lowering of the pasting temperature and the synergistic increase in pseudoplasticity and yield value of the blend pastes. However, carboxymethylated and hydroxypropylated XGs showed a diminished interaction.

In a current rheological study [296], the galactoxyloglucan from *Hymenia courbaril* was mixed with starch containing 66% amylose and with waxy corn starch (amylopectin). The gel mixtures showed, under static rheological conditions, an increase in paste viscosity compared to those of the polysaccharides alone. Dynamic rheometry indicated that the interactions resulted in increased thermal stability of the gel formed in comparison to that of the starch alone.

4.3.4
Functional Properties and Application Potential of Xyloglucans

The tamarind seed XG exerts several biological activities, such as marked inhibitory effect on the binding of BK virus to cells [299] and immunomodulatory effects [300]. Recently, it was reported [301] that XG affects the proliferation of cytokines in various skin-cell lines (Fig. 14) such as HaCaT cells (im-

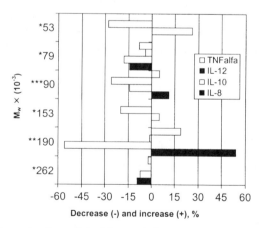

Fig. 14 Effect of the molecular weight of tamarind seed xyloglucan depolymerized by (*) γ-irradiation, (**) ultrasonication, and (***) endo-glucanase treatment on the production of various cytokines (Tumor necrosis factor α, TNF-α; Interleukin 8, IL-8; Interleukin 10, IL-10; and Interleukin 12, IL-12) in HaCaT cells (Immortalized keratinocytes line) [301]

mortalized keratinocytes). The authors suggested the potential application of XG in protection of skin against UV irradiation. XG was successfully applied as a hydrocolloid component of mucoadhesive buccal patches for controlled drug release [302]. Thermoreversible XG gels are useful as vehicles for oral and rectal drug delivery [303–306]. Moreover, the gelling behavior of XG was evaluated for the production of in situ gelling XG formulations for sustained-release ocular drug delivery [307]. Tamarind seed XG is also applicable as a binder for wet granulation and direct compression tableting methods [308]. Due to the adhesive properties of seed storage hemicelluloses, their application as wet-end additives in paper-making was recently proposed [309]. Structurally well-defined oligosaccharides from purified tamarind XG were supposed to be useful in the identification of reaction products from various enzyme assays and as possible acceptor molecules for these enzymes [287].

5
Mixed-linkage β-D-glucans

5.1
Occurrence and Structure

5.1.1
Sources

The mixed-linkage $(1 \rightarrow 3,1 \rightarrow 4)$-$\beta$-D-glucans ($\beta$-glucans) are unique to the Poales, the taxonomic order that includes the cereal grasses. β-Glucans are hemicellulose components of cereal grains, where they are located in the subaleurone and endospermic cell walls [6, 44, 310]. They are the principal molecules associated with cellulose microfibrils during cell growth [311]. Of commercial importance are oat and barley, which usually contain around 3–5% β-glucan, but some oat cultivars contain as much as 6–7% in the groat and some barleys even 12% or more [312–315]. β-Glucans have been reported [6, 316] to be present in non-endospermic tissues of gramineous monocotyl plants as well. A structurally similar, mixed-linkage β-D-glucan (lichenan) is commonly found in the lichen *Cetraria islandica* [317] differing from the cereal β-glucans in the fine structure. It consists of cellotriosyl (78%), cellotetraosyl (4%) and longer cellulose-like (18%) segments.

The scientific interest in cereal β-glucans arose partly from the problems they cause in brewing and animal-feed industries in the case of barley [318] and partly from the health benefits, such as cholesterol reduction [319–321], regulation of postgrandial serum glucose levels in humans and animals [319, 322], and immunostimulatory activity [323, 324]. Some of these activities have been observed with both oat and barley β-glucans [325].

5.1.2
Isolation

The isolation processes for hemicelluloses from cereal tissues need to include steps for the removal of starch, lipids, proteins and/or low-molecular phenolics, and are therefore more complicated than those applied to woody tissues. For extraction of β-glucans from barley and oat, the key methodologies have been developed and the influence of different solvents on the recovery and viscosity of the biopolymers has been investigated [326–328]. The extraction efficiency was 39–59%. Although lowest in yield, fractions isolated under alkaline conditions had the highest β-glucan content. In recent studies on barley β-glucans [329, 330], it was shown how the extraction conditions influenced not only the yield and composition of the polymers but also their molecular weight.

The differences in extractability of the barley β-glucans have been related to the existence of physical interactions between the β-glucan and AX components in the endospermic cereal cell walls [331]. In detailed studies on malting barley, water-extractable hemicelluloses were released from the grist by sequential extraction with water (40 °C), followed by digestion of residual starch at 65 °C using thermostable α-amylase [42]. The total yield of the recovered hemicelluloses from the obtained extracts was low, comprising 2.7% of the dry barley grist. The majority of hemicelluloses was extractable by the subsequent sequential extraction of the treated grist with saturated Ba(OH)$_2$, distilled water, and 1 M NaOH [43], comprising together 5.9% of the barley grist. The hemicellulose fractions obtained contained mainly a mixture of both β-glucan and AX, which could be in part separated by stepwise precipitation at increasing saturation level of the salt (NH$_4$)$_2$SO$_4$ [332]. Although it represents a powerful purification method, some of the isolated β-glucan preparations were still contaminated by other polysaccharides. By a freeze-thaw cycling process, giving gelatinous or fibrous precipitates, highly pure barley β-glucans were prepared [333].

5.1.3
Structure

The structural characteristics of cereal β-D-glucans have been intensively investigated during the last decade using effective analytical methods such as enzymic hydrolysis in combination with HPLC, HPAEC, GC-MS, and 2D-NMR spectroscopy [310, 334–337]. The NMR methods allowed the estimation of the ratio of β-(1 → 3) to β-(1 → 4) linkages as well as the presence of contaminating carbohydrates by nondestructive treatment in a relatively short time. Thus, the primary structure of the β-glucans has been well described. This unbranched homopolymer contains about 70% (1 → 4)-linked and 30% (1 → 3)-linked D-Glc$p$ residues. Blocks of (1 → 4)-linkage sequences (cel-

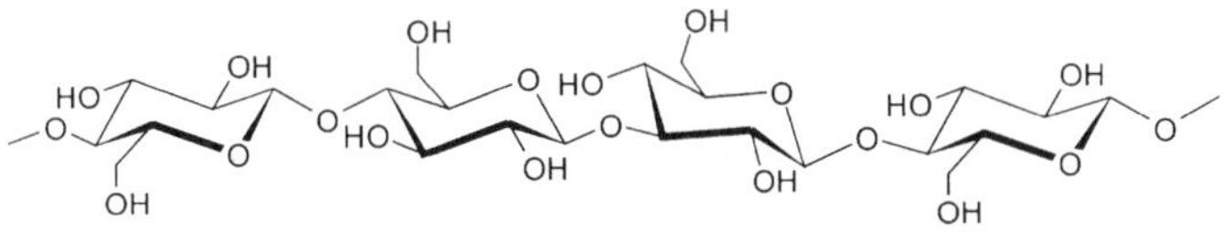

Fig. 15 Primary structure of mixed linkage $(1 \rightarrow 3, 1 \rightarrow 4)$-D-glucan ($\beta$-glucan)

lotriosyl and cellotetraosyl cellulose-like segments) are separated by single $(1 \rightarrow 3)$-linkages (Fig. 15). However, there is evidence for a minor amount of $(1 \rightarrow 4)$-sequences longer than the tetraose type (up to 9–14 Glc$p$ residues were determined), whereas the presence of contiguous β-$(1 \rightarrow 3)$-linked Glc$p$ units has not been detected [335]. The fine structure comprising the arrangement of the cellulose-like segments along the chains was suggested to be random. The molar ratio of cellotriosyl/cellotetraosyl segments have been reported to range between 2.1–2.4 for oat, 2.8–3.3 for barley, and 4.2–4.5 for wheat [317]. Enzyme digestion showed differences in the fine structure among different genera of cereals, but no great variations between samples belonging to the same genera.

The β-glucan-rich fractions from barley [42, 43] showed significant differences in the ratios of $(1 \rightarrow 4)$ to $(1 \rightarrow 3)$ linkages, molecular size, and $[\eta]$ values. In comparison to the water-extractable β-glucan-rich fractions, the alkali-extractable ones were characterized by high ratios of cellotriosyl/cellotetraosyl units and large amounts of long, contiguously linked $(1 \rightarrow 4)$-linkage segments. Such polymers exhibit a tendency for inter-chain aggregation through strong hydrogen bonding along the cellulose-like regions and hence lower solubility. Computer models demonstrated that randomly dispersed $(1 \rightarrow 3)$-β-Glc$p$ units among the cellotriosyl and cellotetraosyl segments confer increased flexibility to the molecular form [310]. The appearance at very high hydrodynamic volumes of molecular components in the SEC chromatograms of some fractions indicated the formation of aggregates. A higher proportion of cellotriosyl segments, suggested to impose some conformational regularity in β-glucans [338], may consequently result in a higher degree of organization in the cell walls.

5.2
Physicochemical Properties

5.2.1
Molecular Weight

The technological troubles of barley β-glucan and the health benefits of cereal β-glucans result, in general, from their ability to form highly viscous solutions and gels. The gel formation is controlled by the molecular weight as well as the solubility of the polysaccharides in aqueous media, which is governed by the structure, i.e., size, proportion, and distribution of the cellulose-like oligomeric blocks [335]. The M_w values reported in the literature for the cereal β-glucans ranged between $0.065-3 \times 10^6$ g/mol for oat, $0.15-2.5 \times 10^6$ g/mol for barley, and $0.25-0.7 \times 10^6$ g/mol for wheat [335 and references therein, 339, 340]. In contrast to the structural features, substantial differences were observed in the molecular weight and viscosity of β-glucans isolated from several barley genotypes [341]. The large variations reflect the diversity of botanic origin, but might also result from the methodology of molecular-weight determination.

The usually applied SEC techniques using pullulan standards yield overestimated values due to the differences in the hydrodynamic properties of both polysaccharides. More acceptable for calibration are β-glucan standards with M_w values determined by light-scattering techniques [342, 343]. For a rapid and simple characterization of the molecular weight of β-glucans, the determination based on $[\eta]$ measurements is advisable as the Mark–Houwink relation affords the M_w estimate. However, $[\eta]$ is affected by the solvent type and temperature, as shown for barley β-glucan [318, 344, 345]. Recently, the calculation methods of $[\eta]$ for a series of commercial β-glucan standards and various barley β-glucans were evaluated [346]. The Mark–Houwink relationships obtained were found to be different from previously reported findings.

A basic requirement in the molecular-weight determination is the full solubility of polysaccharides in appropriate solvents and preparation of solutions, which after dilution contain separated macromolecular chains. This is a crucial problem in the case of the cereal β-glucans. Various reports have described the tendency of β-glucans to associate and form aggregates [317, 347], which were detected by the fluorimetric Calcofluor-FIA method [348]. The aggregates of dissolved β-glucan extracted from beer, observed by light scattering and viscometry, were proposed to be fringed micelles, formed by side-to-side aggregation of chains of M_w about 175 000 g/mol [318].

The problem of characterizing the solution properties of polysaccharides is long-standing. To overcome this fact, the pressure-cell solubilization method was developed for starch solutions [349, 350] and applied with success to hemicellulosic polysaccharides, namely, GaMs [228, 229] and XGs [103, 104]. Suitable conditions of autoclaving were elaborated, based on the effect of the treatment on the dispersability and stability of various polysaccharides in dilute aqueous solutions [351]. However, the problem in autoclaving is that the heating and cooling cycles are relatively long and the precise time of treatment is difficult to control. Good dispersability of a high-molecular-weight β-glucan without degradation was achieved within 4–10 min by microwave heating in a high-pressure vessel [342]. The sample recovery increased from $75 \pm 5\%$ (classical heating under stirring) to $98 \pm 2\%$. However, longer times and higher temperatures induced polymer degradation.

5.2.2
Rheological and Gelling Properties

Knowledge on the flow behavior and potential gelling properties of cereal β-glucans is of special interest in the brewing process, but also important in the elucidation of their physiological effects. Several studies concerning these properties have appeared during the last years [335, 352–356]. The viscoelastic and flow behavior of barley β-glucans [357] indicated the formation of network-like structures under certain circumstances and supported the assumed aggregation tendency of these polymers even in the dilute solution regime, where no overlapping of the chains is expected. Interestingly, the studies of the viscoelastic behavior of various oat β-glucans revealed [358] that only for samples of low molecular weight the solutions display gel-like properties owing to self-association through cellulose-like sequences. It is to be stressed that the preparation of samples and their thermal history have proved to be important factors in the development of associated structures, both affecting the results.

Most recently, Vaikousi et al. [340] reported a more detailed study on the flow behavior and gelling properties of water-soluble barley β-glucans varying in molecular size. The results indicated that, on storage, the low-molecular-weight samples showed unusual shear-thinning, which was explained by aggregation tendency. The β-glucan solutions could undergo sol $\rightarrow$ gel transitions and the gelation rates were inversely proportional to the molecular size. Extremely rapid gelation was observed with samples of high cellotriosyl/cellotetraosyl ratios and very low molecular weight, implying that these features govern the formation of ordered structures. In contrast to small-deformation mechanical tests, the compressing testing showed stronger gels with increasing molecular weight. At the same time, Lazaridou et al. [335] published the effects of fine structure and molecular size on the rheological properties in the solution and gel state of cereal β-glucans in comparison to lichenan (the β-glucan from lichens). With increasing ratio of cellotriosyl/cellotetraosyl segments of the cereal β-glucans (2.12–3.66) and molecular size, the strength of the gels increased, while its brittleness decreased. As shown in Fig. 16, lichenan gel (lic100), compared to the gels prepared from oat, barley, and wheat β-glucans, gives a sharper transition, implying a more cooperative process as expected from its fine structure (ratio of cellotriosyl/cellotetraosyl moieties = 24/49) [317].

An undervalued factor is the microstructure, which the polysaccharides receive during the various isolation and drying processes. Depending on the methods, which are most frequently freeze-drying or spray-drying of aqueous dispersions and solvent exchange from aqueous media at various values of pH, interparticle hydrogen bonds of different strength are formed [359]. These might resist redissolution of the polysaccharide and some of them also resist treatment with aprotic solvents. As the microparticles are not fully sol-

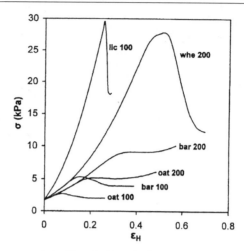

Fig. 16 Compression stress (σ) –"Hencky" strain (ε_H) curves for gels of β-glucan samples from oat: oat100 (2.12) and oat200 (2.13), barley: bar100 (2.80) and bar200 (3.04), wheat: whe200 (3.66), and lichenan lic100 (24.49) with cellotriosyl/cellotetraosyl ratios given in parenthesis. Polymer concentration, 8% w/v and gel curing temperature 25 °C

vated, supramolecular aggregated species remain present even in very dilute solutions.

5.3
Functional Properties and Application Potential of Cereal β-glucans

The increasing interest in β-glucans during the last two decades is due to their physical and physiological properties, which are of commercial as well as of nutritional importance. They have been accepted as functional, bioactive ingredients. In a recent review [314], the relationships between the solution properties of cereal β-glucans and observed physiological effects have been discussed. The author concluded that the efficacy of various soluble-fiber preparations could not simply be related only to the amount of β-glucan, as there are presently insufficient data to allow adequate evaluation. In accord with these facts, a more detailed investigation on the dilute and semi-dilute solution properties of β-glucans derived from oat bran has been carried out [354].

Interactions of β-glucans with various polysaccharides have been investigated during the last decade, very frequently with a wide range of cereal flour starch [326, 360–363] and also with amylodextrin [364]. Substitution of 5% wheat starch with β-glucan resulted in reduced enthalpy of starch gelatinization and a reduced degree of starch granule swelling [326]. The studies have revealed that variations in starch-pasting characteristics (both positive and negative) are dependent on the hydrocolloid type, starch source, concentration and method of measurement.

Due to the viscosity-enhancing effect of the cereal β-glucans as well as the fact that they are able to gel under certain conditions [364], they are applicable as food hydrocolloids. Cereal β-glucans can be utilized as thickening agents to modify texture and appearance in gravies, salad dressings, and ice cream formulations [335], and for the stabilization of emulsions and foams [365]. Most recently, the influence of various β-glucan preparations differing in molecular weight on the perception of mouth-feel and on rheological properties in beverage prototypes was tested [366]. They were also found to be applicable as fat mimetics in the development of low-calorie food [335, 367–369]. A suitable extraction method was developed [370] to produce stable antioxidants from oat fiber exploiting its hemicellulose component as an encapsulating and protective hydrocolloid material. The film-forming properties of barley and oat β-glucan extracts are of interest in the production of new biomaterials [371].

6
L-Arabino-D-galactans

6.1
Occurrence

Arabinogalactans (AGs) are widely spread throughout the plant kingdom. Many edible and inedible plants are rich sources of these polysaccharides. AGs occur in two structurally different forms described as type I and type II, associated with the pectin cell-wall component by physical bonds and some of them are covalently linked to the complex pectin molecule as neutral side chains. Commercial pectins always contain AG ($\sim$ 10–15%). AG of type I has a linear (1 → 4)-β-D-Gal*p* backbone, bearing 20–40% of α-L-Ara*f* residues (1 → 5)-linked in short chains, in general at position 3. It is commonly found in pectins from citrus, apple and potato [6]. Recently, this AG type has been isolated from the skin of *Opuntia ficus indica* pear fruits [372].

AG type II, known as arabino-3,6-galactan, has a (1 → 3)-β-D-Gal*p* backbone (Fig. 17) heavily substituted at position 6 by mono- and oligosaccha-

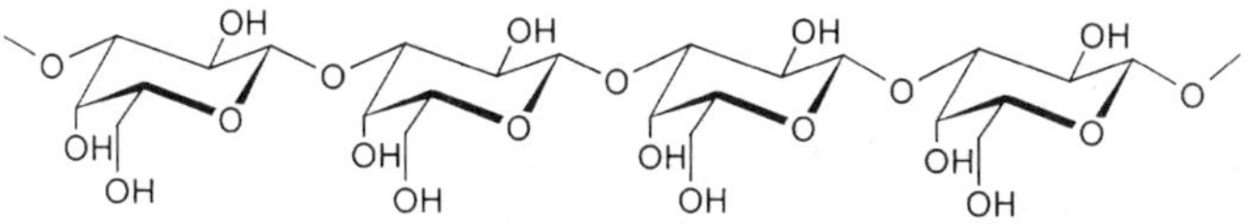

Fig. 17 Structure of the main chain of (1 → 3, 6)-D-galactan (AG type-II)

ride side chains composed of arabinosyl and galactosyl units. It is more widespread than the AG type I and occurs in cell walls of dicots and cereals often linked to proteins (known as arabinogalactan proteins) [373 and references therein]. In some case it contains a small amount of β-GlcA. AG type II comprises, together with GaMs and cellulose, the predominating polysaccharides of green arabica coffee beans [374]. As a component of pectin, it has been very frequently extracted from herbal plants [375, 376]. The acidic AG type II has uronic acid residues incorporated in side chains and belong to the groups of gum exudates [377].

AG type II is most abundant in the heartwood of the genus *Larix* and occurs as minor, water-soluble components in softwoods. Certain tree parts of western larch (*L. occidentalis*) were reported to contain up to 35% AG [378]. The polysaccharide is located in the lumen of the tracheids and ray cells. Consequently, it is not a cell-wall component and, by definition, not a true hemicellulose. However, it is commonly classified as such in the field of wood and pulping research. This motivated us to include the larch AG in the review.

During the last decade, the larch AG, known as larch gum in food applications, has produced emerging commercial and scientific interest, which follows closely upon recent reports related to the beneficial physiological effects of the commercial larch AG and its immunomodulatory properties [379, 380].

6.2
Structural and Physicochemical Properties of Larch Arabinogalactan

Although the general structural features of larch AG have been extensively investigated in the past [6], many uncertainties regarding the molecular weight and fine structure concerning the location, composition, and length of side chains, remained unresolved. These facts have justified reinvestigations, which started about ten years ago.

First attention was paid to the molecular properties. In earlier papers, two molecular weight components, one with M_w values in the range 7500–18 000 g/mol and the second with M_w in the range 37 000–100 000 g/mol, have been reported for AGs from western larch (*L. occidentalis*) [6]. However, only the smaller component was found in AGs from *L. laricina* (tamarack), *L. sibirica* (Siberian) [377] and *L. dahurica* (Mongolian larch) [381]. These variations in molecular weight might be attributed to the different isolation procedures and analytical methodology. Gel filtration as well as light scattering has been used to determine the molecular weight. The first method gave apparent molecular weight values (M_{app}), which depend on the calibration standards used, usually pullulans and dextrans with hydrodynamic properties differing greatly from those of the highly branched AG. A further problem was shown [382] to be the presence of a small proportion of uronic acid units in AG from larch species, which causes non-size-exclusion behavior of the polysaccharide when eluents of low ion strength are used. The resulting chromatograms may contain shoulders or several peaks instead of one peak. Eremeeva and Bykova [383] described the consequences of this effect in their SEC study on Siberian larch AG, but speculated that it was due to intramolecular electrostatic effects rather than to ion exclusion.

Gel filtration of the crude larch arabinogalactan on Sephadex G-75 yielded a series of fractions with M_{app} ranging between 3000 and 93 000 g/mol [384]. Interestingly, these fractions show a decrease of the Gal/Ara molar ratio from 6.9 : 1 to 2.3 : 1, whereas the Gal/Ara ratios of two nearly monodisperse fractions, AG37 (37 000 g/mol) and AG9 (9000 g/mol), isolated by gel-filtration techniques from the crude AG, were about the same (5.4 and 5.7, respectively) [385]. The third fraction AG3 (3000 g/mol) was proofed by a sensitive immunoassay to be an AG, however, with a higher amount of arabinose containing side chains [384]. The fraction AG37, subjected either to autoclaving at 121 °C or exposure to alkali in the presence of sodium borohydride, was reported [385] to break down into fragments of M_{app} ~ 9000 g/mol with a uniform molecular-weight distribution (M_w/M_n of 1.08–1.17). The extent of conversion was strongly dependent on the degradation conditions. The M_w values determined by GFC/LLS, sedimentation equilibrium, and mass spectrometry were in close agreement. Sugar composition, linkage analysis as well as ^{13}C-NMR spectroscopy revealed that the primary structural features of the parent and degraded AG samples were nearly identical.

The fragmentation process of larch AG, when exposed to mild alkali or heat, was investigated by NMR spectroscopy [386]. SEC analysis showed that smaller AG species undergo an order–disorder transition in preference to larger ones, and that the transition can be reversed by drying from or freezing an aqueous solution of the fragments [387]. The authors suggested that the fragmentation was a dissociation of the molecular assembly (multiplex) accompanied by an order–disorder transition, analogous to the known triple helix–single chain random coil transition of $(1 \rightarrow 6)$-branched $(1 \rightarrow 3)$-β-glucans such as scleroglucan [388]. A molecular modeling study [389] revealed that the $(1 \rightarrow 3)$-β-D-galactan can adopt a triple helical structure similar to that of the corresponding $(1 \rightarrow 3)$-β-D-glucan [388] and can accommodate a highly flexible β-D-Gal$p$-$(1 \rightarrow 6)$-β-D-Gal$p$ disaccharide moiety as a side group, which is 6-linked to nearly every Gal$p$ unit in the main chain. The resulting triple helix was reported to be applicable to western larch arabinogalactan. Depending on the side groups quite different morphologies can be assumed. Preliminary X-ray fiber diffraction data supported these suggestions.

To elucidate structural discrepancies of larch AG, the degradation in mild alkali, known as the peeling-off reaction, was applied [390]. The method is well suited to structural analysis of this polysaccharide, because of the rapid peeling of the $(1 \rightarrow 3)$-galactan chain from the reducing end by sequential β-alkoxycarbonyl elimination followed by rearrangements of the former reducing end group into 3-deoxyaldonic acid (metasaccharinic acid). The separated products were characterized by compositional and methylation analyses as well as by means of NMR spectroscopy. The results confirmed the earlier determined structural features of the larch AG:

(i)   most main-chain Gal$p$ residues carry a side chain on O-6,
(ii) about half of the side chains are β-(1 → 6)-linked Gal$p$ dimers,
(iii) about a quarter are single Gal$p$ residues,
(iv) the rest contain three or more Gal$p$ residues and include most of the arabinose present,
(v)   arabinose occur as single α-Ara$f$ units and dimeric side chains (14).

$$\beta\text{-L-Ara}p\text{-}(1 \to 3)\text{-}\alpha\text{-L-Ara}f\text{-}(1 \to \tag{14}$$

The authors suggested that larch arabinogalactan is composed of repeating units of similar molecular weight and composition, joined through a yet undetermined linkage, which is susceptible to cleavage at low alkali concentrations and moderate temperatures. These suggestions as well as that about the location of the Ara$f$ residues and side chains (14) in the proposed structure of the AG molecule (15) are rather speculative and need further experimental proof.

$$\beta\text{-D-Gal}p\text{-}(1 \to 3)\text{-}\beta\text{-D-Gal}p\text{-}(1 \to 3)\text{-}\beta\text{-D-Gal}p\text{-}(1 \to 3)\text{-}\beta\text{-D-Gal}p\text{-}(1 \to$$

```
β-D-Galp-(1 → 3)-β-D-Galp-(1 → 3)-β-D-Galp-(1 → 3)-β-D-Galp-(1 →
6                           6                 6
↑                           ↑                 ↑
R              R-(1 → 3)-β-D-Galp        β-D-Galp
                            6                 6
                            ↑                 ↑
                            R              β-D-Galp
                                                              (15)
```

R: H or β-D-Gal$p$(1 →
or α-L-Ara$f$-(1 →
or β-L-Ara$p$-(1 → 3)-α-L-Ara$f$-(1 →)

In another paper of the series, devoted to renewed chemical investigation of the larch arabinogalactan, the polymeric products of the partial acid hydrolysis were studied using compositional, linkage and ^{1}H-NMR spectroscopy in combination with SEC in low- and high-ionic strength solvents [391]. As well as confirming the earlier known structural features of AG, interesting results were presented. As nearly all arabinosyl units were removed by the acidic treatment, very pure galactans with different structures were obtained, depending on whether the starting material was in the ordered or disordered form. From the ordered form, AG contains relatively long main chains that bear a high proportion of non-reducing Gal$p$ end-groups. Such products could have potential in medical and biomedical applications involving interaction with galactose-specific binding sites in the human body [386, 392, 393].

The polymeric products from disordered AG have much smaller molecular weights and reduced tendency to self-associate in solution. Such properties may be desirable in applications requiring membrane permeability.

6.3
Functional Properties and Application Potential

A preliminary study of the use of larch AGs in aqueous two-phase systems [394] revealed that this polysaccharide provides a low-cost alternative to fractionated dextrans for use in aqueous two-phase, two-polymer systems with polyethylene glycol (PEG). The narrow molecular-weight distribution (M_w/M_n of 1–2) and low viscosity at high concentration of AG can be exploited for reproducible separations of proteins under a variety of conditions. The AG/PEG systems were used with success for batch extractive bioconversions of cornstarch to cyclodextrin and glucose.

In addition to thickening and the effect on perceived texture, AGs from various plant sources are gaining attention as potential drug carriers [387, 395–397]. Purified AG from *L. occidentalis* as well as its low-molecular-weight fractions might be suitable carriers for targeted drug delivery [385]. It can also be used as a coating material for the preparation of superparamagnetic iron oxide particles used in liver magnetic resonance imaging [398].

7
Hemicellulose Derivatives and their Applications

7.1
Xylan Derivatives

Due to the lack of a commercial supply, as well as their usually low molecular weight and poor solubility, xylans have found little industrial utility and interest in their modification has been rather low in comparison to commercially available polysaccharides such as cellulose or starch. With the aim of improving the functional properties of xylans and/or imparting new functionalities to them, various chemical modifications have been investigated during the past decade. Most of them were presented in recent reviews [3, 399].

Partial etherification of the beech wood MGX with *p*-carboxybenzyl bromide in aqueous alkali yielded fully water-soluble xylan ethers with DS up to 0.25 without significant depolymerization; the M_w determined by sedimentation velocity was 27 000 g/mol [400, 401]. By combination of endo-β-xylanase digestion and various 1D- and 2D-NMR techniques, the distribution of the substituents was suggested to be blockwise rather than uniform. The derivatives exhibited remarkable emulsifying and protein foam-stabilizing activities. By reaction of the beech wood xylan and its sulphoethyl derivative with 1-bromododecane in DMSO at moderate reaction temperatures, further amphiphilic derivatives with remarkable tensioactive properties were prepared indicating their potential as biosurfactants [402].

Cationic groups are usually introduced into xylan, similarly as to cellulose or starch, by the formation of ether bonds applying various reagents (Fig. 18). In previous studies [399], investigations were directed towards xylan-rich waste materials such as hardwood sawdust, corncobs, and sugar cane bagasse by reaction with 3-chloro-2-hydroxypropyltrimethylammonium chloride (CHTMAC). Subsequent extraction steps with water and dilute alkali lead to fractionation into trimethylammonium-2-hydroxypropyl (TMAHP) cellulose, TMAHP-hemicellulose, and lignin. More recently, the cationization of xylans, isolated from beech wood, corncobs, rye bran and the viscose spent liquor was investigated [403]. The results indicated that the DS depended on the molar ratios of CHTMAC/xylan and NaOH/CHMTAC as well as on the xylan type used. The functionalization pattern of the cationized xylans with DS 0.25–0.98 was characterized by ^{13}C-NMR spectroscopy after the hydrolytic chain degradation [404]. Monosubstitution of xylose units was found mainly to occur at lower DS, where position 2 is preferred, whereas at higher DS the regioselectivity is lost.

The TMAHP-MGX isolated from cationized aspen sawdust was reported to be applicable as a beater additive; it significantly increased the tear strength of bleached spruce organosolv pulp [3]. The TMAHP derivatives prepared from isolated xylans were shown to improve the paper-making properties and

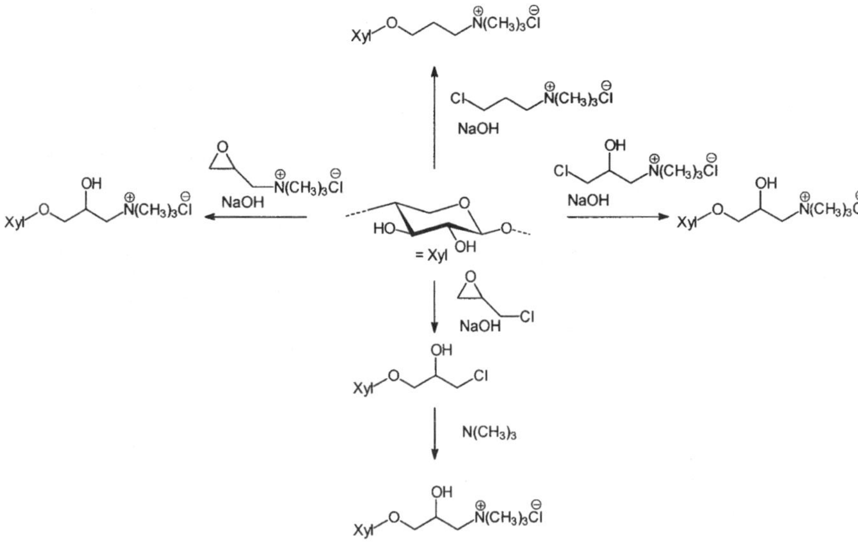

Fig. 18 Reaction pathways for the introduction of cationic moieties into xylan

act as flocculants for pulp fibers at very low additions (∼ 0.25%), very probably due irreversible adsorption onto cellulose fibers [405]. The derivatives exhibit antimicrobial activity against some Gram-negative and Gram-positive bacteria, depending on the DS and xylan type [406].

Carboxylic acid esters of xylan are prepared under typical conditions used for polysaccharide esterification, i.e., activated carboxylic acid derivatives are allowed to react with the polymer both heterogeneously and homogeneously (Fig. 19). Heterogeneous esterification of oak wood sawdust and wheat bran hemicelluloses with excess octanoyl chloride (without solvent) was described [407]. The separated liquid fraction contained, as well as lignin, degraded esterified hemicelluloses. Homogeneous acylation of xylan-rich hemicelluloses has mostly been carried out in DMF in combination with LiCl [408]. Under moderate reaction conditions, acylated derivatives with DS ranging from 0.18 to 1.71 were prepared from MGX of poplar wood [409] and wheat straw AGX [410]. The DMF/LiCl solvent was also used for acetylation of hemicellulose fractions isolated from wheat straw and poplar also in the presence of dimethylaminopyridine, DMAP, as a catalyst [411, 412]. SEC measurements revealed that polymer degradation was low at reaction temperatures below 80 °C. The conversion of wheat straw and bagasse hemicelluloses with succinic anhydride in aqueous alkaline solutions yielded carboxyl groups containing derivatives with DS < 0.26 [413]. Application as thickening agents and metal-ion binders were proposed for these derivatives. Various catalysts have been used to accelerate acetylation, succinylation and oleoylation of wheat straw and bagasse hemicelluloses [414, 415].

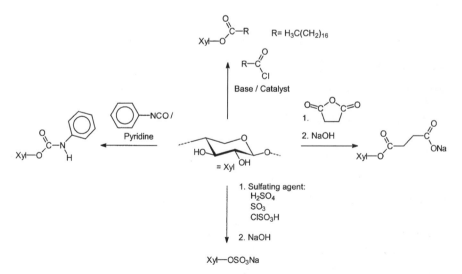

Fig. 19 Introduction of ester groups into xylan

By the treatment of oat spelt xylan with phenyl or tolyl isocyanate in pyridine the fully functionalized corresponding carbamates were prepared [416]. Xylan 3,5-dimethylphenylcarbamate showed higher recognition ability for chiral drugs compared to that of the same cellulose derivative [417].

In a most recent paper [418] the preparation of corn fiber arabinoxylan esters by reaction of the polymer with C_2-C_4 anhydrides using methanesulfonic acid as a catalyst is described. The water-insoluble derivatives with high molecular weight showed glass-transition temperatures from 61 to 138 °C, depending on the DS and substituent type. The products were thermally stable up to 200 °C. Above this temperature their stability rapidly decreased.

Xylan sulphates, known also as pentosan polysulphates (PPS), are permanently studied with regard to their biological activities [3, 419–422]. Usually, sulphuric acid, sulphur trioxide, or chlorsulphonic acid are employed as sulphating agents alone or in combination with alcohols, amines or chlorinated hydrocarbons as reaction media [423].

Thermoplastic xylan derivatives have been prepared by in-line modification with propylene oxide of the xylan present in the alkaline extract of barley husks [424, 425]. Following peracetylation of the hydroxypropylated xylan in formamide solution yielded the water-insoluble acetoxypropyl xylan. The thermal properties of the derivative qualify this material as a potential biodegradable and thermoplastic additive to melt-processed plastics. Xylan from oat spelts was oxidized to 2,3-dicarboxylic derivatives in a two-step procedure using $HIO_4/NaClO_2$ as oxidants [426].

The neutral X_m-type homoxylan from the seaweed *Palmaria decipiens* was oxidized with bromine solution [427]. Oxidation occurs preferentially at C-2, as shown by means of gel-permeation chromatography, GPC, after reductive cleavage, reduction, and peracetylation. Periodate oxidation of xylan-introduced dialdehyde functions, which gave ligands for the coordination of Cu(II) after reaction with *p*-chloroaniline [428]. Hydrophobic films were prepared from corn bran xylan in a two-step process [429]. The dialdehyde of the polymer prepared by periodate oxidation under controlled reaction conditions was subjected to reductive amination using laurylamine in aqueous medium. The DS of the dodecylamine-grafted xylan films ranged between 0.5 and 1.1, depending on the reaction conditions. The plastic behavior at ambient temperature correlated with the glass-transition temperature around – 30 °C. The product indicates potential application as a bioplastic [429].

7.2
Modification of Mannan-Type Hemicelluloses

The widely commercially exploited guar GaM has been the subject of some studies dealing with chemical or enzymic modifications aimed to extend the application range of this polysaccharide. Specific oxidation on the C-6 position of the Gal*p* side chain units was performed by β-galactosidase [241, 430].

Grafting of polyacrylamide onto guar gum [431] and *Ipomoea* gum [178] in aqueous medium initiated by the potassium persulphate/ascorbic acid redox system was performed in the presence of atmospheric oxygen and Ag^+ ions. After grafting, a tremendous increase of the viscosity of both gum solutions was achieved, and the grafted gums were found to be thermally more stable.

The hydroxypropyl derivative of guar GaM (HPG) was prepared with propylene oxide in the presence of an alkaline catalyst. HPG was subsequently etherified as such with docosylglycidyl ether in isopropanol and presence of an alkaline catalyst [432]. The peculiar features of the long-chain hydrophobic derivatives were ascribed to a balance between inter- and intramolecular interactions, which is mainly governed by the local stress field.

Due to the low solubility of the *Cassia tora* gum, composed mainly of GaM [181], the polymer was modified by graft copolymerization of acrylamide onto the gum [433, 434], cyanoethylation [435], and carboxymethylation [436]. The modified GaMs from *Cassia tora* seeds have been applied as a beater additive in paper-making [437]. The hydroxypropyl derivative of guar gum has already found application as a water-blocking agent in the formulation of explosive cartridge, as a processing aid in the mining and mineral industry, water-based paints, etc., as reviewed in [432].

From the GaM of the seed endosperm of *Adenanthera pavonina*, films were prepared by casting solutions of GaM/collagen blends and subsequent cross-linking with glutaraldehyde [438]. The thermal, dielectric, and piezoelectric properties of the films have been studied to develop new materials for electronic devices used for various biological applications. IR spectroscopy of the films indicated that no interaction or binding occurred between both polymers as the typical spectral bands showed no shift. The cross-linking reaction was found to decrease the swelling of the films as a function of the GaM concentration, which also influenced the dielectric properties.

The GGM-rich hemicelluloses, isolated from water-impregnated spruce chips by heat-fractionation [218], has been used as pre-polymers after modification with methacrylic functions [439]. Radical polymerization of the modified hemicelullose with 2-hydroxyethyl methacrylate in water yielded elastic, soft, transparent, and easily swollen hydrogels.

7.3
Xyloglucan Derivatives

From tamarind seed xyloglucan, carboxymethyl derivatives with different levels of DS were prepared in isopropanol medium [440]. Swelling power, solubility and tolerance to organic solvents of the derivatives increased with increasing DS. The interaction properties of the unmodified xyloglucan with calcium chloride and sodium tetraborate were found to be reversed upon carboxymethylation.

A range of derivatives of tamarind seed polysaccharide has been prepared, characterized and selected solution properties examined [259]. Following oxidation of the terminal Gal*p* residues with galactose-oxidase, subsequent oxidation and reductive amination have been used to prepare a range of carboxylated and alkylaminated derivatives, respectively. Sulfated derivatives have been prepared by reaction with a sulphur trioxide-pyridine complex in DMF. Based on the dependence of [η] on ionic strength, carboxylated and sulphated derivatives were found to have characteristically stiffened backbones as found previously for the native polysaccharide [266]. Binding of divalent cations to carboxylated derivatives is shown to be relatively weak, although polymer precipitation was noted in the presence of Pb^{2+}. Alkylaminated polysaccharides show only a modest decrease in surface and interfacial tension compared with the native polysaccharide, although significant foam formation and stabilization was found for a nonylaminated sample. Following enzymic depolymerization, this material showed a marked decrease in surface and interfacial tension suggesting that interfacial activity in alkylaminated tamarind polysaccharide is only apparent under disruptive solution conditions. The results of ^{1}H-NMR line-width and T1 measurements before and after depolymerization indicates that this is due to solution viscosity rather than specific interaction effects.

8
Concluding Remarks

The number of reports about hemicelluloses that have been covered by this review indicates the significantly increased importance of all types of hemicelluloses as plant constituents and isolated polymers during the last decade. Attention has been paid not only to known hemicelluloses but also to the primary structure, physicochemical, physical, and various functional properties of hemicelluloses isolated from hitherto uninvestigated plants. The efforts to exploit a variety of plant as potential sources of hemicelluloses were pointed out particularly for agricultural crops, wood wastes, as well as for by-products of pulp and rayon fiber technologies. Many studies were devoted to characterize seed-storage hemicelluloses from plants that have been traditionally applied in food and medicine of many underdeveloped countries to find substitutes for imported commercial food gums.

The structural varieties of hemicelluloses offer a number of possibilities for specific chemical, physical, and enzymic modifications. Future advancements will be based on the synthesis of hemicellulose-based polymers with new functionalities and with a well-defined and preset primary structure both on the level of the repeating unit and the polymer chain. Hemicelluloses have also started to be attractive to synthetic polymer chemists as sources for the production of new bioactive and biocompatible polymers and other biomaterials, and this interest is expected to increase in the future. The usefulness of hemicellulose-based or derived materials in an industrial and biomedical context is now beyond dispute. This will stimulate further research into the development of isolation and purification processes from natural sources, and the development of analytical tools for the structural characterization of the isolated hemicelluloses and their derivatives.

Acknowledgements The authors acknowledge the financial support of the research in the field of hemicellulose by the Slovak grant agency VEGA (project no. 2/7138), by the German BWMA/AiF–DGfH (project no. 13698 BR) as well as by the European Community program COST D29 (project no. D29/0008/03), and COST D28 (project no. D28/006/03). The authors are indebted to Dr. Katrin Petzold for proofreading of the manuscript.

References

1. Schulze E (1891) Ber Dtsch Chem Ges 24:2277
2. Fry SS (1989) J Exp Bot 40:1
3. Ebringerová A, Hromádková Z (1999) In: Harding SE (ed) Biotechnology and Genetic Engineering Reviews. Intercept, England, vol. 16, p. 325
4. Ebringerová A, Heinze T (2000) Macromol Rapid Comm 21:542
5. Gatenholm P, Tenkanen M (2004) (eds) ACS Symposium Series 864, Science and Technology. American Chemical Society, Washington
6. Stephen AM (1983) In: Aspinall GO (ed) The Polysaccharides, vol. 2. Academic, New York, p. 97
7. Yamagaki T, Maeda M, Kanazawa K, Ishizuka Y, Nakanishi H (1997) Biosci Biotech Bioch 61:1077
8. Yamagaki T, Maeda M, Kanazawa K, Ishizuka Y, Nakanishi H (1997) Biosci Biotech Bioch 61:1281
9. Deniaud E, Quemener B, Fleurence J, Lahaye M (2003) Int J Biol Macromol 33:9
10. Deniaud E, Fleurence J, Lahaye M (2003) J Physiol 39:74
11. Deniaud E, Fleurence J, Lahaye M (2003) Bot Mar 46:366
12. Shatalov AA, Evtuguin DV, Neto CP (1999) Carbohydr Res 320:93
13. Evtuguin DV, Tomás JL, Silva AMS, Neto CP (2003) Carbohydr Res 338:597
14. Teleman A, Tenkanen M, Jacobs A, Dahlman O (2002) Carbohydr Res 337:373
15. Teleman A, Lundqvist J, Tjerneld F, Stålbrand H, Dahlman O (2000) Carbohydr Res 329:807
16. Jacobs A, Palm M, Zacchi G, Dahlman O (2003) Carbohydr Res 338:1869
17. Habibi Y, Mahrouz M, Vignon MR (2002) Carbohyd Res 337:1593
18. Vignon MR, Gey C (1998) Carbohydr Res 307:107
19. Haq ON, Gomes J (1997) Bangladesh J Sci Ind Res 12:76
20. Ishurd O, Ali Y, Wei W, Bashir F, Ali A, Ashour A, Pan Y (2003) Carbohydr Res 338:1609
21. Dinand E, Vignon M (2001) Carbohyd Res 330:285
22. Igartuburu JM, Pando E, Luis FR, Gil-Serrano A (1998) J Nat Prod 61:876
23. Igartuburu JM, Pando E, Luis FR, Gil-Serrano A (2001) J Nat Prod 64:1174
24. Watanabe T, Mitsuishi Y, Kato Y (1999) J Appl Glycosci 46:281

25. Vierhuis E, Schols HA, Beldman G, Voragen AGJ (2000) Carbohydr Polym 43:11
26. Mafra I, Lanza B, Reis A, Marsilio V, Campestre C, De Angelis M, Coimbra MA (2001) Physiol Plantarum 111:439
27. Ray B, Loutelier-Bourhis C, Lange C, Condamine E, Driouich A, Lerouge P (2004) Carbohydr Res 339:201
28. Mais U, Sixta H (2004) In: Gatenholm P, Tenkanen M (eds), ACS Symposium Series 864, Science and Technology. American Chemical Society, Washington, p. 94
29. Bochicchio R, Reicher F (2004) Carbohydr Polym 53:127
30. Ebringerová A, Kardošová A, Hromádková Z, Hříbalová V (2003) Fitoterapia 74:52
31. Ebringerová A, Hromádková Z, Alföldi J, Hříbalová V (1998) Carbohyd Polym 37:231
32. Ishii T (1997) Plant Sci 127:11
33. Hromádková Z, Kovačiková J, Ebringerová A (1999) Ind Crop Prod 9:101
34. Schooneveld-Bergmans MEF, Beldman G, Voragen AGJ (1999) J Cereal Sci 29:63
35. Schooneveld-Bergmans MEF, Hopman AMCP, Beldman G, Voragen AGJ (1998) Carbohydr Polym 35:39
36. Huisman MMH, Schols HA, Voragen AGJ (2000) Carbohydr Polym 43:269
37. Delcour JA, Win HV, Grobet PJ (1999) J Agr Food Chem 47:271
38. Dervilly G, Leclercq C, Zimmermann D, Roue C, Thibault J-F, Saulnier L (2002) Carbohydr Polym 47:143
39. Vinkx CJA, Delcour JA (1996) J Cereal Sci 24:1
40. Voragen AGJ, Gruppen H, Verbruggen MA, Vietor RJ (1992) In: Visser J, Beldmann G, Kusters van Someren AS, Voragen AGJ (eds) Xylans and Xylanases. Elseveir Science, Amsterdam, p. 51
41. Izydorczyk MS, Biliaderis CG (1995) Carbohydr Polym 28:33
42. Izydorczyk MS, Macri LJ, MacGregor AW (1998) Carbohydr Polym 35:249
43. Izydorczyk MS, Macri LJ, MacGregor AW (1998) Carbohydr Polym 35:259
44. Roubroeks JP, Andersson R, Åman P (2000) Carbohydr Polym 42:3
45. Han J-Y (2000) Food Chem 70:131
46. Nandini CD, Salimath PV (2001) Food Chem 74:417
47. Rao MVSSTS, Muralikrishna G (2001) Food Chem 72:187
48. Nilsson M, Saulnier L, Andersson R, Åman PM (1996) Carbohydr Polym 30:229
49. Cleemput G, van Oort M, Hessing M, Bergmans MEF, Gruppen H, Grobet PJ, Delcour JA (1995) J Cereal Sci 22:73
50. Nilsson M, Andersson R, Andersson RE, Autio K, Åman PM (2000) Carbohydr Polym 41:397
51. Vinkx CJA, Stevens I, Gruppen H, Grobet PJ, Delcour JA (1995) Cereal Chem 72:411
52. Saulnier L, Marot C, Chanliaud E, Thibault J-F (1995) Carbohydr Polym 26:279
53. Samuelson AB, Lund I, Djahromi JM, Paulsen BS, Wold JK, Knutsen SH (1999) Carbohydr Polym 38:133
54. Fischer MH, Yu N, Gray GR, Ralph JR, Anderson L, Marlett JA (2004) Carbohydr Res 339:2009
55. Samuelsen AB, Cohen EH, Paulsen BS, Brull LP, Thomas-Oates JE (1999) Carbohydr Res 315:312
56. Kardošová A, Malovíková A, Rosík J, Capek P (1990) Chem Pap-Chem Zvesti 44:111
57. Kardošová A, Matulová M, Malovíková A (1998) Carbohydr Res 308:99
58. Kardošová A, Malovíková A, Pätoprstý V, Nosáľová G, Matáková T (2002) Carbohydr Polym 4:27
59. Jacobs A, Larsson PT, Dahlman O (2001) Biomacromolecules 2:979
60. Gruppen H, Komerlink FJM, Voragen AGJ (1993) J Cereal Sci 18:111

61. Viëtor RJ, Komerlink FJM, Angelino SAGF, Voragen AGJ (1994) Carbohydr Polym 24:113
62. Hoffman RA, Geijtenbeek T, Kamerling JP, Vliegenthart JFG (1992) Carbohydr Res 223:19
63. Puls J (1997) Macromol Symp 120:196
64. Dervilly-Pinel G, Tran V, Saulnier L (2004) Carbohydr Polym 55:171
65. Atkins EDT (1992) In: Visser J, Beldmann G, Kusters van Someren AS, Voragen AGJ (eds) Xylans and Xylanases. Elsevier Science, Amsterdam, p. 39
66. Yui T, Imada K, Shibuya N, Ogawa K (1995) Biosci Biotech Bioch 59:965
67. Attala RH, Hackney JM, Uhlin I, Thompson NS (1993) Int J Biol Macromol 15:109
68. van Hazendonk JM, Reinerink EJM, de Waard P, van Dam JEG (1996) Carbohydr Res 291:53
69. Sun RC, Fang JM, Goodwin A, Lawther JM, Bolton AJ (1998) Carbohydr Polym 37:351
70. Lawther JM, Sun RC, Banks WB (1996) J Appl Polym Sci 60:1827
71. Sun R, Lawther JM, Banks WB (1996) Carbohydr Polym 29:325
72. Sun RC, Hughes S (1998) Carbohydr Polym 36:293
73. Sun J-X, Sun RC, Sun X-F, Su YQ (2004) Carbohydr Res 339:291
74. Sun RC, Tomkinson J, Ma PL, Liang SF (2000) Carbohydr Polym 42:111
75. Bataillon M, Mathaly P, Cardinali A-PN, Duchiron F (1998) Ind Crop Prod 8:37
76. Methacanon P, Chaikumpollert O, Thavorniti P, Suchiva K (2003) Carbohydr Polym 54:335
77. Eriksson O, Lindgren BO (1997) Svensk Papperstidn 80:59
78. Wallace G, Russel WR, Lomax JA, Jarvis MC, Lapierre C, Chesson A (1995) Carbohydr Res 272:41
79. Lapierre C, Pollet B, Ralet M-C, Saulnier L (2001) Phytochemistry 57:765
80. Lam TBT, Kadoya K, Iiyama K (2001) Phytochemistry 57:987
81. Saulnier L, Vigouroux J, Thibault J-F (1995) Carbohydr Res 272:241
82. Ebringerová A, Hromádková Z, Berth G (1994) Carbohydr Res 264:9
83. Saulnier L, Andersson R, Åman P (1997) J Cereal Sci 25:121
84. Sun RC, Sun VF (2002) Carbohydr Polym 49:415
85. Gustavsson M, Bengtsson M, Gatenholm P, Glasser W, Teleman A, Dahlman O (2001) In: Chiellini E, Gil H, Braunegg G, Burchert J, Gatenholm P, van der Zee M (eds), Biorelated Polymers: Sustainable Polymer Science and Technology. Kluver Academic Plenum, Dordrecht, p. 41
86. Hromádková Z, Ebringerová A (2003) Ultrason Sonochem 10:127
87. Sun RC, Tomkinson J (2003) Eur Polym J 39:751
88. Sun RC, Tomkinson J (2002) Carbohydr Polym 50:263
89. Sun RC, Sun XF, Ma XH (2002) Ultrason Sonochem 9:95
90. Sun RC, Wang XY, Sun XF, Sun JX (2002) Polym Degrad Stabil 78:295
91. Glasser WG, Kaar WE, Jain RK, Sealey JE (2000) Cellulose 7:299
92. Shimizu K, Sudo K, Ono H, Ishihara M, Fujii T, Hishiyama S (1998) Biomass Bioenerg 14:195
93. Schooneveld-Bergmans MEF, Dignum MJW, Grabber JH, Beldman G, Voragen AGJ (1999) Carbohydr Polym 38:309
94. Jacobs A, Lundqvist J, Stålbrand H, Tjerneld F, Dahlman O (2002) Carbohydr Res 337:711
95. Faurot A-L, Saulnier L, Berot S, Popineau Y, Petit M-D, Rouau X, Thibault J-F (1995) Food Sci Technol 28:436
96. Delcour JA, Rouseu N, Vanhaesendonck IP (1999) Cereal Chem 76:1

97. Hespell RB (1998) J Agr Food Chem 46:2615
98. Saake B, Erasmy N, Kruse Th, Schmekal E, Puls J (2004) In: Gatenholm P, Tenkanen M (eds) Hemicelluloses Science and Technology. ACS Symposium Series 864, p. 53
99. N'Diaye S, Rigal L (2000) Bioresource Technol 75:13
100. Maréchal P, Jorda J, Pontalier P-Y, Rigal L (2004) In: Gatenholm P, Tenkanen M (eds) Science and Technology, ACS Symposium Series 864. Washington, p. 38
101. Roubroeks JP, Saake B, Glasser W, Gatenholm P (2004) In: Gatenholm P, Tenkanen M (eds) Hemicelluloses Science and Technology. ACS Symposium Series 864, p. 167
102. Ebringerová A, Hromádková Z, Burchard W, Dolega R, Vorwerg W (1994) Carbohydr Polym 24:161
103. Picout DR, Ross-Murphy SB, Errington N, Harding SE (2003) Biomacromolecules 4:799
104. Wang Q, Ellis PR, Ross-Murphy SB, Burchard W (1997) Carbohydr Polym 33:115
105. Chanliaud E, Roger P, Saulnier L, Thibault J-F (1996) Carbohydr Polym 31:41
106. Dhami R, Harding SE, Elizabeth NJ, Ebringerová A (1995) Carbohydr Polym 28:113
107. Ebringerová A, Hromádková Z, Alföldi J, Berth G (1992) Carbohydr Polym 19:99
108. Chanliaud E, Saulnier L, Thibault J-F (1997) Carbohydr Polym 32:315
109. Schooneveld-Bergmans MEF, Van Dijk YM, Beldman G, Voragen AGJ (1999) J Cereal Sci 29:49
110. Dervilly-Pinel G, Thibault J-F, Saulnier L (2001) Carbohydr Res 330:365
111. Eremeeva TE, Khinoverova OE (1990) Cell Chem Technol 24:439
112. Eremeeva TE, Bykova TO (1993) J Chromatogr 639:159
113. Bykova T, Arnis T (2002) Plant Physiol Bioch 40:347
114. Rattan O, Izydorczyk MS, Biliaderis CG (1994) Food Sci Technol 27:550
115. Girhammar U, Nair BM (1995) Food Hydrocolloid 9:133
116. Picout DR, Ross-Murphy SB (2003) Carbohydr Res 337:1781
117. Garcia RB, Ganter JLMS, Carvalho RR (2003) Eur Polym J 36:783
118. Biliaderis CG, Izydorczyk MS, Rattan O (1995) Food Chem 53:165
119. Courtin CM, Delcour JA (2002) J Cereal Sci 35:225
120. Sasaki T, Yasui T, Matsuki J (2000) Food Hydrocolloid 14:295
121. Gudmundsson M, Eliasson A-C, Bengtsson S, Åman P (1991) Starch/Stärke 43:5
122. Courtin CM, Delcour JA (1998) J Agr Food Chem 46:4066
123. Haque A, Morris ER, Richardson RK (1994) Carbohydr Polym 25:337
124. Williams PA, Phillips GO (1995) In: Stephen AM (ed), Food Polysaccharides and Their Applications. Marcel Dekker, New York, p. 463
125. Larrson PT (2004) In: Gatenholm P, Tenkanen M (eds) Science and Technology, ACS Symposium Series 864. American Chemical Society, Washington, p. 254
126. Paananen A, Osterberg M, Rutland M, Tammellin T, Saarinen T, Tappura K, Stenius P (2004) In: Gatenholm P, Tenkanen M (eds) Science and Technology, ACS Symposium Series 864. American Chemical Society, Washington, p. 269
127. Linder A, Gatenholm P (2004) In: Gatenholm P, Tenkanen M (eds) Science and Technology, ACS Symposium Series 864. American Chemical Society, Washington, p. 236
128. Ebringerová A, Sroková I, Talába P, Hromádková Z (1998) In: Praznik W, Huber A (eds) Carbohydrate as Organic Raw Materials IV. WUV-Universitätsverlag, Wien, Austria, p. 118
129. Glicksman M (1983) Food Hydrocolloids, vol II. CRC, Boca Raton, Florida
130. Sarker DK, Wilde PJ, Clark DC (1998) Cereal Chem 75:493
131. Gabrielii I, Gatenholm P (1998) J Appl Polym Sci 69:1661

132. Gabrielii I, Gatenholm P, Glasser WG, Jain RK, Kenne L (2000) Carbohydr Polym 43:367
133. Kayserilioglu BS, Bakir U, Yilamz L, Akkas N (2003) Bioresource Technol 87:239
134. Sun RC, Fang JM, Rowlands P, Bolton J (1998) J Agr Food Chem 46:2804
135. Xiao B, Sun XF, Sun RC (2001) Polym Degrad Stabil 74:307
136. Rauschenberg N, Dhara K, Palmer J, Glasser W (1990) Polymer Progr (Am Chem Soc Div Polym Chem) 31:650
137. Fang JM, Sun RC, Tomkinson J, Fowler P (2000) Carbohydr Polym 41:379
138. Irvine GM (1984) Tappi J 67:118
139. Marchessault RH (2004) In: Gatenholm P, Tenkanen M (eds) Science and Technology, ACS Symposium Series 864. American Chemical Society, Washington, p. 158
140. Ohta T, Yamasaki S, Egashira Y, Sanada H (1994) J Agr Food Chem 52:653
141. Belicová A, Ebringer L, Krajčovič J, Hromádková Z, Ebringerová A (2001) World J Microb Biot 17:293
142. Bukovský M, Kardošová A, Koščová H, Kočálová D (1998) Biologia (Bratislava) 53:771
143. Ghoneum MH (1998) Biochem Bioph Res Co 243:25
144. Ghoneum M, Jewett A (2000) Cancer Detect Prev 24:314
145. Ebringerová A, Hromádková Z, Hříbalová V (1995) Int J Biol Macromol 17:327
146. Ebringerová A, Kardošová A, Hromádková Z, Malovíková A, Hříbalová V (2002) Int J Biol Macromol 30:1
147. Nosáľová G, Kardošová A, Fraňová S (2000) Pharmazie 55:65
148. Lloyd LL, Kennedy JF, Methacanon P, Paterson M, Knill CJ (1998) Carbohydr Polym 37:315
149. Miraftab M, Qiao Q, Kennedy JF, Anand SC, Collyer GJ (2001) In: Anand SC (ed) Medical Textiles. Woodhead, Cambridge, p. 164
150. Garcia RB, Nagashima T, Praxedes AKC, Raffin FN, Moura TFAL, do Egito EST (2001) Polym Bull 46:371
151. Crittenden RG, Plyne MJ (1996) Trends Food Sci Tech 7:353
152. Fooks LJ, Fuller R, Gibson GR (1999) Int Dairy J 9:53
153. Chirstakopolulos P, Katapodis P, Kalogeris E, Kekos D, Macris BJ, Stamatis H, Skaltsa H (2003) Int J Biol Macromol 31:171
154. de Vries RP, Visser J (2001) Microbiol Mol Biol Rev 65:497
155. Gregory ACE, O'Connell AP, Bolwell GP (1998) Biotechnol Genet Eng Rev 15:439
156. Biely P (2003) In: Whitaker P, Voragwen AGJ, Wong H (eds) Handbook of Food Enzymology. Marcel Dekker, New York, p. 879
157. Lappalainen A, Tenkanen M, Pere J (2004) In: Gatenholm P, Tenkanen M (eds) Hemicelluloses Science and Technology. ACS Symposium Series 864, p. 140
158. Kabel MA, Schols HA, Voragen AGJ (2002) Carbohydr Polym 50:191
159. Tanczos I, Schwarzinger C, Schmidt H, Balla J (2003) J Anal Appl Pyrolysis 68:151
160. Tenkanen M, Siika-aho M (2000) J Biotechnol 178:149
161. Kabel MA, de Waard P, Schols HA, Voragen AGJ (2003) Carbohydr Res 337:719
162. Kačuráková M, Wilson RH (2001) Carbohydr Polym 44:291
163. Reis A, Coimbra MA, Domingues P, Ferrer-Correia AJ, Domingues MRM (2004) Carbohydr Polym 55:401
164. Rydlund A, Dahlman O (1997) Carbohydr Res 300:95
165. Sun H-J, Yoshida S, Park N-H, Kusakabe I (2002) Carbohydrate Res 337:657
166. Carvalheiro F, Esteves MP, Parajo JC, Pereira H (2004) Bioresource Technol 91:93
167. Palm M, Zacchi G (2003) Biomacromolecules 4:617
168. Garrrote G, Dominguez H, Parajo JC (1999) J Chem Technol Biotechnol 74:1101

169. Koukios EG, Pastou A, Koullas DP, Sereti V, Kolosis F (1999) In: Overend RP, Chornet E (eds), Biomass: a growth opportunity in green energy and value-added products. Pergamon, Oxford, p. 641
170. Kabel MA, Carvalheiro F, Garotte G, Avgerinos E, Koukios E, Parajo JC, Girio FM, Schols HA, Voragen AGJ (2002) Carbohydr Polym 50:47
171. Varma AJ, Kennedy JF, Galgali P (2004) Carbohydr Polym 56:429
172. Lindblad MS, Liu Y, Albertsson A-C, Ranucci E, Karlsson S (2002) Adv Polym Sci 157:139
173. Lindblad MS, Ranucci E, Albertsson A-C (2001) Macromol Rapid Comm 22:962
174. Ranucci E, Spagnoli H, Ferruti P (1999) Macromol Rapid Comm 20:1
175. Reid JSG, Edwards ME (1995) In: Stephen AM (ed), Food polysaccharides and their applications. Marcel Dekker, New York, p. 155
176. Navarini L, Gilli R, Gombac V, Abatangelo A, Bosco M, Toffanin R (1999) Carbohydr Polym 40:71
177. Scherbukhin VD, Anulov OV (1999) Appl Biochem Microbiol 35:299
178. Singh V, Srivastava V, Pandey M, Esthi R, Sanghi R (2003) Carbohydr Polym 51:357
179. Chaubey M, Kapoor VP (2001) Carbohydr Res 332:439
180. Kapoor VP, Taravel FR, Joseleau J-P, Milas M, Chanzy H, Rinaudo M (1998) Carbohydr Res 306:231
181. Soni PL, Pal R (1996) Trends Carbohydr Chem 2:33
182. Ishrud O, Zahid M, Zhou H, Pan Y (2001) Carbohydr Res 335:297
183. Ganter JLMS, Heyraud A, Petkowicz CLOM, Rinaudo M, Reicher F (1995) Int J Biol Macromol 17:13
184. Corsaro MM, Giudicianni I, Lanzetta R, Marciano CE, Monaco P, Parrilli M (1995) Phytochemistry 39:1377
185. Adinolfi M, Corsaro MM, Lanzetta R, Parrilli M, Folkard G, Grant W, Sutherland J (1994) Carbohydr Res 263:103
186. Ganter JLMS, Cardoso ATM, Kaminsk M, Reicher F (1997) Int J Biol Macromol 21:137
187. Oosterveld A, Harmsen JS, Voragen AGJ, Schols HA (2003) Carbohydr Polym 52:285
188. Bradbury AGW, Halliday DJ (1990) J Agric Food Chem 38:389
189. Rinaudo M (2001) Food Hydrocolloid 15:433
190. Daas PJH, Schols HA, de Jongh HHJ (2000) Carbohydr Res 329:609
191. Kapoor VP, Chanzy H, Taravel FR (1995) Carbohydr Polym 27:229
192. Thomas J, Darvill AG, Albersheim P (1983) Plant Physiol 72:59
193. Bacic A, Harris PJ, Stone BA (1988) In: Priess J (ed), The Biochemistry of Plants, A Comprehensive Treatise, vol. 14. Academic, New York, p. 297
194. Whitney SE, Brigham JE, Darke AH, Reid JSG, Gidley MJ (1998) Carbohydr Res 307:299
195. Sjöström W (1993) Wood Chemistry—Fundamentals and Applications, 2nd ed. Academic, New York
196. Schröder R, Nicolas P, Vincent SJF, Fischer M, Reymond S, Redgwell RJ (2001) Carbohydr Res 331:291
197. Kubačková M, Karacsonyi S, Bilisics L (1992) Carbohydr Polym 19:125
198. Capek P, Alföldi J, Lišková D (2002) Carbohydr Res 337:1033
199. Capek P, Kubačková M, Alföldi J, Bilisics L, Lišková D, Kákoniová D (2000) Carbohydr Res 329:635
200. Teleman A, Nordström M, Tenkanen M, Jacobs A, Dahlman O (2003) Carbohydr Res 338:525
201. Ishrud O, Zahid M, Ahmad VU, Pan YJ (2001) J Agric Food Chem 49:3772

202. Lee JK, Lee MK, Yun YO, Kim Y, Kim JS, Kim YS, Kim K, Han SS, Lee CK (2001) Int Immunopharmacol 1:1275
203. Reynolds T, Dweck AA (1999) J Ethnopharmacol 68:3
204. Manna S, McAnalley BH (1993) Carbohydr Res 241:317
205. Wozniewski T, Blaschek W, Franz G (1990) Carbohydr Res 198:387
206. Femenia A, Sánchez ES, Simal S, Rosseló (1999) Carbohydr Polym 39:109
207. Kardošová A, Ebringerová A, Pinkas P, Machová E, Hříbalová V (2002) Chem Listy 97:767
208. Femenia A, Garcia-Pasual P, Simal S, Rosselo C (2003) Carbohydr Polym 51:397
209. Nishinari K, Williams PA, Phillips G (1992) Food Hydrocolloid 6:199
210. Gidley MJ, McArthur AJ, Underwood DE (1991) Food Hydrocolloid 5:129
211. Zhang H, Yoshimura M, Nishinari K, Williams MAK, Foster TJ, Norton IT (2001) Biopolymers 59:38
212. Katsuraya K, Okuyama K, Hatanaka K, Oshima R, Sato T, Matsuzaki K (2003) Carbohydr Polym 53:180
213. Hongshu Z, Jinggan Y, Yan Z (2002) Carbohydr Polym 47:83
214. Kim KH, Tucker MP, Keller FA, Aden A, Nguen QQ (2001) Appl Biochem Biotech 91–93:253
215. Puls J, Poutanen K, Korner H-U, Viikari L (1985) Appl Microb Biotechnol 22:416
216. Schultz TP, McGinnis GD, Biermann CJ (1984) Energ Biomass Wastes 8:171
217. Koshijima T (1986) Wood Res 72:1
218. Lundqvist J, Teleman A, Junel L, Zacchi G, Dahlman O, Tjerneld F, Stålbrand H (2002) Carbohydr Polym 48:29
219. Lundqvist JL, Jacobs A, Palm M, Guido Zacchi, Dahlman O, Stålbrand H (2003) Carbohydr Polym 51:203
220. Paszner L, Jeong C, Quinde A, Awardel-Karim S (1993) Energy Biomass Wastes 16:629
221. Willför S, Sjöholm R, Laine C, Roslund M, Hemming J, Holmbom B (2003) Carbohydr Polym 52:175
222. Harvey DJ (1996) J Chromatogr A 720:429
223. Harvey DJ (1999) Mass Spectrom Rev 18:349
224. Dahlman OB, Jacobs A, Nordström M (2004) In: Gatenholm P, Tenkanen M (eds) Science and Technology, ACS Symposium Series 864. American Chemical Society, Washington, p. 80
225. Richardson RK, Ross-Murphy S (1987) Int J Biol Macromol 9:249
226. Wang Q, Ellis PR, Ross-Murphy SB (2000) Food Hydrocolloid 14:129
227. Cheng Y, Brown KM, Prud'homme RK (2002) Int J Biol Macromol 31:29
228. Picout DR, Ross-Murphy SB, Errington N, Harding SE (1997) Carbohydr Polym 33:115
229. Picout DR, Ross-Murphy SB, Jumel K, Harding SE (2002) Biomacromolecules 3:761
230. Jacon SA, Rao MA, Cooley HJ, Walter RH (1993) Carbohydr Polym 20:35
231. Rayment P, Ross-Murphy SB, Ellis PR (2000) Carbohydr Polym 43:1
232. Rayment P, Ross-Murphy SB, Ellis PR (1995) Carbohydr Polym 28:121
233. Kohyama K, Iida H, Nishinari K (1993) Food Hydrocolloid 7:213
234. Yousaria AB, William MB (1994) Starch/Stärke 46:134
235. Annabel P, Fitto MG, Harris B, Phillips GP, Williams PA (1994) Food Hydrocolloid 8:351
236. Yoshimura M, Takaya T, Nishinari K (1996) J Agric Food Chem 44:2970
237. Yoshimura M, Takaya T, Nishinari K (1998) Carbohydr Polym 35:71
238. Lo CT, Ramsden L (2000) Nahrung 44:211

239. Chandrasekaran R, Janaswamy S, Morris VJ (2003) Carbohydr Res 338:2889
240. Seifert A, Heinevetter L, Cölfen H, Harding SE (1995) Carbohydr Polym 28:325
241. Yamauchi F, Suetsuna K (1993) J Nutr Biochem 4:450
242. Kasapis S, Al-Marhoobi IMA, Khan AJ (2000) Int J Biol Macromol 27:13
243. Tizard IR, Carpenter RH, McAnalley BH, Kemp MC (1989) Mol Biother 1:290
244. Ebihara K, Masuhara R, Kiriyama S, Manabe M (1981) Nutr Rep Int 23:577
245. Ramesh HP, Yamaki K, Tsushida T (2003) Carbohydr Polym 50:79
246. Schiermeier S, Schmidt PC (2002) Eur J Pharm Sci 15:295
247. Vincken JP, York WS, Beldman G, Voragen AGJ (1997) Plant Physiol 114:9
248. Pauly M, Albersheim P, Darvill A, York WS (1999) Plant J 20:629
249. Carpita NC, Gibesut DM (1992) Plant J 98:357
250. McCann MC, Roberts KJ (1994) J Exp Bot 45:1683
251. Ha MA, Apperly DC, Jarvis M (1997) Plant Physiol 115:593
252. Hayashi T, Ogawa K, Mitsuishi Y (1994) Plant Cell Physiol 35:1199
253. Mishima T, Hisamatsu M, York WS, Teranishi K, Yamada T (1998) Carbohydr Res 308:389
254. Hayashi T, Takeda T, Ogawa K, Mitsuishi Y (1994) Plant Cell Physiol 35:893
255. Lima DU, Buckeridge MS (2001) Carbohydr Polym 46:157
256. Lima DU, Loh W, Buckeridge MS (2004) Plant Physiol Biochem 42:389
257. Renard CMGC, Lemunier C, Thibault J-F (1995) Carbohydr Polym 28:209
258. Watt DK, Brasch DJ, Larsen DS, Melton LD (1999) Carbohydr Polym 39:165
259. Lang P, Masci G, Dentini M, Crescenzi V, Cooke D, Gidley MJ, Fanutti C, Reid JSG (1992) Carbohydr Polym 17:185
260. Spronk BA, Rademaker GJ, Haverkamp J, Thomas-Oates JE, Vincken J-P, Voragen AGJ, Kamerling JP, Vliegenthart JFG (1998) Carbohydr Res 305:233
261. Femenia A, Rigby NM, Selvendran RR, Waldron KW (1999) Carbohydr Polym 39:151
262. Vierhuis E, York WS, Kolli VSK, Vincken J-P, Schols HA, Van Albeeck G-JWM, Voragen AGJ (2001) Carbohydr Res 332:285
263. Huisman MMH, Weel KGC, Schols HA, Voragen AGJ (2000) Carbohydr Polym 42:185
264. Hantus S, Pauly M, Darvill AG, Albersheim P, York WS (1997) Carbohydr Res 304:11
265. Niemann C, Carpita NC, Whistler RL (1997) Starch/Stärke 49:154
266. Gidley MJ, Lillford PJ, Rowlands DW, Lang P, Dentini M, Creszenzi V, Edwards M, Fanutti C, Reid JSG (1991) Carbohydr Res 214:299
267. Thompson JE, Fry SC (2000) Planta 211:275
268. Buckeridge MS, Crombie HJ, Mendes CJM, Reid JSG, Gidley MJ, Vieira CCJ (1997) Carbohydr Res 303:233
269. Busato AP, Vargas-Rechia CG, Reicher F (2001) Phytochemistry 58:525
270. Cutillas-Iturralde A, Peña MJ, Zarra I, Lorences EP (1998) Phytochemistry 48:607
271. Doco T, Williams P, Pauly M, O'Neill MA, Pellerin P (2003) Carbohydr Polym 53:253
272. Igartuburu JM, Pando E, Rodríguez Luis F, Gil-Serrano A (1997) Phytochemistry 46:1307
273. Vierhuis E, Schols HA, Beldman G, Voragen AGJ (2001) Carbohydr Polym 44:51
274. Onweluzo JC, Ramesh HP, Tharanathan RN (2002) Carbohydr Polym 47:253
275. Vincken J-P, Wijsman AJM, Beldman G, Niessen WMA, Voragen AGJ (1996) Carbohydr Res 288:219
276. York WS, Kolli VSK, Orlando R, Albersheim P, Darvill AG (1996) Carbohydr Res 285:99
277. Jia Z, Qin Q, Darvill AG, York WS (2003) Carbohydr Res 338:1197
278. Maruyama K, Goto C, Numata M, Suzuki T, NakagawaY, Hoshino T, Uchiyama T (1996) Phytochemistry 41:1309

279. Sims I, Munro SLA, Currie G, Craik D, Bacic A (1996) Carbohydr Res 293:147
280. Pauly M, Andersen LN, Kaupinnen S, Kofod LV, York WS, Albersheim P, Darvill AG (1999) Glycobiology 9:93
281. Vincken J-P, Beldman G, Voragen AGJ (1997) Carbohydr Res 298:299
282. de Alcântara PHN, Dietrich SMC, Buckeridge MS (1999) Plant Physiol Biochem 37:653
283. Shirakawa M, Yamatoya K, Nishinari K (1998) Food Hydrocolloid 12:25
284. Yamagaki T, Mitsuishi Y, Nakanishi H (1997) Biosci Biotech Bioch 61:1411
285. Yamagaki T, Mitsuishi Y, Nakanishi H (1998) Rapid Commun Mass Sp 12:307
286. Yamagaki T, Mitsuishi Y, Nakanishi H (1998) Tetrahedron Lett 39:4051
287. Marry M, Cavalier DM, Schnurr JK, Netland J, Yang Z, Pezeshk V, York WS, Pauly M, White AR (2003) Carbohydr Polym 51:347
288. Shankaracharya NB (1998) J Food Sci Technol 35:193
289. Lang P, Burchard W (1993) Makromol Chem 194:3157
290. Lang P, Kajiwara K (1993) J Biomat Sci Polym Ed 4:517
291. Wang Q, Ellis PR, Ross-Murphy SB, Reid JS (1996) Carbohydr Res 284:229
292. Sims IM, Gane AM, Dunstan D, Allan GC, Boger DV, Melton LD, Bacic A (1998) Carbohydr Polym 37:61
293. Reid JSG, Edwards ME, Dea ICM (1988) In: Phillips GO, Williams PA, Wedlock DJ (eds), Gums and Stabilisers for the Food Industry, vol. 4. IRL Press, Oxford, p. 391
294. Yamanaka S, Yuguchi Y, Urakawa H, Kajiwara K, Shirakawa M, Yamatoya K (2000) Food Hydrocolloid 14:125
295. Ikeda S, Nitta Y, Kim BS, Temsiripong T, Pongsawatmanit R, Nishinari K (2004) Food Hydrocolloid 18:669
296. Freitas RA, Gorin PAJ, Neves J, Sierakowski M-R (2003) Carbohydr Polym 51:25
297. Yoshimura M, Takaya T, Nishinari K (1999) Food Hydrocolloid 13:101
298. Prabhanjan H, Ali SZ (1995) Carbohydr Polym 28:245
299. Sinibaldi L, Pietropaolo V, Goldoni P, Ditaranto C, Orsi N (2002) J Chemother 4:16
300. Lanhers MC, Fleurentin J, Guillemni F (1996) Ethnopharmacol 18:42
301. Ebringerová A, Hromádková Z, Vodeničárová M, Dvořáková D, Gregušová K, Velebný V (2002) XXIst International Carbohydrate Symposium, Cairns, Australia, p P237
302. Burgalassi S, Panichi L, Saettone MF, Jacobsen J, Rassing MR (1996) Int J Pharm 133:1
303. Renard D, Rober P, Lavenant L, Melcion D, Popineau Y, Guéguen J, Duclairoir C, Nakache E, Sanchez C, Schmitt C (2002) Int J Pharm 242:163
304. Suisha F, Kawasaki N, Miyazaki S, Shirakawa M, Yamatoya K, Sasaki M, Attwood D (1998) Int J Pharm 172:27
305. Kawasaki N, Ohkura R, Miyazaki S, Uno Y, Sugimoto S, Attwood D (1999) Int J Pharm 181:227
306. Sumathi S, Ray AR (2002) Sci J Pharm Pharm 5:12
307. Miyazaki S, Suzuki S, Kawasaki N, Endo K, Takahashi A, Attwood D (2001) Int J Pharm 229:29
308. Kulkarni D, Dwivedi AK, Singh S (1998) Ind J Pharm Sci 60:50
309. Lima DU, Oliveira RC, Buckeridge MS (2003) Carbohydr Polym 52:367
310. Wood PJ, Weisz J, Beer MU, Newman CW, Newman RK (2003) Cereal Chem 80:329
311. Buckeridge MS, Rayon C, Urbanowicz B, Tiné MAS, Carpita NC (2004) Cereal Chem 81:115
312. Colleoni-Sirghie M, Fulton BD, White PJ (2003) Carbohydr Polym 54:237

313. Colleoni-Sirghie M, Kovalenko IV, Briggs JL, Fulton B, White PJ (2003) Carbohydr Polym 52:439
314. Wood PJ (2004) Trends Food Sci Tech 15:313
315. Cervantes-Martinez CT, Frey KJ, White PJ, Wesenberg DM, Holland JB (2002) Crop Sci 42:730
316. Ramesh HP, Tharanathan RN (1999) Food Chem 64:345
317. Wood PJ, Weisz J, Blackwell BA (1994) Cereal Chem 71:301
318. Grimm A, Krüger E, Burchard W (1995) Carbohydr Polym 27:205
319. Bhatty RS (1999) Cereal Chem 76:73
320. Kahlon TS, Woodruff CL (2003) Cereal Chem 80:260
321. Braaten JT, Wood PJ, Scott FW, Wolynetz MS, Lowe MK, Bradley-White P, Colins MW (1994) Europ J Clin Nutr 48:465
322. Wood PJ (1994) Carbohydr Polym 25:331
323. Estrada A, Yun CH, Van Kessel A, Li B, Hauta S, Larveld B (1997) Microbiol Immunol 41:991
324. Yun CH, Estrada A, Van Kessel A, Park B, Laarveld B (2003) FEMS Immunol Med Mic 35:67
325. Delaney B, Nicolosi J, Wilson TS, Carlson T, Fraser S, Zheng G, Hess R, Ostergren K, Haworth J, Knutson N (2003) J Nutr 133:468
326. Symons LJ, Brennan CS (2004) J Food Sci 69:257
327. Bhatty RS (1995) J Cereal Sci 22:163
328. Beer MU, Arrigoni E, Amado R (1996) Cereal Chem 73:58
329. Temelli F (1997) J Food Sci 62:1194
330. Burkus Z, Temelli F (1998) Cereal Chem 75:805
331. Izydorczyk MS, MacGregor AW (2000) Carbohydr Polym 41:417
332. Izydorczyk MS, Biliaderis CG, Macri LJ, MacGregor AW (1998) J Cereal Sci 27:321
333. Morgan KR, Ofman DJ (1998) Cereal Chem 75:879
334. Johansson L, Virkki L, Maunu S, Letho M, Ekholm P, Varo P (2000) Carbohydr Polym 42:143
335. Lazaridou A, Biliaderis CG, Micha-Screttas M, Steele BR (2004) Food Hydrocolloid 18:837
336. Roubroeks JP, Mastromauro DI, Andersson R, Christensen BE, Åman P (2000) Biomacromolecules 1:584
337. Storsley JM, Izydorczyk MS, You S, Biliaderis CG, Rossnagel B (2003) Food Hydrocolloid 17:831
338. Tvaroška I, Ogawa K, Deslandes Y, Marchessault RH (1983) Can J Chem 61:1608
339. Wood PJ, Weisz J, Mahn W (1991) Cereal Chem 68:530
340. Vaikousi H, Biliaderis CG, Izydorczyk MS (2004) J Cereal Sci 39:119
341. Izydorczyk MS, Storsley J, Labossiere D, MacGregor AW, Rossnagel BG (2000) J Agric Food Chem 48:982
342. Wang Q, Wood PJ, Cui W (2002) Carbohydr Polym 47:35
343. Wang Q, Wood PJ, Huang X, Cui W (2003) Food Hydrocolloid 17:845
344. Gomez C, Navarro A, Manzanares P, Horta A, Carbonell JV (1997) Carbohydr Polym 32:17
345. Linemann A, Kruger E (1998) Brauwelt Int 3:214
346. Burkus Z, Temelli F (2003) Carbohydr Polym 54:51
347. Varum KM, Smidsrod O, Brants DA (1992) Food Hydrocolloid 5:497
348. Gomez C, Navarro A, Manzanares P, Horta A, Carbonell JV (1997) Carbohydr Polym 32:7
349. Aberle T, Burchard W, Vorwerg W, Radosta S (1994) Starch/Stärke 46:329

350. Vorwerg W, Radosta S (1995) Macromol Symp 99:71
351. Wang Q, Wood PJ, Cui W, Ross-Murphy SB (2001) Carbohydr Polym 45:355
352. Dawkins NL, Nnanna IA (1995) Food Hydrocolloid 79:1
353. Skendi A, Biliaderis CG, Lazaridou A, Izydorczyk MS (2003) J Cereal Sci 38:15
354. Ren Y, Ellis PR, Ross-Murphy SB, Wang Q, Wood PJ (2003) Carbohydr Polym 53:401
355. Böhm N, Kulicke WM (1999) Carbohydr Res 315:302
356. Böhm N, Kulicke WM (1999) Carbohydr Res 315:293
357. Gomez C, Navarro A, Manzanares P, Horta A, Carbonell JV (1997) Carbohydr Polym 34:141
358. Doublier JL, Wood PJ (1995) Cereal Chem 72:335
359. Hromádková Z, Ebringerová A, Sasinková V, Šandula J, Hříbalova V, Omelková J (2003) Carbohydr Polym 51:9
360. Bahnassy YA, Breene WM (1994) Starch/Stärke 46:134
361. Biliaderis CG, Arvanitoyannis I, Izydorczyk MS, Prokopowich F (1997) Starch/Stärke 49:278
362. Shi XH, BeMiller JN (2002) Carbohydr Polym 50:7
363. Carriere JC, Inglett GE (1999) Carbohydr Polym 40:9
364. Lazaridou A, Biliaderis CG, Izydorczyk MS (2003) Food Hydrocollid 17:693
365. Burkus Z, Temelli F (2000) Food Res Int 33:27
366. Lyly M, Salmenkallio-Marttila M, Suorti T, Autio K, Poutanen K, Lähteenmäki L (2004) Cereal Chem 80:536
367. Volakis P, Biliaderis CG, Vamvakas C, Zerfiridis GK (2004) Food Res Int 37:83
368. Konuklar G, Inglett GWE, Warner K, Carriere CJ (2004) Food Hydrocolloid 18:535
369. Inglett GE, Carriere CJ, Maneepun S, Tungtrakul P (2004) Int J Food Sci Technol 39:1
370. Lehtinen P, Laakso S (1998) J Agric Food Chem 46:4842
371. Tejinder S (2003) Cereal Chem 80:728
372. Habibi Y, Mahrouz M, Marais MF, Vignon MR (2004) Carbohydr Res 339:1201
373. Serpe MD, Nothnagel EA (1995) Plant Physiol 109:1007
374. Fischer M, Reimann S, Trovato V, Redgwell RJ (2001) Carbohydr Res 330:93
375. Yamada H (1994) Carbohydr Polym 25:269
376. Paulsen BS (2001) Curr Org Chem 5:939
377. Stephen AM, Churms SC (1995) In: Stephen AM (ed), Food Polysaccharides and Their Applications. Marcel Dekker, New York, p. 377
378. Whistler RL (1993) In: Whistler RL, BeMiller JN (eds), Industrial Gums, 3rd edn. Academic Press, San Diego, p. 295
379. Kim LS, Waters RF, Burkholder PM (2002) Altern Med Rev 7:138
380. Hauer J, Anderer FA (1993) Cancer Immunol Immun 38:237
381. Odonmazhig P, Ebringerová A, Machová E, Alföldi J (1994) Carbohydr Res 252:317
382. Ponder GR, Richards GN (1997) J Carbohydr Chem 16:181
383. Eremeeva TE, Bykova TO (1992) Carbohydr Polym 18:217
384. Prescott JH, Groman EV, Gulyas G (1997) Carbohydr Res 301:89
385. Prescott JH, Enriques P, Jung C, Menz E, Groman EV (1995) Carbohydr Res 278:113
386. Manley-Harris M (1997) Carbohydr Polym 34:243
387. Ponder GR, Richards GN (1997) J Carbohydr Chem 16:195
388. Brigand G (1993) In: Whistler RL, BeMiller JN (eds), Industrial Gums, Polysaccharides and Their Derivatives, 3rd edn. Academic, San Diego p. 461
389. Chandrasekaran R, Janaswamy R (2002) Carbohydr Res 337:2211
390. Ponder GR, Richards GN (1997) Carbohydr Polym 34:251
391. Ponder GR (1998) Carbohydr Polym 36:1

392. Gallez B, Lacour V, Demeure R, Debuyst R, Dejehet F, DeKeyser JL, Dumont P (1994) Magn Reson Imaging 12:61
393. Hagmar B, Erkell LJ, Ryd W, Skomedal H (1994) Cell Pharmacol 1:87
394. Christian TJ, Manley-Harris M, Richards GN (1998) Carbohydr Polym 35:7
395. Dong Q, Fang J (2001) Carbohydr Res 332:109
396. Huisman MM, Brüll LP, Thomas-Oates JE, Haverkamp J, Schols HA, Voragen AGJ (2001) Carbohydr Res 330:103
397. Loosveld A-M, Grobet PJ, Delcour JA (1997) J Agric Food Chem 45:1998
398. Wisner ER, Amparo EG, Vera DR, Brock JM, Barlow TW, Griffey SM, Drake C, Katzberg RW (1995) J Comput Assist Tomogr 19:211
399. Heinze Th, Koschella A, Ebringerová A (2004) In: Gatenholm P, Tenkanen M (eds) Hemicelluloses: Science and Technology. ACS Symposium Series 864. American Chemical Society, Washington, p. 312
400. Ebringerová A, Hromádková Z, Malovíková A, Sasinková V, Hirsch J, Sroková I (2000) J Appl Polym Sci 78:1191
401. Ebringerová A, Alföldi J, Hromádková Z, Pavlov GM, Harding SE (2000) Carbohydr Polym 42:123
402. Ebringerová A, Sroková I, Talába P, Kačuráková M, Hromádková Z (1998) J Appl Polym Sci, 67:1523
403. Ebringerová A, Hromádková Z, Kačuráková M, Antal M (1994) Carbohydr Polym 24:301
404. Ebringerová A, Hromádková Z (1996) Angew Makromol Chem 24:97
405. Antal M, Ebringerová A, Hromádková Z, Pikulík I, Laleg M, Micko MM (1997) Papier 51:223
406. Ebringerová A, Belicová A, Ebringer L (1994) J Microbiol Biotechnol 10:640
407. Thiebaud S, Borredon ME (1998) Bioresource Technol 63:139
408. Fang JM, Sun RC, Fowler P, Tomkinson, Hills CAS (1999) J Appl Polym Sci 74:2301
409. Sun RC, Fang JM, Tomkinson J, Hill CAS (1999) J Wood Chem Technol 19:287
410. Sun RC, Fang JM, Tomkinson J (2000) Polym Degrad Stabil 67:345
411. Sun RC, Fang JM, Tomkinson J, Jones GL (1999) Ind Crop Prod 10:209
412. Sun RC, Fang JM, Tomkinson J, Geng ZC, Liu JC (2001) Carbohydr Polym 44:29
413. Sun RC, Sun XF, Bing XJ (2002) Appl Polym Sci 83:757
414. Sun X-F, Sun RC, Tomkinson J, Baird MS (2003) Carbohydr Polym 53:483
415. Sun XF, Sun RC, Sun JX (2004) J Sci Food Agr 84:800
416. Vincendon M (1993) Macromol Chem 194:321
417. Okamoto Y, Noguchi J, Yashima E (1998) React Funct Polym 37:183
418. Buchanan CM, Buchanan NL, Debenham JS, Gatenholm P, Jacobsson M, Shelton MC, Watterson TL, Wood MD (2003) Carbohydr Polym 52:345
419. Giedrojc J, Radziwon P, Klimiuk M, Bielawiec M, Breddin HK, Kloczko J (1999) J Physiol Pharmacol 50:111
420. Figg WD, Pluda JM, Sartor O (1999) In: Teicher BA (ed), Antiangiogenic Agents in Cancer Therapy. Humana, Totowa, NJ, p. 371
421. Ghosh P (1999) Semin Arthritis Rheum 28:211
422. Elliot SJ, Striker LJ, Stetler-Stevenson WG, Jacot TA, Striker GE (1999) J Am Soc Nephrol 10:62
423. Klemm D, Philipp B, Heinze T, Heinze U, Wagenknecht W (1998) Comprehensive Cellulose Chemistry. Wiley-VCH, Weinheim
424. Jain RK, Sjöstedt MA, Glasser WG (2001) Cellulose 7:319
425. Glasser WG, Jain RK, Sjöstedt MA (1996) Biotechnol Adv 14:605

426. Matsumura S, Nishioka M, Yoshikawa S, Yoshikawa S (1991) Macromol Rapid Comm 12:89
427. Jerez JR, Matsuhiro B, Urzua CC (1993) Carbohydr Polym 32:155
428. Barroso NP, Costamagna J, Matsuhiro B, Villagran M (1997) Bol Soc Chil Quim 42:301
429. Fredon E, Granet R, Zerrouki R, Krausz P, Saulnier L, Thibault J-F, Rosier J, Petit C (2002) Carbohydr Polym 49:1
430. Sierakowski MR, Milas M, Desbrieres J, Rinaudo M (2000) Carbohydr Polym 42:51
431. Bajpai UDN, Alka J, Sandeep R (1990) J Appl Polym Sci 39:2187
432. Lapasin E, de Lorenzi L, Pridl S, Torriano G (1995) Carbohydr Polym 28:195
433. Sharma BR, Kumar V, Soni PL (2003) Carbohydr Polym 54:143
434. Sharma BR, Kumar V, Soni PL (2002) J Appl Polym Sci 86:3250
435. Sharma BR, Kumar V, Soni PL (2003) Starch/Stärke 55:38
436. Sharma BR, Kumar V, Soni PL, Sharma PJ (2003) J Appl Polym Sci 89:3216
437. Soni PL, Singh SV, Naithani S (2000) Paper Intern 5:14
438. Figueiro SD, Goes JC, Moreira RA, Sombra ASB (2004) Carbohydr Polym 56:313
439. Lindblad MS, Albertsson AC, Ranucci E (2004) In: Gatenholm P, Tenkanen M (eds) Science and Technology, ACS Symposium Series 864. American Chemical Society, Washington, p 347
440. Prabhanjan H (1989) Starch/Stärke 41:409

Adv Polym Sci (2005) 186: 69–101
DOI 10.1007/b136817
© Springer-Verlag Berlin Heidelberg 2005
Published online: 30 August 2005

Bioactive Pectic Polysaccharides

Berit Smestad Paulsen (✉) · Hilde Barsett

School of Pharmacy, Department of Pharmaceutical Chemistry, Section Pharmacognosy,
P.O.box 1068 Blindern, 0316 Oslo, Norway
b.s.paulsen@farmasi.uio.no, hilde.barsett@farmasi.uio.no

Abstract Polysaccharides from plants have been the subject of studies for a very long time, mainly focussed on their physical properties, their chemical and physical modification, and their application. Over the last 20 years there has been increasing interest in the biological activity of the natural polysaccharide polymers. These studies became possible as a result of the scientific development of isolation, purification and characterisation methods concomitant with the development of fairly simple in vitro tests for effects especially on the immune system. The growing acceptance of the knowledge to be gained by people still using so-called traditional medicine in finding sources worthy of study has led to new sources for interesting bioactive plant polysaccharides. This chapter contains only the knowledge on bioactive plant polysaccharides of the pectic type gained over approximately the last ten years and the focus is on those papers where structural characterisation has been performed. For this reason, the reader may not find all of the plants studied during this period within the chapter. Discussions concerning the structural aspects of polysaccharides that may be responsible for activity, are included where relevant.

Keywords Arabinogalactans · Bioactivity · Medicinal plants · Rhamnogalacturonans · Pectins · Structure-activity relations

Abbreviations

AceA	aceric acid
AFM	atomic force microscopy
AG-I	arabinogalactan type I
AG-II	arabinogalactan type II
Api	apiose
Ara$f$	arabinofuranose
DHA	3-deoxy-D-lyxo-2-heptulosaric acid
DP	Degree of polymerisation
Gal$p$	galactopyranose
Gal$p$A	galactopyruronic acid
GlcA	glucuronic acid
Glc$p$	glucopyranose
IL-6	interleucine 6
KDO	3-deoxy-D-manno-octulosonic acid
NK cells	natural killer cells
p.o.	per os
RG-I	rhamnogalacturonan type I
RG-II	rhamnogalacturonan type II
Rha$p$	rhamnopyranose
Xyl	xylose

1
Introduction

In traditional medicine, plants have been used to treat various types of illnesses, including wounds, both external and internal. The use of plants can be found as part of traditional medicine on all continents, and plants are still in use even in the Western countries as so-called "traditional remedies". Modern science has shown that many of these plants contain polysaccharides that exhibit biological activity of different kinds. In addition, many of these polysaccharides are able to form gels or viscous solutions that are of great industrial value. Research to form a scientific basis for rational, traditional use of polysaccharides isolated from plants as immunostimulatory, antitumour or anti-inflammatory agents was not possible before approximately 20 years ago. This was according to Wagner and Kraus [1] mainly due to following reasons:

- The isolation and purification of sufficient amounts of pure bioactive polysaccharides in a reproducible manner was difficult due to the lack of isolation and purification methods.
- Unknown structure activity relationships, mainly because methods for activity testing were poorly available.
- Lack of information concerning the exact mechanism of the action for the immunomodulatory pharmacological activities.
- Lack of information concerning pharmacokinetics and bioavailability after p.o. and parenteral administration of the polymers.

The groups of Wagner in München, Germany, and Yamada in Tokyo, Japan, have made important contributions towards the development of methods that could be used for testing the biological activity of polysaccharides in vitro and in vivo, which are well documented in various reviews in the field [2–7]. Yamada [4] reports that 1984 was the first year when studies on a pure, complex polysaccharide with biological activity were cited in the literature. This was isolated from the upper part of *Echinacea purpurea*, a plant that had a long traditional use against cold and influenza. The use of this plant was first reported from the Native Americans of the old North America, and another traditional use was as a remedy for woundhealing [8]. Soon after this report, further studies of bioactive polysaccharides were published and most of the studies performed on biologically active polysaccharides took the traditional information on the use of plants in woundhealing or related areas as the starting point for choosing what plants were to be studied. Especially plants used for the treatment of external wounds and dermal ailments, as well as those used against ulcer and tumours, were early in focus.

Plants, lichens and algae have all been tested for the content of polysaccharides with biological effect in various systems. Polysaccharides appear in many different forms and in different locations in plants. They are present in all organs, both inside cells, and intercellularly, and they are important as strengthening substances, i.e. as fibres and as the material forming the matrix of the cells. Polysaccharides are structurally a heterogeneous group of compounds, they are neutral or acidic, they may consist of only one type of monosaccharide, or of two and up to approximately ten different types, some of which may be in repeating units; they can be linear or branched and be substituted with different types of organic groups like methyl and acetyl groups. Often the polysaccharides with biological activity are charged, that is they contain uronic acids, e.g. D-galacturonic acid as in the pectic type polymers. Different types of polysaccharides isolated from plants used in traditional medicine are identified for their activities on the complement system; e.g. arabinans, arabinogalactans and rhamnogalacturonans [4]. Similar types of polymers have also been shown to have effects on macrophages, T-lymphocytes and NK-cells. The majority of these polysaccharides exhibit also a mucoadhesive effect. They bind to the surface of cells, and can then be the cause of local effects seen in certain experiments [9].

This review will focus on pectic substances that have been shown to exhibit biological activity. Their structure, bioactivity and possible structure activity relations will be discussed.

2
Chemical Structure of Pectic Type Polymers

Pectins can generally be divided into neutral and acidic polymers, but certain structural features are common between the different types of pectic substances. These will be described below. Concerning the general chemistry of carbohydrate types of linkages, the reader can find details on this in general textbooks of chemistry and biochemistry and is for this reason not included.

2.1
Arabinans

The arabinans found in plants are basically composed of L-arabinofurano-sides. Depending on the source they may be linear or branched, and primarily linked through positions 3 or 5, Fig. 1 [10, 11]. Linkages through C-2 are also observed, but generally they are less frequent than the 3-linkage. It is generally accepted that the core linkage of the arabinans are the 5-linkages and the branches occur at C-3 or C-2. But it is not obvious that the arabinans exist as such in nature. They are most probably linked to the galactans in the pectic complex and released either via enzymatic action or weak acid hydrolysis during the extraction process. The cell-walls where the arabinans are found are also rich both in *exo-* and *endo*-glycanases.

$$\rightarrow 5)\alpha\text{-L-Ara}f(1\rightarrow 5)\alpha\text{-L-Ara}f(1\rightarrow 5)\alpha\text{-L-Ara}f(1\rightarrow 5)\alpha\text{-L-Ara}f(1\rightarrow$$

$$\begin{array}{cc} 3 & 2\text{or}3 \\ \uparrow & \uparrow \\ 1 & 1 \\ \alpha\text{-L-Ara}f & \alpha\text{-L-Ara}f \end{array}$$

Fig. 1 Proposed structure of a part of an arabinan

A few examples of pure arabinans with effect in the complement system have been isolated from, for example, the fruits of *Ziziphus jujuba* (α-2,5-arabinofuranan) and from the roots of *Bupleurum falcatum* (α-3,5-arabino-furanan-α-1,4-glucan complex) [12, 13].

2.2
Arabinogalactans Type I and II (AG-I and AG-II)

The arabinogalactans have more frequently been reported for activity in various biological systems. Arabinogalactans are often classified in three groups: arabino-4-galactans (Type I), arabino-3,6-galactans (Type II) and polysaccharides with arabinogalactan side chains (Type III) [14]. The latter type are also called the real pectins [10, 11]. Only types I and II will be dealt with in this chapter, as Type III are equal to the pectins discussed below.

AG-I is found to variable degrees in the cell wall and is composed of a β-1,4 linked galactan backbone with side chains of arabinans basically linked through position 3 of the galactose units. The AG-I structures are mainly found as a constituent of the pectic complex RG-I that will be described below. AG-II has as its main core a galactan that can have either 3 or 6 linkages in the main chain and is highly branched with the 1,3,6-linked galactose units at the branching points. Also this type of arabinogalactan is frequently found bound to RG-I. Both types of arabinogalactans are found to be linked through position 4 of the rhamnose units of the pectic chain (see below). One easy method to distinguish between the two arabinogalactans is their ability to precipitate the so called Yariv reagent [15, 16]. Only AG-II has the ability to form a red precipitate with the Yariv reagent, and this is frequently used to show the presence of AG-II in bioactive polymers, and can also be used for a quantitative assessment of the amount of this type of polymer in the total pectic complex [6].

2.3
Pectic Acid, Pectins

Pectic acids or pectins are common words for describing polymers containing galacturonic acid. These were earlier thought to mainly consist of the acid only, but it has now long been recognised that pectins are a very complex group of polysaccharides. Long sequences of polygalacturonans (homogalacturonans) can be found in the pectins, and it has been shown by, for example, Nothnagel et al. [17] and Samuelsen et al. [18] that the galacturonan part of the pectins may consist of up to at least ten units. Several neutral monosaccharides are normally also present in these polymers. Pectins are found mainly between the cells and in the primary cell wall in most plants [10, 11], and have during the last 20 years been shown to be responsible for different types of bioactivity when used as traditional medicines in different cultures [1–4, 19]. The pectic type of polymer possess complex structures, as mentioned, and today they are divided into two main types, Rhamnogalacturonan I and Rhamnogalacturonan II. They differ so much in structure that it is feasible to deal with them separately [10].

2.4
Rhamnogalacturonan I (RhaGalA-I, RG-I)

Rhamnogalacturonan I or RG-I was first used by Albersheim's group [11, 20, 21]. It was observed that by treating the polysaccharides obtained from a suspension of cultured sycamore cell walls with a α-1,4-*endo*-polygalacturonase, a polymer that had a core of alternating α-1,4-linked D-galacturonic acid and α-1,2-L-rhamnose units could be isolated. It also turned out [11], when studying the pectin polymer from different sources that they all had a striking similarity. The rhamnose units in the alternating core were frequently found as branch points, primarily on position 4, carrying galactan and arabinan side chains of varying structure. Rhamnose was occasionally branched on position 3 as well. The arabinogalactans attached to the rhamnose units are frequently found to be of the arabinogalactan type II (precipitates with the Yariv reagent), although AG-I occasionally also may be present. Both galactans with the main chain being 1,6-linked and 1,3-linked may be found for AG-II. This re-

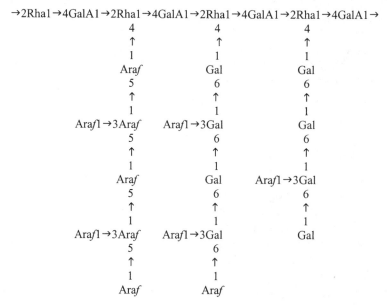

Fig. 2 Average structure of the "hairy" or "ramified" region of an apectic substance, with a rhamnogalacturonan I backbone substituted at position 4 of the rhamnose units with arabinan and arabinogalactan type II side chains

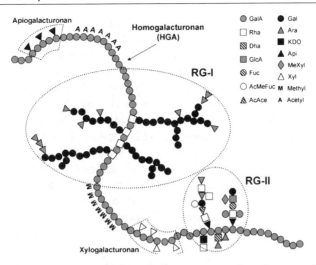

Fig. 3 A schematic presentation of the primary structure of pectins according to Perez et al. [23]

gion is now termed the "hairy" or the "ramified" region of the pectic polymer and an average image of RG-I is given in Fig. 2.

In addition to the hairy region, it was found that the pectic polymer contained a so called smooth region only composed of α-1,4-galacturonic acid residues. This smooth region can carry methyl ester groups and also be acetylated at positions 2 and/or 3 [22]. It has also been reported that the "homogalacturonans" can be substituted, often with single xylose residues. The position of these is schematically shown in the model of the pectic polymer as proposed by Perez et al. [23] in Fig. 3. This chapter will deal with bioactive pectins, and detailed structures of those containing the RG-I sequences will be discussed where relevant.

2.5
Rhamnogalacturonan II (RhaGalA-II, RG-II)

Rhamnogalacturonan II is a part of the pectin complex that cell wall polysaccharides are composed of, and comprise only a minor part of the total amount of pectins present. They have a so-called "homogalacturonan" backbone composed of 9–10 D-galacturonic acid units that are α-1,4-linked, and four different oligosaccharide chains are attached via positions 3 or 4 of the uronic acid backbone. As the backbone consists only of galacturonic acid, RG-II is not really an appropriate name for this polymer, but it has been kept as it was in use for a long time before the real structure of this polymer was discovered. The most characteristic part of RG-II is the presence of the rare sugars 2-O-methylfucose, 2-O-methylxylose, apiose, aceric acid, 2-keto-3-deoxy-D-manno-octulosonic acid (KDO) and 3-deoxy-D-lyxo-2-heptulosaric acid (DHA) [11, 24]. The real structure, as it probably exists, as a three dimensional polymer was proposed by the group of Perez [25]. Figure 4 shows the primary structure of RG-II. This is the basis for the three dimensional orientation proposed that is composed of 2 units that appear to be linked via two apiosyl residues being cross-linked by a 1 : 2 borate diol ester. Apiose in the A chain (Fig. 4), is involved in the linkage although it is not in the centre of the molecule. This location leaves space for borate to be incorporated and as also shown by AFM in the same paper, creates a dimer structure that may be the basic structure for a possible extension to form a larger complex (Fig. 5). The data for the 3D-structure is basically deducted from NMR studies. RG-II is, as seen, a complex structure that appears to be present in the primary cell wall in most plants. Discussions on the biosynthesis and the high number of enzymes involved in producing this well conserved, complex oligomer are focussed on by Perez et al. [23]. The structure of RG-II that was isolated from red wine after the fermentation process, showed the presence of both the monomeric and dimeric substance. The reason for finding this high molecular weight oligosaccharide present after the fermentation, may be explained by the fact that RG-II is rich in the atypical monomers and oligomers mentioned above that are of such a nature that enzymes having those as substrates do not exist. Twenty-nine different plants

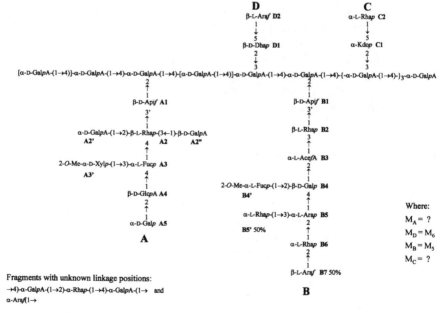

Fig. 4 The complete structure of rhamnogalacturonan II as described by Rodrigues-Carvajal et al. [25]

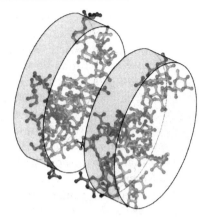

Fig. 5 Presentation of the 3D organisation of the RG-II dimer in which two apioses are cross-linked by a 1 : 2 borate-diol ester, as proposed by Perez et al. [23]

have been reported in the literature up to 2003 to contain RG-II with more or less the same structure.

Various pectin containing plants that are used in traditional medicine have been shown to contain RG-II. The pectic substance Bupleuran 2IIc from *Bupleurum falcatum*, having an effect on the gastric ulcer, and studied in great detail by the group of Yamada contains a minor region of RG-II. *Atractylodes lancea* pectin, ALR-b, and leaf pectin from *Panax ginseng*, both contain RG-II in the bioactive fraction. So does *Angelica acutiloba* and *Glycyrrhiza uralensis* [26–28]. The bioactivity of these polymers will be discussed below, as will their structure activity relations.

3
Bioactive Pectic Polysaccharides Isolated from Plants

As mentioned in the introduction, various reviews over the last ten years show that many plants contain bioactive polysaccharides. Most of the plants studied were chosen due to their traditional use for different kinds of illnesses where the immune system could be involved. The following section will describe the pectic type polymers from the plants most studied for their structure, and activities related to the structure where possible.

3.1
Acanthus ebracteatus Vahl, Acanthaceae

Acanthus ebracteatus is a plant traditionally used for various ailments, amongst those skin diseases in Thai traditional medicine. The stem of the plant was shown to contain neutral and acidic polysaccharides with effect in the complement system that was quite high compared to the normal standard used. The neutral polymer is composed of galactose, 3-O-methylgalactose and arabinose in the ratio 3 : 4 : 1. Both galactose and 3-O-methylgalactose are mainly β-1,4-linked, arabinose α-1,5, in addition they are also terminal groups and branch points, for the galactoses mainly on position 6. Fractionation by gelfiltration of the slightly more active acidic fraction gave six different subfractions, of which the highest Mw fraction (~ 1500 kDa), A1002a, had the lowest concentration of galacturonic acid and the highest of 3-O-methylgalactose of the subfractions obtained. It also had the highest activity of all, significantly higher than those with an Mw in the region 30–60 kDa. These latter had comparable activity with the standard (PMII from *Plantago major*) used, and had also similar Mw, monosaccharide composition and structural features as PMII. Both the high amount of the unusual sugar 3-O-methylgalactose, and the high molecular weight of the most active polymer from *A. ebracteatus* may be important factors for the activity observed [29, 30].

3.2
Angelica acutiloba Kitagawa, Umbelliferae

Immunostimulating polysaccharides were already in 1982 observed in the water extract of roots from *Angelica acutiloba* [31]. This paper was the basis for further studies on the polysaccharides from this plant. One of the first arabinogalactans for which an activity on the complement system was shown, was in fact an arabinogalactan from a hot water extract of the roots of this plant [32]. The polymer was called AGII and was, although only a minor component of the total amount of polysaccharides present in the root, the most potent one. The polymer was shown to be an arabino β-galactan, the backbone most probably composed of an α-1,6-linked galactan with α-1,5-arabinofuranosyl residues linked through C-3 of the galactan backbone. This polymer activates the complement system both via the alternative and classical pathways and contains structures similar to the general AG-II polymers. The polymer was subjected to degradation both by mild acid hydrolysis and enzymatic digestions. Linkage analyses of the obtained oligomers gave the following basic structures: a 1,6-linked galactan with unbranched short side chains of Ara*f* on position 3; 1,4-linked gal was also observed, as well as highly branched Ara*f*-oligomers with linkages both on positions 3 and 5. Base catalysed degradation gave rise to rhamnogalacturonan I sections having substitutions on position 4 of the rhamnose units [33]. Other pectins with effect on the complement system isolated from the roots of *A. acutiloba* were composed of over 90% of a galacturonan region with a small amount of the ramified region [34]. The ramified region, isolated after degradation of the polysaccharide with pectinase and pectinesterase, contained the rhamnogalacturonan core possessing side chains rich in neutral carbohydrate chains, which were directly attached to position 4 of rhamnose [35]. The ramified region from each pectin had a more potent complement-activating effect than the corresponding original pectins, and the oligogalacturonides had weak or negligible activities. These facts suggest that the complement-activating potency of these pectins is expressed mainly by their ramified regions [34]. The total structure of this polymer has been proposed by Kiyohara and Yamada [36]. The relationship between the structure and the effect on the complement system of this polymer is discussed in great detail [37]. Removal of the external Ara$f$ units of the molecule resulted in higher activity than that of the "mother-molecule", the arabinogalactan side chains, AG-I type, showed the most potent activity of the side chains prepared. Degradation of the rhamnogalacturonan core decreased the activity slightly, the two acidic arabinogalactan units comprised of highly active arabinogalactan and galactan side chains. It was concluded that the 1,3,6-β-galactan moiety of the arabinogalactan side chains contributed the most to the effect after the degradation. When the 1,6 D-galactosyl side chains were removed, the activity was not altered, intact 1,6 galactan chains had negligible activity, and it was thus concluded that the 1,3-β-D galactan backbone was essential for the activity. The "mother-molecule" AGIIb-1 mainly expresses the effect on the complement system via the classical pathway, while the one obtained after enzymatic removal of the external arabinose moieties expressed its activity via both classical and alternative pathways. The suggestions from these findings were that the 1,3,6-β-D-galactan moiety is involved in the expression of activity via both pathways, while the Ara$f$ side chains may inhibit the expression of the activity through the alternative pathway. In other studies a reduction of the activity was seen after removal of the galactose units by the exo-β-1,3-galactanase, but the digested product was still active. These results also showed that 1,6-linked galactose side chains are important for the activity, and that the attachment of those to a 1,3 galactan backbone is necessary for the optimum activity [38].

As most of the complement activating polysaccharides present in *Angelica acutiloba* contain pectin with "ramified" regions also being important for the bioactivity of the polymers, Kiyohara et al. [39] studied in detail the relationship between structure and activity of these regions. They found that digestion of the polymer called AR-2IIa with $endo$-α-D-1,4-polygalacturonase gave rise to a ramified region called PG-1a, being a rhamnogalacturonan with neutral side chains, and oligogalacturonides. The resistant product, E-PG-1a, obtained after degradation with exo-α-L-arabinofuranosidase and exo-β-D-galactosidase had the same effect on the complement system as the ramified region itself. This core contained both long and short galactosyl chains consisting of a non-reducing terminal, 1,6-linked and 1,3,6-linked units for the long chains and the short ones contained basically 1,6-linked units. Degradation of the GalA moieties in PG-1a decreased markedly the effect on the complement system, while the long and short galactosyl chains expressed $\sim$ 50 and $\sim$ 20%, respectively, of the activity of E-PG-1a.

It is interesting to note that another root pectic polysaccharide from the same plant also expresses antitumour activity against an ascitic form of Sarcoma-180, IMC carcinoma, a Meth A fibrosarcoma, as well as the solid form of a MM-46 tumour [40]. Structural studies on this polymer showed that it consists of a rhamnogalacturonan moiety with branches on C4 of rhamnose, and also contained a highly branched 3,5 arabinan and a 1,4-linked galactan, indicating an arabinogalactan type I (AG-I) structure. It also contains 1,3, 1,6 and 1,3,6-linked galactose units, which were also present in the pectic polymer with effect of the complement system. It was interesting, though, that the antitumour polysaccharide did not have significant effect on the complement system, indicating that the neutral 1,4-linked galactose chains may be of importance for the anti-tumour activity. Early studies on pectins from Japanese medicinal herbs indicated that *A. acutiloba* polysaccharides contained RG-II, but activities related to this structural element have not been pursued further [41].

3.3
Atractylodes lancea DC Asteraceae

The rhizomes of the plant *Atractylodes lancea* were shown to contain three polysaccharides that contributed to the expression of the immunomodulating activity that was found from a preparation of traditional Japanese Kampo medicine [42]. The test system for the immunomodulating activity was based on the ability of the polymers studied to express a stimulating effect on the cytokine production of the Peyer's patch cells. One of the polysaccharides was characterised as an arabino-3,6-galactan (AG II). After removal of the outer Ara*f* units and treatment of the remaining polymer with an *endo*-β-D-1,6-galactanase, the activity was remarkably decreased. Structural analyses showed that the removed side chains were mainly composed of β-D-1,6-galactopyranosyl oligosaccharides with a DP ranging from 1 to 8 units. Degradation of the β-D-1,3 galactan backbone also reduced significantly the activity, indicating that some of the side chains attached to the backbone were also responsible for the activity. The basic structure of the polymer is a β-D-1,3 galactan backbone with β-D-1,6-galactopyranosyl side chains attached, and these are again decorated with arabinofuranoside residues, the major part being terminal units [43, 44].

Two pectic polysaccharides (ALR-a and ALR-b) were also responsible for intestinal immune system modulating activity [45]. The pectins were degraded with an *endo*-α-D-1,4-polygalacturonase, that on gelfiltration gave three fractions. The highest Mw fraction of ALR-b was shown to consist of the hairy region, RG-I, and the LMW fraction of oligogalacturonans, primarily. None of these fractions showed any intestinal immune system modulating activity, while the intermediate molecular weight fraction, called PG-2, showed a potent effect in the same system. Chemical analysis of this showed that it contained several of the sugars normally found in rhamnogalacturonan II (RG-II), but did show differences to this one on the linkage type, and also by lacking some of the sugars normally found in RG-II. Gelfiltration and anion exchange chromatography of the enzyme degraded ALR-a resulted in a fraction called ALR-a-Bb with potent intestinal immune system modulating activity through the Peyer's patch cells. This fraction resembled RG-II based on the monosaccharide composition in the same way as PG-2 did. A review of these bioactive polymers has also been written [46].

3.4
Bupleurum falcatum L. Umbelliferae

The roots of the Sino-Japanese medicinal herb have a long tradition for use in the treatment of chronic hepatitis, nephrotic syndrome and various auto-immuno diseases. Yamada's group have extensively studied the polysaccharide responsible for the anti-ulcer and mitogenic effect shown for the active principle in the roots. The polysaccharide fraction responsible was denominated Bupleuran 2IIc, and was shown to be a pectic type polymer. It consists of approx. 70% α-1,4-linked galacturonic acid, of which 30% are methyl-esterified. Parts of the galacturonic residues present are also branchpoints. Bupleuran 2IIc also contains ramified or hairy regions consisting of a rhamnogalacturonan core having neutral side chains of mainly galactose and arabinose units, attached to the 2-linked rhamnose units in the main core or to the 4-linked galacturonic acid residues. This region has structural elements typical of the rhamnogalacturonan type I (RG-I) pectins. The Bupleuran 2IIc also contains a minor region with similarities to the pectic rhamnogalacturonan II (RG-II) that contain the rare sugars KDO, DHA, Apiose and Aceric acid [46, 47] (and refs. cited therein). Bupleuran 2IIc was shown to have potent complement activating and antiulcer activities. The mechanism behind the mucosal protection was suggested to be due to its anti-secretory activity on acid and pepsin, its increased protective coating and its radical scavenging effect, but was not involved in the action of endogenous prostaglandins and mucus synthesis [46] (and refs. cited therein). As the traditional medicines are mainly taken orally, they also studied the possible uptake and tissue distribution of the Bupleuran 2IIc by means of a polyclonal antibody raised in rabbits by injection of the hairy region obtained after *endo*-polygalacturonase treatment of the native polymer [48]. When Bupleuran 2IIc was intravenously injected into mice, the polysaccharide disappeared from circulation within 24h, and was mainly detected in the liver by an ELIZA method based on the antibody raised. When a crude mixture containing mainly Bupleuran 2IIc was administered orally to the mice, the polysaccharide was detected in the liver and in the Peyer's patches. It was found that the outer part of the hairy region consisting of two different oligosaccharide chains having GlcA or 4-O-MeGlcA at the non-reducing end bound to 6-linked galactose units were the antigenic epitopes of the anti-ulcer polysaccharide Bupleuran 2IIc, Fig. 6 [49]. This verifies that the polymer can have an effect when administered orally.

The other effect that was substantially studied was the mitogenic effect of the polymer. When mice were fed the polysaccharide for 7 days, proliferative responses of the spleen cells were enhanced. In vitro studies showed that Bupleuran 2IIc proliferates B-cells in the absence of macrophages, and the activated B-cells are induced into anti-body forming cells in the presence of IL-6. Bupleuran 2IIc in the presence of the antibody raised against the hairy region gave a reduced mitogenic activity. The mitogenic activity was

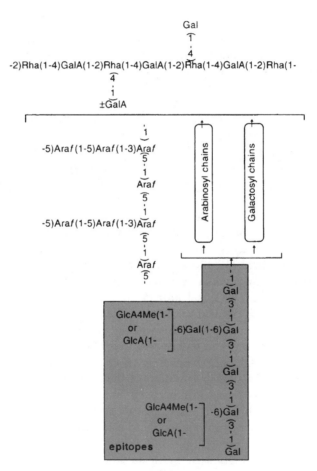

Fig. 6 Proposed structure of the antigenic epitopes in the "ramified" region of bupleuran 2IIc for anti-bupleuran 2IIc/PG-1-IgG as proposed by Sakurai et al. [49]

also reduced in the presence of β-D-Glc$p$A-1,6-β-D-Gal$p$-β-D-1,6-β-D-Gal$p$ or the dimer with one gal$p$ unit, showing that the epitope on Bupleuran 2IIc that was recognised as the antibody binding part of the molecule also acts as the active site of the polysaccharide for the mitogenic activity. Bupleuran 2IIc also enhances the IgM secretion from highly purified murine normal B-cells. The hairy or ramified region of the polysaccharide showed potent IL-6 secretion-enhancing activity, indicating that the same active site as above may partly contribute to the enhancement of IgM secretion as an autocrine and/or paracrine mechanism [50].

3.5
Cistanche deserticola Y.C.Ma. Scrophulariaceae

This organism is a holoparasite that grows on the roots of the hardwood *Haloxylon ammodendron* that is widely distributed in the Gobi desert, Mongolia, and has a wide use as a traditional medicinal remedy. The drug was found to contain various polysaccharides of pectic nature with mitogenic and comitogenic effect using Zymosan as a positive control. Methylation and NMR spectroscopic studies show that the pectic type polymers in *C. deserticola* most probably are of the rhamnogalacturonan type I, having side chains both of arabinogalactan type I and type II. Methyl-esters and acetyl-groups are identified by carbon-NMR spectroscopy [51]. Further studies lead to the isolation, after enzymatic degradation for removal of starch and protein, different precipitation methods, ion-exchange chromatography and gelfiltration, of a pectic arabinogalactan called Cistan A. This polymer can be characterised as an immunomodulating pectic substance based on the dose-dependant mitogenic and comitogenic effect found that is higher than the standard used. The methylation analysis of Cistan A shows that this molecule does not contain the common features of the Rhamnogalacturonan I type polymers as the rhamnose basically is 1,4-linked, not 1,2 as is common in the RG-I type. The arabinose and galactose linkages confirm that both AG-I and AG-II polymers are attached to the main core of Cistan A. Other bioactive fractions were also obtained, their activity was lower than that of the Cistan A. Their monomeric composition were fairly similar apart from one having a higher content of xylose than the other [52].

3.6
Cuscuta chinensis Lam. Convolvulaceae

Polysaccharides isolated from the seeds of *C. chinensis* have effects both as immunostimulants and as antioxidants. The polysaccharide CS-A-3-β has a backbone of α-D-1,4-linked Gal$p$A and β-L-1,2-Rha$p$ units with branches at C-4 of the Rha$p$ residues and at C-3 of Gal$p$A residues that are composed of an arabinogalactan and glucobiose. The Ara$f$ units are terminal and 1,5-

linked, while the Gal units are terminal, 1,6- and 1,3,6-linked, typical for the AG-II type polymer chains as part of the rhamnogalacturonan type I (RG-I) polysaccharide. The glucobiose is α-1,4-linked and may be linked to the GalA units. The effect of the polysaccharide fractions on hydrogen peroxide induced cell lesion of rat pheochromocytoma line PC 12 was tested. Pretreatment of the cells prior to exposure with hydrogen peroxide gave an enhanced cell survival when the pure CS-A-3b was used, showing that this polysaccharide has a protective effect against hydrogen peroxide induced cell toxicity on PC 12. The glucobiose on the GalA units is not common for pectins and is a characteristic feature for this polymer that may be partly responsible for the effect seen. This polysaccharide was probably the first reported to protect resting nerve cells from free radical-induced injury. Further structural studies will reveal the part of the pectin that is responsible for the bioactivity shown [53].

3.7
Diospyros kaki L. Ebenaceae

D. kaki leaves contain a polysaccharide that has a backbone of alternating α-1,4 GalpA units and α-1,2 Rhap units. Most of the Rha units are branchpoints with branches on C4, consisting of β-1,4-linked xylose units, and others consisting of β-1,3 and β-1,6-linked galactose units. These side chains are also substituted with Araf on O-2 of the xylose units and O-3 on the β-1,6 Gal units. The structure proposed is not common amongst the rhamnogalacturonans as the side chain on the xylose units appear to be rare. The polysaccharide stimulates the LPS-induced B-lymphocyte proliferation, but not ConA-induced T-lymphocyte proliferation. It is proposed that the labile Araf units not are important for the expression of the enhancement of the immunological activity, the presence of GalA in the backbone appears to have an important, but not a crucial effect on the expression of the activity. Further studies will have to be performed on this polymer in order to verify this new type of pectic polymer [54].

3.8
Entada africana Guill. et Perr. Mimosaceae

The polysaccharides from Entada africana were isolated by water extraction and further separated by anion exchange chromatography. The fraction called EA 100 Acidic 1, had an effect in the complement system equivalent to the standard polymer used. This polymer was shown to basically be an arabinogalactan protein with a relatively high degree of xylose. The ratio ara : gal : xyl was 3 : 2,5 : 2 and the protein part was rich in hydroxyprolin. Linkage analyses as well as precipitation with the Yariv reagent showed that this polymer belongs to the AG-II type. The xylose present was 1,4-linked. Removal of Araf units by weak acid hydrolysis converted most of the 1,4-xylose units into terminal ends, indicating that araf was linked on position 4 of single xylose units. It was also obvious that the Ara*f* units were linked through both positions 3 and 4 of galactose, and in addition to the presence of 1,3,6-linked galactose, 1,4-linkages were also present, indicating that part of the fraction contains AG-I structures. Removal of the Ara*f* units by acid hydrolysis decreased the activity in the complement system. This may be caused either by the denaturing of the protein part or the fact that Ara*f* units influence the activity. The polysaccharide fractions extracted from *E. africana* also contained polymers of the RG-I type, but the activity in the complement system for these was minor compared to that of EA 100 Acidic 1 [55].

3.9
Glinus oppositifolius (L.) Aug. DC. Aizoaceae

The West-African plant *Glinus oppositifolius* has been used, amongst other ailments, in the treatment of wounds. The plant was shown to contain polysaccharide fractions that gave a high effect in the complement system. Two pectic type polysaccharides, GOA1 and GOA2, was isolated by different chromatographic methods. The polysaccharide GOA1 contains terminal, 1,3- and 1,5-linked Ara*f* and 1,4-, 1,3- and 1,3,6-linked Gal*p*, suggesting the presence of both AG-I and AG-II type arabinogalactans. The GOA2 is supposed to be an RG-I type pectic polymer with side chains of AG-II structures attached to position 4 of the rhamnose units in the main core. In addition to the potent complement fixing activity, both polymers were shown to have chemotactic properties towards macrophages, T-cells and NK-cells, showing that both polymers have immunomodulating abilities [56].

3.10
Glycyrrhiza uralensis Fisch ex DC. Fabaceae

The first paper on the bioactive polysaccharides from *Glycyrrhiza uralensis* roots was published in 1996 by Kiyohara et al. [57]. They isolated a pectic type polymer with anti-complementary and mitogenic activity that was an acidic pectin, possibly containing rhamnogalacturonan type I as part of the total structure. Degradation of the uronic acid part of the molecule decreased both types of bioactivities. The neutral oligosaccharide chains were shown to retain some of the activities of the native polymer, but it was suggested that they should be attached to the acidic core to retain maximum activity.

It was also shown that the pectic type polymers from *G. uralensis* contain a minor part of the RG-II type structure [26].

In a more recent study it was shown that the roots from *G. uralensis*, after isolation, fractionation and purification, contain two bioactive polysaccharides of the pectin family termed GU-3IIa-2 and 3IIb-1.

They are both immunologically active, but they did react differently in the different test systems used. GU-3IIa-2 was a potent bone marrow cell proliferating activity enhancer, while GU-3IIb-1 was a potent NK cell-mediated tumour cytotoxicity enhancer. 3IIa-2 is mainly composed of arabinose (35%), galactose (25%) and galacturonic acid (15%). The polymer contains arabinogalactan type II as it forms a precipitate with the Yariv reagent, and is also composed of terminal, 1,5 and 1,3,5-linked arabinose units, and a high degree of 1,3,6-linked galactose units in addition to terminal, 1,3-, 1,4-, and 1,6-linked units. The galacturonic acid is mainly 1,4-linked, with some terminal and 1,2,4-linked units.

3IIb-1 is a more complex pectic type polymer. It contains the most common types of linkages for acidic pectins and it does not precipitate with the Yariv reagent and methylation studies shows that the galactan moieties are rich in 1,4 linkages, typical for the arabinogalactan type I polymer. This bioactive polymer is most probably a typical rhamnogalacturonan type I containing AG-I structures. Detailed structure activity relations of both of these polymers await further studies [58].

3.11
Larix occidentalis Nutt, Pinaceae

Larch arabinogalactan has a structure (Fig. 7) that can be classified as an arabinogalactan type II, but it differs from other AG-II's so far found in the plant kingdom as the larch arabinogalactan will not precipitate with the Yariv-reagent, and it does not have an effect on the complement system, as reported by Yamada et al. [4]. The larch arabinogalactan has been shown to have various other interesting biological activities making the polymer an excellent source as a dietary fiber, and as such has been approved by the FDA. It increases the production of short fatty acids, and decreases the generation and absorption of ammonia. It also appears to have a significant effect on enhancing beneficial gut flora, especially the *Bifidobacteria* and *Lactobacillus*. Experimentally it has been shown that the larch arabinogalactan can stimulate natural killer cell cytotoxicity and other functional aspects of the immune system, and also inhibit the metastasis of tumour cells to the liver [59].

A monograph on the larch arabinogalactan has been published, containing pharmacokinetics, clinical indications, the lack of side-effects and dosage [60]. In a report from 2003, the effect over time of in-vivo administration of the larch arabinogalactan on the immune and hemopoietic cell

Fig. 7 Structure of the Arabinogalactan type II polymer from Larch

lineages in murine spleen and bone marrow is discussed. Results show that 7–14 days after administration to mice, lymphoid cells in the bone marrow were significantly decreased relative to control, but remained unchanged at both time intervals in the spleen. NK cells were also, after 7 days exposure, decreased significantly in the bone marrow, but not in the spleen. After 14 days the NK cells in the bone marrow returned to a normal level, but were increased in the spleen.

A vast cascade of cytokines appear to be induced by the presence of this polysaccharide, and immunopoiesis- and hemopoiesis-inhibition are probably the most prevalent during the first two weeks of daily exposure [61]. Studies relating structure to the biological activity have not been performed.

3.12
Lycium barbarum L. Solanaceae

The fruit of the medicinal plant *Lycium barbarum* contains an arabinogalactan protein, for which the carbohydrate part was classified as an arabinogalactan type I polymer. The structural features of the carbohydrate part are somewhat unusual as it is proposed, from the data obtained, to consist of a backbone of β-1,4-linked galactose units, and all units are branched at position 3 with chains of different compositions, see Fig. 8. The native polymer was shown to promote splenocyte proliferation directly in normal mice, and the carbohydrate part of the polymer had a stronger effect than the glycoconjugate, showing that the immuno-modulating effect of the glycoconjugate resides in the arabinogalactan moiety. Experiments also showed that the most likely target cells were the B-lymphocytes that appear to carry receptor binding sites acting with the polymer [62].

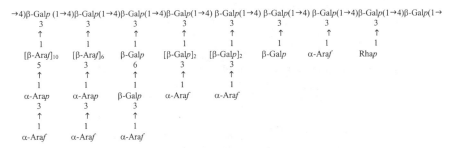

Fig. 8 Arabinogalactan from *Lycium barbarum* as proposed by Peng et al. [63]

3.13
Melocactus depressus Hook, Cactaceae

The pulp of this cactus contains an arabinogalactan type I polymer with the ability to stimulate phagosytosis. The galactose units are 1,4-linked with branches on position 3 on some of the galactose units and on position 6 on others. The side chains are proposed to be trisaccharides consisting of an Ara*f* unit that has terminal Ara*f* substituted on positions 2 and 5 in some cases and 3 and 5 in others [63].

3.14
Panax ginseng C.A.Meyer, Araliaceae

Ginsenan S-IIA, a polysaccharide fraction from the roots of *P. ginseng* is a potent inducer of IL-8 production by human monocytes and THP-1 cells, and this induction is accompanied by increased IL-8 mRNA expression. The polysaccharide appears from the structural feature to be a mixture of arabino-galactan type I and type II, based on the presence of 1,3-, 1,6-, 1,3,6-, 1,4-, and 1,4,6-galactose units as well as terminal arabinose and 1,5-, 1,3,5-, and 1,2,5-linked units. It also contains 1,4,6-linked glucose units that together with the 1,2,5-linked arabinose units are different from the units found in other ginseng polysaccharides and may thus be of importance for the activity [64].

Already in 1988 and 1991, Gao et al. [65,66] detected four different polysaccharides present in the leaves of *Panax ginseng* that had an effect on the complement system, but only two of them, the neutral, GL-NIa, and one of the acidic ones, GL-AIa, had potent activities at low concentrations. GL-NIa was found to be mainly an arabinigalactan type II polymer. GL-AIa was a polysaccharide with a rhamnogalacturonan core with neutral side chains of the AG-II type, confirmed by a strong reaction with the Yariv reagent and the methylation results. It was shown that the crude polysaccharide fraction contained KDO and DHA, suggesting the presence of Rhamnogalacturonan II in

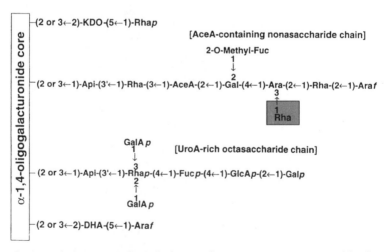

Fig. 9 Rhamnogalacturonan II from the leaves of *Panax ginseng* as proposed by the group of Yamada [3]

the fraction. Three different RG-II polymers were isolated without the digestion of the native polymer with *endo*-polygalacturonase [67, 68]. One of these, denominated GL-4IIb-2, was a macrophage Fc receptor expression enhancing polysaccharide, and also had the most potent IL-6 production enhancing activity of macrophages of the polymers isolated [67, 69]. The structure of this RG-II (Fig. 9) was shown to be slightly different in the structure than the RG-II shown in Fig. 4. The active polymer was present as a dimer with the borate linking the apiose of the two units together. Dissociation led to diminished activity, while redimerisation recovered the activity [68]. As other RG-II dimers were tested for the same activity giving no effect, it was suggested that the specific structure of the GL-4IIb-2 dimer was necessary for the activity [68]. Yamada's group has also found that this polymer shows a potent secretion enhancing activity of the nerve growth factor (NGF) which is known to play a role as a neurotrophic factor for the survival of neuronal cells and preventing aging and dementia, indicating that the polysaccharides of ginseng may have several interesting biological activities [3].

3.15
Piper nigrum L. Piperaceae

Black pepper contains several polysaccharides of which one shows a strong effect as an immune enhancer based on the fact that the polymer is an anticomplementary polysaccharide. The polysaccharide has an Mw of approx. 40 kD. It is composed basically of rhamnose, arabinose, galactose and galacturonic acid, and shows a high binding capacity for the Yariv reagent. This indicates that the side chain of the polymer is of the arabinogalactan type II, which is a common structure for several polysaccharides with an effect on the complement system [70].

3.16
Plantago major L. Plantaginaceae

The leaves of *Plantago major* have a long tradition in most areas where the plant grows as a woundhealing agent [71–73]. Two different polymers of the pectic type with high activity in the complement system are present in the leaves of *P. major*. PMIa was shown to be an arabinogalactan type II polysaccharide, it gave a positive Yariv reaction, and is composed of a galactan 1,3 backbone, heavily branched with 1,6-linked side chains linked through O-6 of the galactan core. To these side chains, terminally and 1,5-linked arabinofuranoside-residues are linked through position O-3 of the 1,6-linked galactan side chains. The polymer contains a protein part that is rich in hydroxyproline, serine and alanine, being typical of AG-II polymers [71].

The other polymer denominated PMII [18, 72, 73], is a pectic type polymer with mainly a galacturonic acid backbone. The structure of PMII was found to be similar to the general type described by Voragen et al. [22] with a smooth region, consisting of α-D-galacturonosyl units 1,4-linked and with hairy regions. Two different types of hairy regions were isolated after degradation of the polymer with pectinesterase and pectinase. The main core in both regions were alternate 1,4-linked galacturonic acid units and 1,2-linked rhamnose units. The main difference structurally between the two types of hairy regions were the molecular weight, the GalA/Rha ratio, being higher for the compound with the lowest molecular weight and a lower degree of substitution of the rhamnose units for the same fraction. The one with the largest Mw was called PVa, the smallest PVb. The first was more active in the complement system than the latter. The structure of PVa resembled the RG-1 structure shown above.

Arabinofuranosides were removed from PVa, which resulted in an increase in the activity; this was in contrast to the findings of Kiyohara et al. [35] who found no change in activity after removal of similar units from *A. acutiloba* pectin. For this polymer it was suggested that the minimum requirement for complement activation via the classical pathway was β-1,6-linked galactan attached to the rhamnogalacturonan backbone, which also appears to be an important part of the backbone for PVa.

More detailed studies on the interaction between human complement and PMII [74] showed that PMII appears to be an activator both on the classical and the alternative pathway. The complement activation was performed using serum from ten different individuals as the complement source. The surprising result was that the activation differed considerably depending on the complement source, a 200-fold difference in ICH_{50} value was observed. The levels of antibodies against PMII detected in the different sera did not correlate with the levels seen for ICH_{50} activity of PMII. PMII appears to be as potent an activator of complement as aggregated human immunoglobulin, (IgG), and these results might be related to the reported wound-healing effect of the leaves of *Plantago major*. An in vivo study was also performed, and the results demonstrated that PMII protects against pneumococcal infection in mice when administered systemically prechallenge, and also that the protective effect was owing to stimulation of the innate and not the adaptive immune system [75, 76].

3.17
Salvia officinalis L. Lamiaceae

The aerial part of sage was successively extracted with water, potassium oxalate, DMSO and KOH, and gave rise to different fractions with bioactivity. Partial purification of the water extract gave rise to a polydisperse fraction called A, that based on the monosaccharide composition, IR, and NMR, was thought to be composed of arabinogalactans associated with the highly ramified rhamnogalacturonans core. Fraction B, extracted with oxalate, was characterised (based on the same information as for A) as a typical pectin material with arabinan side chains. Both of these polysaccharide fractions are thought to be partly responsible for the immunomodulatory effect of the sage extract. The effect was shown by the mitogenic and comitogenic activity the polymers exhibited [77].

3.18
Tinospora cordifolia Miers, Menispermaceae

The stems of the tree were found to contain polysaccharides consisting of arabinose, galactose and galacturonic acid and only minor amounts of rhamnose. Structural studies indicate that the polymeric material consists of 1,4-linked galacturonic acid residues, terminal, 1,4-, 1,6- and 1,3,6 galactose units and terminal and 1,5-linked arabinofuranose residues. Further studies must be performed on this in order to determine what type of pectin it can be classified as. The linkage data indicate that both AG-I and AG-II are present. This polymer was shown to activate polyclonal B-cells [78].

The antioxidant properties of the polysaccharides against iron-mediated lipid damage and γ-ray induced protein damage was studied. The polysaccharide was shown to be a good protector against the iron-mediated lipid peroxidation of rat brain homogenate revealed by the thiobarbituric acid reactive substances and the lipid hydroperoxide assays. The polysaccharide provides a significant protection to proteins against γ-ray induced damage. The protective effect was explained by the high reactivity towards DPPH, superoxide radicals and the most damaging of the radicals, the hydroxyl radical [79].

3.19
Trichilia emetica subsp. suberosa JJ de Wilde, Meliaceae

The leaves of the tree *T. emetica* have a traditional use as a wound healer, and is especially useful against old wounds, both infested and cancerous ones. The polysaccharide content of the leaves was shown to have an effect on the complement system [80] and was studied further. Both extractions with water (50 and 100 °C) gave an active fraction. Anion exchange chromatography of the material extracted at 100 °C leads to four different fractions with varying activities. Structurally, all acidic polymers have similar features with 1,4-linked galacturonic acid as the main constituent and rhamnose both 1,2 and 1,2,4-linked, indicate that the polymer is of the RG-I type with side chains on C-4 of rhamnose. The side chains appear to be of the AG-II type having all the relevant structural features of this as described above. Removal of the Ara*f* units by weak acid hydrolysis reduced the effect on the complement system, indicating that these units may play a role in the activity. It appears that the arabinose units primarily are linked through position 3 on 1,6-linked galactose units, as the 1,3,6-linked units were substantially reduced for all polymers with a concomitant increase of the 1,6-linked units. No alteration of the amount of 1,3-linked galactose was seen indicating a backbone of 1,3-linked units with side chains of 1,6-linked galactoses that are decorated with ara*f* units [80].

3.20
Vernonia kotschyana Sch. Bip. Ex Walp.
Baccharoides adoensis var. *kotschyana* (Sch. Bip. Ex Walp.), Asteraceae

The roots of this plant are extensively used in Mali for the treatment of gastrointestinal disorders and wound healing and are part of a registered improved traditional medicine (ITM) called Gastrocedal. Water extracts of the roots gave rise to acidic polysaccharide fractions that showed a dose dependant activity in the complement system. The monosaccharide composition revealed those typical for pectic substances in addition to a high content of fructose, which was shown to be an inulin that was inactive in the test systems used in this paper. The Yariv reagent revealed that all fractions contain AG-II type polymers. The acidic fractions denominated Vk50A2 and Vk100A2 showed mitogenic activity as they proliferated spleen cells in a dose-dependant manner. When the cell population of the spleen cells responding to the stimulation by the two polysaccharides was investigated by flow cytometry analysis, the population of B-cell positive cells increased during the experimental period. An increase in apoptotic cells was not observed suggesting that the increase of B-cells was not due to a decrease of T-cells. Further, it was found that the polysaccharide fractions could be characterised as B-cell mitogens, and that the T-cells and macrophages were not involved in the stimulation. The induction of the lysosomal enzyme activity in macrophages was dose dependant, but the activity somewhat lower than the positive control used [81]. Further studies of the Vk100A2 fraction lead to the isolation of two fractions by size exclusion chromatography, Vk100A2a and Vk100A2b. The latter, containing mainly galacturonic acid, ~ 85%, and smaller amounts of the neutral sugars arabinose, galactose and rhamnose, showed a complement fixing ability, but no activity on the proliferation of B- and T-cells. Vk100A2a showed a dose-dependant complement fixing ability and a T-cell independent induction of B-cell proliferation. Both polymers induced chemotaxis of human macrophages, T-cells and NK-cells. Enzymatic degradation and methylation studies of Vk100A2a showed that the polymer was a typical arabinogalactan pectin consisting of a highly branched rhamnogalacturonan core with approximately 50% of the rhamnose units as branch points on position 4. The side chains are composed of arabinose as terminal, 1,3-linked and 1,3,5-linked units, the galactose mainly as 1,4-linked units, but also as terminally, 1,6-, and 1,3,6-linked units. The arabinose units are mainly attached to galactose in position 3. Even after enzymatic treatment of Vk100A2a with arabinofuranosidase and/or galactosidase, the resulting core polymer still showed high activity in the complement system suggesting that the complement fixing ability may at least in part be expressed by the carbohydrate structures present in the inner portions of Vk100A2a. The structural studies show that this polysaccharide can be classified as a RG-I type polymer with side chains both of the AG-I and AG-II types. It is interesting to note that after arabinofuranosidase treatment, the complement fixing ability is somewhat higher than for the native polymer indicating that the arabinose is modulating (reducing) the activity. The treatment involving galactanase show a reduced activity indicating that the part removed, being mainly the structural elements consisting of the 1,4-linked galactose units (AG-I) also play a positive role for the complement fixing ability [82].

4
Structure Activity Relations

As said in the introduction, this chapter focusses only on pectic type polysaccharides that have shown activity in different biological systems. The majority have shown effect on the immune system in one way or another, but a few other effects have also been observed. Table 1 gives an overview of the polysaccharide structures found for the active polysaccharides and as can be seen, a majority of the polysaccharides contain a rhamnogalacturonan I backbone. Most of these polysaccharides have attached arabinogalactan II side chains, and a few have arabinogalactan type I chains attached. Examples of pure arabinogalactan II polymers as well as rhamnogalacturonan II polymers have been found. The basic structures of all these polymers are, as described in the section "Chemical structure of pectic type polysaccharides", but as not all pectic type polymers exhibit biological activity, it is obvious that certain specific structural aspects must be present in those that show a bioactivity in the different test systems. Although structural details of the polysaccharides have been presented, only in a few cases have studies been carried out to ascertain which parts or structural details are really responsible for the bioactivity, and the group headed by Professor Haruki Yamada, Tokyo, has done most of this work.

Bupleurum falcatum pectins have been studied in great detail [46–50] as they were shown to have an effect on the complement system, anti-ulcer activity and macrophage Fc-receptor up-regulating activity to enhance immune complex clearance. The most potent fraction was the Bupleuran 2IIc that also showed a potent mitogen effect against mouse spleen cells and Peyer's patch cells of the small intestines in vitro. Detailed structural studies revealed that the ramified region contained the bioactive parts, and on this section of the molecule the oligosaccharide β-D-4-O-methyl-Glc$p$A- or β-D-Glc$p$A-1,6-β-

Table 1 An overview of the pectic type structures and bioactivities of the polysaccharides presented in this review

Plant reviewed	Type of structure				Type of activity shown	Ref.
	AG-I	AG-II	RG-I	RG-II		
Acanthus ebracteatus			×		Effect on the complement system	[29]
Angelica acutiloba	×		×		Antitumour activity	[40]
Angelica acutiloba	×	×	×		Effect on the complement system	[32–36]
Atractylodes lancea		×	×	×	Intestinal immune system modulating activity	[42–45]
Bupleurum falcatum		×	×	×	Antiulcer	[46–50]
Bupleurum falcatum		×	×	×	Effect on the complement system	[46–50]
Cistanche deserticola	×	×	×		Mitogenic and comitogenic activity	[51, 52]
Cuscuta chinensis		×	×		Immunostimulating	[53]
Cuscuta chinensis		×	×		Antioxidant	[53]
Diospyros kaki		×	×		Immunostimulating	[54]
Entada africana	×	×			Effect on the complement system	[55]
Glinus oppositifolius	×	×	×		Effect on the complement system	[56]
Glinus oppositifolius	×	×	×		Chemotactic properties towards Macrophages, T- and NK Cells	[56]
Glycyrrhiza uralensis		×			Effect on the complement system	[57]
Glycyrrhiza uralensis		×			Mitogenic activity	[57]
Glycyrrhiza uralensis		×			NK cell-mediated tumour cytotoxicity enhancer	[58]
Glycyrrhiza uralensis		×			Bone marrow cell proliferating activity	[59]
Larix spp		×			Various effects on the immunesystem and as a dietary fiber	[50–61]
Lycium barbarium	×				Immunomodulating	[62]
Melocactus depressus	×				Phagocytosis stimulating	[63]
Panax ginseng	×	×			Stimulation of production of IL-8 (Immunomodulating)	[64]

Table 1 continued

Plant reviewed	Type of structure				Type of activity shown	Ref.
	AG-I	AG-II	RG-I	RG-II		
Panax ginseng				×	Intestinal immune system modulating activity	[67–69]
Panax ginseng				×	Macrophage Fc receptor expression enhancer, IL-6 production enhancer, nerve growth factor secretion enhancer	[3, 67–69]
Panax ginseng		×	×		Effect on the complement system	[65, 66]
Piper nigrum		×			Effect on the complement system	[70]
Plantago major		×	×		Effect on the complement system	[72, 73]
Plantago major			×		Pneumococcal infection-protector	[75, 76]
Salvia officinalis			×		Mitogenic and comitogenic effect	[77]
Tinospora cordifolia			×		Activation of B-cells	[78]
Tinospora cordifolia			×		Antioxidant	[79]
Trichilia emetica		×	×		Effect on the complement system	[80]
Vernonia kotschyana	×	×	×		Effect on the complement system	[81, 82]
Vernonia kotschyana	×	×	×		Effects in different systems involved in immunomodulation	[81, 82]

D-Gal$p$-β-D-1,6-β-D-Gal$p$ was shown to be a possible structural unit for the recognition of the carbohydrate receptors on the B-cells. This has not yet been published for other polysaccharides, but the group of the authors has in collaboration with Yamadas group found similar oligosaccharide structures in polysaccharides from plants that traditionally have been used against ulcers. The polyclonal antibodies prepared having affinity towards this oligomeric structure were used to show that Bupleuran 2IIc was indeed taken up by the body when given orally [48, 49].

The structure that appears to be important for the effect on the complement system is most probably the complex galactan oligomer being composed of separate 1,3- and 1,6-linked galactose chains with branch points being of 1,3,6 nature. These units have been proposed as the active sites for this activity for the pectins from *Angelica acutiloba* [36–39], *Glinus oppositifolius* [56], *Glycyrrhiza uralensis* [26, 57, 58] and *Vernonia kotschyana* [81, 82], but they are also found in most of the polysaccharides that have an effect on the complement system. This is the so-called AG-II structure. But as not all polysaccharides containing this have an effect on the complement system, for example the larix AG-II [4], other factors related to the position of these structural units on the polymer must also be important, which has not been clarified yet. The size of these structures may be important [30] as well as the possibility that it has to have more than one binding site to be active.

Although the ramified region appears to be the most important structural feature for the effect on the complement system, it has been shown that the presence of the homogalacturonan region in the native polymers are responsible for a modulating effect, i.e. they have a down-regulating effect on the activity. This has been shown for pectins from both *Angelica acutiloba*, *Glycyrrhiza uralensis*, *Plantago major* and for various polysaccharides that are under study in the laboratory of the authors of this chapter. The Ara*f* units may also play a modulating role on the complement effect. It is also interesting to note that the pectic type polymer from *Acanthus ebracteatus* that is rich in 3-O-methyl galactose, coupled with an Mw higher than 1 mill., shows an extremely high activity compared to those pectins from the same source almost devoid of this structural feature [29].

One of the pectic fractions from *Angelica acutiloba* showed potent antitumour activity, and this polymer was rich in the AG-I type structure, indicating that the 4-linked galactose units are important for this activity [40].

Pectin polymers containing the well conserved rhamnogalacuronan II have been shown to have different types of effects in immunological test systems, but only a few of those polymers have undergone structural studies. Most of the polymers have only been characterised for the presence of some of the unusual monosaccharides that are normally found in RG-II. The one most thoroughly studied was isolated from the leaves of *Panax ginseng*, and this is according to Kiyohara [83] the only RG-II polymer that has been shown to have biological activity, and it was also shown that only the dimer form of the ginseng RG-II is bioactive [68]. How the active structural features of *P. ginseng* differ from the structural features of RG-II's isolated from other plants is not known.

Bioactivities found for some of the polysaccharides described in this chapter have been assigned to certain structural features. The antioxidant effect of the *Cuscuta chinensis* pectin was proposed to be caused by the presence of a glucobiose unit linked via a GalA unit on the RG-I polymer [53], but this structural feature was not found for the anti-oxidant polysaccharide from *Tinospora cordifolia* [78, 79].

Sufficient scientific data is still lacking to really pinpoint the bioactive sites of the pectic type polymers described in this chapter, but on the basis of the work of the group headed by Yamada over the last ten years, a better understanding of the importance of some of the structural features has been obtained. There appears to be more than one active site to accommodate all of the different activities seen, and from this review certain features have emerged as being more important than others.

5
Conclusion

From the research reviewed in this chapter it is obvious that pectins may be an important source of biologically active substances that can be used to improve the health of mankind, and especially the health of those from less favoured regions of the Earth. It is also interesting to note that most of the plants used for the isolation of the bioactive pectic type polymers reviewed have a long tradition in the use against various ailments for which the immune system may be involved in one way or another. For this reason, ethnopharmacological surveys carried out amongst people still using plants as an important part of their healthcare may be an important source for finding new bioactive plant pectins. A few surveys have been reported over the last decade. Yamada [84] has given an overview up to approx. 1995 of the contribution of pectins to healthcare. He shows that pectins have a variety of pharmacological effects, such as immunostimulating, anti-metastatic, anti-ulcer, anti-nephrosis activities, and cholesterol-reducing effects amongst others. He also describes the use of pectins in drug delivery, as adjuvant in vaccines, and also how the fine structure of each polymer is important for the activity shown of the specific polymer and concludes with observations on how pectins can be used in many ways for the benefit for human beings.

Yamadas group [85, 86] has also taken a Japanese Kampo medicine consisting of many different plants as a starting point for identifying bioactive plant polysaccharides. They found in the Kampo medicine Juzen-Taiho-To, composed of many plants, several bioactive polysaccharides with effects in different test systems that may influence the immune system. A study like this can lead to the identification of the best possible source of the plants in the mixture that contain bioactive polysaccharides.

It is not only the traditional use of a plant that may lead us to interesting new plants for bioactive pectins. From the far east, i.e. the Maritime territory and the Amur Region of Russia, Tomshich et al. [87] studied ten species from six families and found that they contained polysaccharides with possible pectic structure, and five of these plants were shown to have either immunostimulatory or antitumoural activity, or both.

Four different European herbaceous plants have also been tested for possible mitogenic and comitogenic activities. Using Zymosan as a positive control, all polysaccharide fractions tested had higher activity in the bioassays than the Zymosan. The monosaccharide composition of the polymers investigated vary, but indicate that polysaccharides of the pectic type may be present. Further fractionation and purification of the fractions obtained may lead to the identification of pectic type polysaccharides with immunomodulatory effects [88].

Various pectins from plants traditionally used in Europe were tested for antimutagenic activity against nitroaromatic compounds [89]. Of those studied, the following pectic type polymers were found to be active: Araban from sugar beet, weakly positive; acidic pectin from apple, effective; pectin from *Cichorium sp.*, *Citrus sp.*, pectins from sugar beet were all weakly effective, while the rhamnogalacturonan from *Althaea officinalis* roots was found to be strongly effective. The latter has a rhamnogalacturanan type I structure, but is unusual as it contains side chains of glucuronic acid.

Recently, the focus has also been on wound healing plants traditionally used in the West-African country Mali. Two surveys are reported; one from the region around the capital Bamako [90], and the other from a more rural area, Dogonland [91]. The survey from the Bamako region reports the use of 123 different species belonging to 50 different plant families. Those most frequently used, were analysed for their content of polysaccharides. The monosaccharide compositions as well as their effects on the complement system were determined, and lead to the conclusion that several of the plants studied are important sources for future studies of bioactive pectic type polysaccharides. The survey from Dogonland reported 73 plants from 34 families used for wound healing that may be of interest for further investigation in order to obtain biologically active pectins.

From the science performed mainly over the last ten years it is obvious that the role of pectic substances in health care has been substantiated. For some of the pectic substances, parts of the structure of the bioactive sites have been determined, but further studies of the relevant structures for the individual active polymers must be performed in order to find a possible common structure for the activities observed. It also appears that there are special structural features present in some of the polymers, which are not found in others, and which are important for their activity, and this may explain the different behaviour of the polymers in the same system.

Acknowledgements The authors are grateful for permission to reproduce the following figures:

Figures 3 and 5 from Biochimie, vol 85, Perez S, Rodrigues-Carvajal MA, Doco T (2003) "A complex plant cell wall polysaccharide: rhamnogalacturonan II. A structure in quest of a function." p 109–p121

Figure 4 from Carbohydrate Research, vol 338, Rodrigues-Carvajal MA, du Penhoat CH, Mazeau K, Doco T, Perez S (2003) "The three dimensional structure of the mega-oligosaccharide rhamnogalacturonan II monomer: a combined molecular modelling and NMR investigation." p 651–p671

Figure 6 from Carbohydrate Research, vol 311, Sakurai MH, Kiyohara H, Matsumoto T, Tsumuraya Y, Hashimoto Y, Yamada H (1998) "Characterization of antigenic epitopes in anti-ulcer pectic polysaccharides from Bupleurum falcatum L. using several carbohydrases." p 219–p229, all with permission from Elsevier

Figure 9 from Paulsen BS (ed) Bioactive Carbohydrate Polymers. Yamada H (2000) "Bioactive plant polysaccharides from Japanese and Chinese traditional herbal medicines", p 15–p24. Kluwer Academic Publishers, with permission from Springer.

References

1. Wagner H, Kraus S (2000) In: Paulsen BS (ed) Bioactive Carbohydrate Polymers. Proc Phytochem Soc of Europe. Klüwer Academic Publishers, Dordrecht, p 1
2. Wagner H, Kraus S, Jurcic K (1999) In: Wagner H (ed) Immunomodulatory agents from plants. Birkhäuser, Basel, p 1
3. Yamada S (2000) In: Paulsen BS (ed) Bioactive Carbohydrate Polymers. Proc Phytochem Soc of Europe. Klüwer Academic Publishers, Dordrecht, p 15
4. Yamada H, Kiyohara H (1999) In: Wagner H (ed) Immunomodulatory agents from plants. Birkhäuser, Basel, p 161
5. Yamada H, Kiyohara H (1989) Abstract of Chinese Medicines 3:104
6. Paulsen BS (2001) Curr Org Chem 5:939
7. Paulsen BS (2002) Phytochemistry Rev 1:379
8. Stimple M, Prolsch A, Wagner H, Lohmann-Matthes ML (1984) Infection and Immunity 46:845
9. Schmidgall J, Schnetz E, Hensel A(2000) Planta Med 66:48
10. Stephen AM (1983) In: Aspinall GO (ed) The Polysaccharides. Academic Press, London, p 97
11. O'Neill M, Albersheim P, Darvill A (1990) In: Dey PM, Harbourne JB (eds) Methods in Plants Biochemistry, vol 2. Academic Press, London p 415
12. Yamada H, Nagai T, Cyong JC, Otsuka Y, Tomoda M, Shimizu N, Shimada K (1985) Carbohydr Res 144:101
13. Yamada H, Ra K-S, Kiyohara H, Cyong JC, Yang HC, Otsuka Y (1988) Phytochemistry 27:3163
14. Clarke AE, Anderson RL, Stone BA (1979) Phytochemistry 18:521
15. Jermyn MA, Yeow YM (1975) J Plant Physiol 2:501
16. Van Holst G-J, Clarke AE (1985) Anal Biochem 148:446
17. Nothnagel E, McNeil M, Albersheim P (1983) Plant Physiol 71:916
18. Samuelsen AB, Cohen EH, Paulsen BS, Wold JK (1996) In: Visser J, Voragen AGJ (eds) Pectins and Pectinases. Elsevier Science, Amsterdam, p 619
19. Yamada H (1996) In: Visser J, Voragen AGJ (eds) Pectins and Pectinases. Elsevier Science, Amsterdam, p 173
20. Talmadge KW, Keegstra K, Bauer WD, Albersheim P (1973) Plant Physiol 51:158
21. McNeil M, Darvill AG, Albersheim P (1980) Plant Physiol 66:1128
22. Voragen AGJ, Daas PJ, Schols HA (2000) In: Paulsen BS (ed) Bioactive Carbohydrate Polymers. Proc Phytochem Soc of Europe. Klüwer Academic Publishers, Dordrecht, p 129
23. Perez S, Rodriguez-Carvajal MA, Doco T (2003) Biochimie 85:109
24. Doco T, Williams P, Vidal S, Pellerin P (1997) Carbohydr Res 297:181

25. Rodriguez-Carvajal MA, Penhoat CHD, Mazeay K, Doco T, Perez S (2003) Carbohydr Res 338:651
26. Hirano M, Kiyohara H, Yamada H (1994) Planta Med 60:450
27. Yamada H (1994) Carbohydr Polym 25:269
28. Yamada H, Kiyohara H, Matsumoto T (2003) In: Voragen F et al. (eds) Advances in Pectin and Pectinase Research. Kluwer, Dordrecht, p 481
29. Hokputsa S, Harding SE, Inngjerdingen K, Jumel K, Michaelsen TE, Heinze T, Koschella A, Paulsen BS (2004) Carbohydr Res 339:753
30. Pangburn MK (1989) J Immunol 142:2766
31. Kumazawa Y, Mizunoe K, Otsuka Y (1982) Immunology 47:75
32. Yamada H, Kiyohara H, Cyong JC, Otsuka Y (1985) Mol Immunol 22:295
33. Kiyohara H, Yamada H, Otsuka Y (1987) Carbohydr Res 167:221
34. Kiyohara H, Cyong JC, Yamada H (1988) Carbohydr Res 182:259
35. Kiyohara H, Yamada H (1989) Carbohydr Res 187:255
36. Kiyohara H, Yamada H (1989) Carbohydr Res 193:173
37. Kiyohara H, Cyong J-C, Yamada H (1989) Carbohydr Res 193:193
38. Kiyohara H, Zhang YW, Yamada H (1997) Carbohydr Polym 323:249
39. Kiyohara H, Cyong J-C, Yamada H (1989) Carbohydr Res 193:201
40. Yamada H, Komiyama K, Kiyohara H, Cyong JC, Hirakawa Y, Otsuka Y (1990) Planta Med 56:209
41. Hirona M, Matsumo T, Kiyohara H, Yamada H (1994) Planta Med 60:248
42. Yu K-W, Kiyohara H, Matsumoto T, Yang H-C, Yamada H (1998) Planta Med 57:555
43. Yu K-W, Kiyohara H, Matsumoto T, Yang H-C, Yamada H (2001) Carbohydr Polym 46:147
44. Taguchi I, Kiyohara H, Matsumoto T, Yamada H (2004) Carbohydr Res 339:763
45. Yu K-W, Kiyohara H, Matsumoto T, Yang H-C, Yamada H (2001) Carbohydr Polym 46:125
46. Yamada H (2000) In: Paulsen BS (ed) Bioactive Carbohydrate Polymers. Kluwer Academic Publishers, Dordrecht, p 15
47. Yamada H (2000) In: Nothnagel et al. (eds) Cell and Developmental Biology of Arabinogalactan Proteins. Kluwer Academic Publishers, Boston p 221
48. Sakurai MH, Matsumoto T, Kiyohara H, Yamada H (1996) Planta Med 62:341
49. Sakurai MH, Kiyohara H, Matsumoto T, Tsumuraya Y, Hashimoto Y, Yamada H (1998) Carbohydr Res 311:219
50. Guo Y, Matsumoto T, Kikuchi Y, Ikejima T, Wang B, Yamada H (2000) Immunopharmacology 49:307
51. Ebringerova A, Hromadkova Z, Machova E, Naran R, Hribalova V (1997) Chem Papers 51:289
52. Ebringerova A, Hromadkova Z, Hribalova V, Hirsch J (2002) Chem Papers 56:320
53. Bao X, Wang Z, Fang J, Li X (2002) Planta Med 68:237
54. Duan J, Wang X, Dong Q, Fanf J-N, Li X (2003) Carbohydr Res 338:1291
55. Diallo D, Paulsen BS, Liljeback THA, Michaelsen TE (2001) J Ethnopharmacol 74:159
56. Inngjerdingen K, Michaelsen TE, Diallo D, Paulsen BS (2003) In: Perez S (ed) Abstracts, 12th European Carbohydrate Symposium, p 210
57. Kiyohara H, Takemoto N, Zhao J-F, Kawamura H, Yamada H (1996) Planta Med 62:14
58. Hwang J-H, Lee K-H, Yu K-W (2003) Neutraceuticals Food 8:29
59. Kelly GS (1999) Altern Med Rev 4:96
60. ANON (2000) Altern Med Rev 5:463
61. Currie NL, Lejtenyi D, Miller SC (2003) Phytomedicine 10:145
62. Peng X-M, Huang L-J, Qi C-H, Zhang Y-X, Tian G-Y (2001) Chinese J Chem 19:1190

63. Da Silva BP, Parente JP (2002) Planta Med 68:74
64. Sonoda Y, Kasahara T, Mukaida N, Shimuzu N, Tomoda M, Takeda T (1998) Immunopharmacology 38:287
65. Gao Q-P, Kiyohara H, Cyong J-C, Yamada H (1988) Carbohydr Res 181:175
66. Gao Q-P, Kiyohara H, Cyong J-C, Yamada H (1991) Planta Med 57:132
67. Shin K-S, Kiyohara H, Matsumoto T, Yamada H (1997) Carbohydr Res 300:239
68. Shin K-S, Kiyohara H, Matsumoto T, Yamada H (1998) Carbohydr Res 307:97
69. Sun X-B, Matsumoto T, Yamada H (1994) Phytomedicine 1:225
70. Chun H, Shin DH, Hong BS, Cho WD, Cho HY, Yang HC (2002) Biol Pharm Bull 225:1203
71. Samuelsen AB, Paulsen BS, Wold JK, Knutsen SH, Yamada H (1998) Carbohydr Polym 35:145
72. Samuelsen AB, Paulsen BS, Wold JK, Otsuka H, Yamada H, Espevik T (1995) Phytother Res 9:211
73. Samuelsen AB, Paulsen BS, Wold JK, Otsuka H, Kiyohara H, Yamada H, Knutsen SH (1996) Carbohydr Polym 30:37
74. Michaelsen TE, Gilje A, Samuelsen AB, Høgåsen K, Paulsen BS (2000) Scand J Immunol 52:483
75. Hetland G, Samuelsen AB, Løvik M, Paulsen BS, Aaberge IS, Groeng E-C, Michaelsen TE (2000) Scand J Immunol 52:348
76. Hetland G (2003) Curr Med Chem 2:135
77. Capek P, Hribalova V, Svandova E, Ebringerova A, Sasinkova V, Masarova J (2003) Int J Biol Macromol 33:113
78. Chintalwar G, Jain A, Sihapimalani A, Banerji A, Sumariwalla P, Ramakrishnan R, Sainis K (1999) Phytochemistry 52:1089
79. Subramanian M, Chintalwar GJ, Chattopadhyay S (2002) Redox Report 7:137
80. Diallo D, Paulsen BS, Liljeback THA, Michaelsen TE (2000) J Ethnopharmacol 84:279
81. Nergard CS, Diallo D, Michaelsen TE, Malterud KE, Kiyohara H, Matsumoto T, Yamada H, Paulsen BS (2004) J Ethnopharmacol 91:141
82. Nergard CS, Matsumoto T, Inngjerdingen M, Inngjerdingen K, Hokputsa S, Harding SE, Michaelsen TE, Diallo D, Kiyohara H, Paulsen BS, Yamada H (2005) Carbohydr Res 340:115
83. Kiyohara H (1998) In: Ageta H, Aimi N, Ebizuka Y, Fujita T, Honda G (eds) Towards Natural Medicine Research in the 21st Century. Elsevier Science, Amsterdam, p 161
84. Yamada H (1996) In: Visser J, Voragen AGJ (eds) Pectins and Pectinases. Elsevier Science, Amsterdam 14:173
85. Kiyohara H, Matsumoto T, Yamada H (2002) Phytomedicine 9:614
86. Yamada H (2003) In: Watanabe H (ed) Pharmacological research on traditional herbal medicines. Taylor and Francis, London, p 170
87. Tomshich SV, Komandrova NA, Kalmykova EN, Prokofeva NG, Momontova VA, Gorovoi PG, Ovodov YS (1997) Chem Nat Compounds 33:146
88. Ebringerova A, Kardosova A, Hromadkova Z, Hribalova V (2003) Fitoterapia 74:52
89. Hensel A, Meier K (1999) Planta Med 65:395
90. Diallo D, Sogn C, Samake FB, Paulsen BS, Michaelsen TE, Keita A (2002) Pharm Biol 40:117
91. Inngjerdingen K, Nergård CS, Diallo D, Mounkoro PP, Paulsen BS (2004) J Ethnopharmacol 92:233

Adv Polym Sci (2005) 186: 103–149
DOI 10.1007/b136818
© Springer-Verlag Berlin Heidelberg 2005
Published online: 31 August 2005

Organic Esters of Cellulose:
New Perspectives for Old Polymers

Omar A. El Seoud[1] (✉) · Thomas Heinze[2]

[1]Instituto de Química, Universidade de São Paulo, C.P. 26077, 05513-970 São Paulo, S.P.,
Brazil
elseoud@iq.usp.br

[2]Friedrich Schiller University of Jena, Humboldtstrasse 10, D-07743 Jena, Germany
Thomas.Heinze@uni-jena.de

Abstract The impetus for the increased interest in the synthesis of functionalized natural polymers, in particular esters of cellulose, is their easy biodegradability and conformity to the principles of green chemistry. This review is concerned with the preparation of cellulose esters under homogeneous reaction conditions, including products that cannot be obtained by the (industrial) heterogeneous reaction. This scheme, which leads to products of reproducible properties, includes three stages: Cellulose activation (by solvent exchange or heat), dissolution (in derivatizing or non-derivatizing solvent systems), and functionalization of the solubilized polymer. Dissolution in non-derivatizing solvent systems, in particular LiCl/DMAc; $(C_4H_9)_4NF$·hydrate/DMSO and ionic liquids (green solvents) is due to the disruption of the H-bonding within the polymer structure. Dissolution in derivatizing solvents, e.g., acid-anhydrides leads to functionalization of cellulose, and may be fruitfully employed in controlling the regioselectivity of polymer substitution. Optimization of each of the above-mentioned reaction stages is a pre-requisite in order to meet the requirements of green chemistry.

Keywords Cellulose, esters of · Cellulose, derivatization of ·
Cellulose esters, properties of · Cellulose, homogeneous reaction ·
Green chemistry, principles of · Green chemistry, application to cellulose

Abbreviations

AGU	anhydroglucose unit of cellulose
CA	cellulose acetate
CDI	N,N-carbonyldiimidazole
Cell	cellulose
Cell-Tos	cellulose tosylate
CMC	carboxymethyl cellulose
DCC	dicyclohexylcarbodiimide
DMAc	N,N-dimethylacetamide
DMF	N,N-dimethylformamide
DMSO	dimethyl sulfoxide
DS	average degree of substitution in the AGU
DSC	differential scanning calorimetry
Et_3N	triethylamine
EWNN	alkaline solution of iron sodium tartarate
HEC	hydroxyethylcellulose
HRC	homogenous reaction conditions
Ic	index of crystallinity of cellulose
IL	ionic liquid
LS	light scattering
Py	pyridine
SEC	size exclusion chromatography
TBAF	tetra-n-butyl ammonium fluoride·$3H_2O$
TFA	trifluoroacetic acid
Tos – Cl	tosyl chloride

1
Introduction—Scope of the Review

In 2002, the world production of polymers (not including synthetic fibers and rubbers) was ca. 190 million metric tons. Of these, the combined production of poly(ethylene terephthalate), low- and high-density polyethyelene, polypropylene, poly(vinyl chloride), polystyrene, and polyurethane was 152.3 million metric tons [1]. These synthetic, petroleum-based polymers are used, *inter alia*, as engineering plastics, for packing, in the construction-, car-, truck- and food-industry. They are chemically very stable, and can be processed by injection molding, and by extrusion from the melt in a variety of forms. These attractive features, however, are associated with two main problems:

- They are derived from a non-renewable source, petroleum;
- They are resistant to chemical, photochemical, and enzyme-mediated biodegradations.

Whereas polyetheylene, polypropylene, and polystyrene are virtually non-biodegradable, only specially modified, not widely employed, polyamides and polyurethanes are susceptible to biodegradation [2]. Consequently, the use of these polymers has generated a serious environmental problem, namely their accumulation in municipal dumping sites (it may take more than 200 years to biodegrade a polyethylene film!). Therefore, biodegradable- and biocompatible-polymers and composites, especially those from renewable sources, namely poly(hydroxy alkanoates), cellulose, starch, and chitin have attracted much attention in recent years. Use of polymeric materials from renewable sources is a viable approach to maintaining sustainable development of economically-, and ecologically attractive technology. The impetus for the current intense interest in this subject includes reduction of the waste disposal problem, reduction of CO_2 release into the atmosphere, and introduction of new possibilities for bio-based business.

Another favorable aspect of these polymers is their conformity with the principles of green chemistry. The latter sets guidelines for the chemical industry in order to secure sustainable development, while increasing process economy [3, 4]. Briefly, green chemistry, and the related green engineering [5] call for an increase in, and/or upgrading of:

- Process economy, by preventing waste generation. This represents a much superior approach to waste treatment;
- Atom economy, by incorporating all reagents employed in the final product. This also contributes to reduction and/or elimination of waste;
- Process safety, e.g., by using non-toxic, non-inflammable solvents and reagents;
- Process efficiency, e.g. by: material recycling into the process; where possible, use of catalytic pathways; use of catalysts that can be regenerated/recycled; rational use of energy, and reduction of the number of intermediate steps.

Green chemistry also calls for design for biodegradable end products, principally, by employing chemicals from renewable sources, and dictates the use of real-time, on-line analysis for better process control.

Several review articles are available on the synthesis, physico-chemical properties, and biodegradability of natural-based polymers, and their composites [6–9]. The same aspects have been the subjects of recent books [10–12]. In the following account, we concentrate on organic esters of cellulose.

Cellulose acetate (CA) has been known, and industrially employed for decades as films, fibers, filters, membranes, tubes, and utensils, as well as other consumer products, including eyewear, fashion accessories, pens, brushes, toys, among others [13]. The market for Filter Tow, which is made from crimped, endless CA filaments, has seen a tremendous growth in the cigarette market, reaching more than 600 thousand metric tons in 2003 [14]. Additionally, cellulose mono-acetates have several potential applications, because they can be made into either water absorbent, or water-soluble polymers [15].

Biodegradation of CA and other cellulose esters has been recently reviewed, and may occur under aerobic or anaerobic conditions [16]. Thus CA sheets with an average degree of substitution, DS, of ≈ 2 are readily degraded in two weeks, or less, if composting is employed. Biodegradation by cellulases produces cellobiose, which is further hydrolyzed to glucose by β-glucosidades. The rate of biodegradation of other cellulose esters decreases as a function of increasing DS in the anhydroglucose unit, AGU, and the length of the ester moiety. Interestingly, it seems that the DS of the acetate ester, not its crystallinity, is the predominant factor that controls its biodegradability [17].

The green chemistry approach, and the surge of biopolymers as candidates for substituting synthetic ones in several applications require detailed understanding of the following aspects, *at the molecular level*:

(i)   The structural chemistry of cellulose;
(ii) Details of its dissolution in derivatizing, and non-derivatizing solvent systems;
(iii) Details of the steps of its reaction with the derivatizing reagents, e.g., activated carboxylic acids, acyl chlorides, acid anhydrides and mixed anhydrides.

Whereas cellulose structure is now known in considerable detail [18, 19], and the introduction of new solvent systems for dissolving cellulose has intensified the interest in studying point (ii), less is known about point (iii). Such knowledge, however, is a pre-requisite to control the properties of these polymers, hence increase their competitiveness as possible substitutes for synthetic polymers.

The present article has been written with the above-mentioned aspects in mind. We show that different derivatization schemes lead to new products, whose properties are well controlled at the molecular level, and can be tailored to the application required. Some cellulose derivatives, in particular esters of long-chain fatty acids cannot be obtained by the (industrial) heterogenous esterification process. They are, however, potentially important because their glass-transition temperatures, T_g, and melting temperatures, T_m, are relatively low. This allows their processing by injection molding, and casting from the melt, a problem that has limited the application of cellulose derivatives, e.g., as engineering plastics. Recently, cellulose has been shown to dissolve in ionic liquids, ILs, the so-called "green" solvents. It can be regenerated from these solutions in different forms, as fibers, films, and rods. This, and the fact that the natural polymer can be derivatized in these media represent new, and exiting possibilities for synthesis and applications of cellulose-based biopolymers.

The main focus of this account is to review some aspects of the chemistry of cellulose esters. Emphasis is placed on the esterification reaction, carried out under the homogenous reaction conditions (HRC) scheme. Unconventional methods for the synthesis of cellulose derivatives, e.g., esters and ethers under homogenous and heterogenous reaction conditions have been recently reviewed [20]. Therefore, we update the literature on the synthesis and properties of the derivatives obtained by the HRC scheme, and we address its conformity with the concepts of green chemistry. Based on the preceding paragraph, a clear understanding of the three steps involved is required. This understanding contributes to process optimization, a pre-requisite for its applicability on a large scale, at least for producing specialty products.

The discussion is organized in the following order: First the advantages of HRC scheme, relative to the industrial (i.e., heterogenous) process are briefly commented on; second, the relevance of cellulose activation and the physical state of its solution to optimization of esterification are discussed. Finally, the use of recently introduced solvent systems and synthetic schemes, designed in order to obtain new, potentially useful cellulose esters with controlled, reproducible properties is reviewed. A comment on the conformity of these methods with the concepts of green chemistry is also included.

2
Importance of the Homogenous Reaction Conditions (HRC) Scheme

The molecular structure of cellulose leads to extensive hydrogen bonding. The presence of two major intra-molecular types has been suggested, namely $(O_3 - H...O_5')$ and $(O_2 - H...O_6')$, where O_5' and O_6' originate from an adjacent glucose residue. The major intermolecular hydrogen bonding occurs between $(O_6 - H...O_3')$ along the b axis [18, 19, 21, 22, 26]. The stretching frequency region of the OH groups of cellulose has been recently probed by coupling dynamic mechanical analysis (to stretch the cellulose sheet periodically) with 2D-FTIR. The infrared beam was polarized parallel and perpendicular to the stretching motion of the sample. Several vibration modes were clearly identified, including $(3335 \text{ cm}^{-1}, O_3 - H...O_5')$, $(3260 \text{ cm}^{-1}, O_6 - H...O_3')$ $(3420 \text{ cm}^{-1}, O_2 - H...O_6')$ [27]. As a result of this hydrogen bonding and van der Waals forces, cellulose molecules could align together in a highly ordered state to form crystalline regions, whereas the less ordered molecules constitute the amorphous part. The proportion of ordered to disordered regions (index of crystallinity, Ic) of cellulose varies considerably with its origin and the extent of treatment, both physical and chemical, to which the cellulose was submitted. Generally it decreases in the order cotton > wood pulp > regenerated cellulose [28].

This extensive hydrogen bonding bears on several aspects of the chemistry and applications of cellulose. For instance, being a semi-crystalline polymer, cellulose cannot be processed by the techniques most frequently employed for synthetic polymers, namely, injection molding and extrusion from the melt. The reason is that its T_m presumably lies above the temperature of its thermal decomposition. Several commercial cellulose derivatives, in particular CA and cellulose nitrate, are soluble in common organic solvents, e.g., acetone, alcohol and chloroform, and can be extruded as fibers, films, rods and sheets. Introduction of an organic ester group into cellulose lowers T_m to values comparable to those of extensively used polymers, as shown by the following examples for synthetic ones (T_m in °C, polymer): 137, polyethylene; 176, polypropylene; 223, Nylon-6; 270, poly(ethylene terephthalate), and for cellulose acetate/hexanoate mixed esters with *total* DS = 3 (T_m in °C, $DS_{hexanoate}$): 306, 0; 279, 0.18; 258, 0.22; 219, 0.52; 141, 2.0; 115, 3.0 [29]. Therefore, cellulose esters can, in principle, be processed similarly to synthetic polymers.

Since the AGU has *three free* OH groups (at C_2, C_3, and C_6, respectively) it is possible, in principle, to obtain derivatives of any DS, by adjusting the ratio of the derivatizing agent/AGU. This runs into the problem of the differences between the accessibilities of the OH groups in the amorphous and crystalline regions, respectively [19–22]. Consequently, it is not feasible to obtain a uniformly substituted cellulose derivative with a DS, say of 1 to 2.5 *directly*, i.e., by the (heterogenous) reaction of a suspension of cellulose in the derivatizing reagent. The reason is that the products obtained will be heterogenous, even if the (average) DS is achieved. The AGU's of the amorphous regions will be more substituted than their counterparts in the crystalline regions. This heterogeneity may lead, for example, to serious solubility problems in solvents that are usually employed, e.g., acetone. Industrially, the above-mentioned products are obtained by a sequence of *two heterogenous reactions*, namely functionalization of the polymer to a DS of ca. 3, followed by partial hydrolysis to the product of desired (average) DS. Problems routinely arise, however, because both reactions are not easily controllable. For example, in principle, the ester group in position-6 (the primary site) should be hydrolyzed first. The ester group in position-2 should be attacked second, followed by the one in position-3, as the $C_2 - OH$ group is more exposed than the $C_3 - OH$ group. A recent study on the acid-catalyzed hydrolysis of cellulose acetates with different DS showed, however, that the picture is more complicated, and may not follow theory. For example, the expected order of rate of hydrolysis of the acetate groups (6 > 2 > 3) was observed for samples of low DS (from 0.7 to 1.5), whereas position-2 was favored for two samples with high DS [30]. This, and similar anomalies lead to industrial products whose properties are poorly reproducible, a problem whose consequences are usually minimized by blending the products of different batches. Additionally, the energy consumption is high, even for the (faster) high-temperature acetylation, high-temperature hydrolysis process. For example, 1 to 2.5 kcal are used to recover 1 g of the solvent employed (acetic acid) [31, 32]. The industrial processes are being continually improved, in view of the increasing demand on cellulose acetates, especially as Filter Tow, film, and coating [14, 32].

Several solvent systems dissolve cellulose, a process that *may, or may not* lead to cellulose derivative formation. Both types of solvent systems will be considered, although the important derivatizing reaction employed in the production of Rayon fibers will not be addressed. In this process, alkali cellulose is reacted with carbon disulphide to produce the corresponding xanthate, soluble in excess aqueous NaOH, followed by extrusion and regeneration of the polymer fibers [23].

At first glance, the HRC scheme appears simple: the polymer is activated, dissolved, and then submitted to derivatization. In a few cases, polymer activation and dissolution is achieved in a single step. This simplicity, however, is deceptive as can be deduced from the following experimental observations: In many cases, provided that the ratio of derivatizing agent/AGU employed is stoichiometric, the targeted DS is not achieved; the reaction conditions required (especially reaction temperature and time) depend on the structural characteristics of cellulose, especially its DP, purity (in terms of α-cellulose content), and Ic. Therefore, it is relevant to discuss the above-mentioned steps separately in order to understand their relative importance to ester formation, as well as the reasons for dependence of reaction conditions on cellulose structural features.

3
Cellulose Activation

The problem of enhancing cellulose reactivity is highly relevant to both laboratory scale and industrial processes, and for homogenous and heterogenous reactions. The term reactivity is somewhat ambiguous as it refers to reaction rate, and final DS of the cellulose derivative, obtained after reaching quasi-equilibrium (e.g., for esterification by carboxylic acid anhydrides). Deuteration, moisture gain, iodine sorption by cellulose, as well as its acid hydrolysis, periodate oxidation, dye absorption, and formylation may be employed to measure cellulose accessibility, hence reactivity. Although the absolute results depend on the method employed, the general conclusions that are arrived at by use of different techniques agree qualitatively. It should be noticed, however, that the (heterogenous) activation step is influenced by cellulose structure and vise versa, i.e., cellulose structure is influenced by the treatment that is being employed [24]. Therefore, it is not possible to discuss a "universal" activation scheme that proceeds to the same extent (i.e., at the same rate) for different celluloses, when submitted to distinct activation schemes, vide infra.

The following brief account is concerned with factors that affect the accessibility of the OH groups of cellulose, since this is the determining factor for its dissolution, hence subsequent derivatization. Electron microscopy, X-ray scattering and porosimetry of cellulose fibers have clearly shown the presence of non-uniform pores, capillaries, voids and interstices in the fiber surface [25]. Consequently, the total surface area of cellulose fibers exceeds by far the geometrical outer surface. Pore structure determines the internal accessible surface area of cellulose, and thus affects its reactivity. The void volume of native cellulose was estimated from density measurements to be between 3–4% [33].

Any physical interaction (e.g., absorption of water vapor), or chemical reaction with cellulose involves these "pinholes", therefore, their study is relevant to the chemistry of cellulose and its derivatization. For example, an electron microscopy investigation of acetylation of cotton linters by acetic anhydride, under heterogenous reaction conditions, showed that the mechanism of reagent penetration into the fiber is noticeably dependent on the conditions employed. At high temperature, 90 °C, the reaction progressed in a linear manner, involving the formation of continuous micro-pores, from the surface into the interior of the supra-molecular structure. The successive layers of acetylated fiber became loose, broke, and dissolved in the solution. In contrast, at lower temperature, 45 °C, the reaction took place only through segregated micro-pores on the surface of cellulose, whose size increased as a function of reaction time [34].

Pore structures could be opened by adsorption of vapors, e.g., of ethanol, acetic acid and water [35]. On the other hand, they are reduced by alkali treatment or drying. This reduction is probably a major factor in the so-called *hornification* process that adversely affects cellulose dissolution, and subsequent reactions. Polymer dehydration is likely to induce the formation of new hydrogen bonding, especially for the chains located in the less ordered surface areas of the fibrils. With regard to the pore size of cellulose fibers, native cellulose, in general, has larger pores, whereas voids of colloidal sizes are dominant in regenerated celluloses [36]. A fraction of the readily accessible cellulose surface is given by the surface of crystallites, whose nature, along with crystallinity, contribute to the reactivity of the polymer. Therefore, cellulose samples of different origin, DP and Ic are expected to require different dissolution and/or reaction conditions. The reason is that they have widely different supra-molecular structures, this strongly affects the accessibility of the hydroxyl groups, the site of attack on cellulose.

The objective of the activation step is to increase the diffusion of reagents into the cellulose supra-structure, by making the crystallite surfaces and the crystalline regions more accessible. This is achieved by inter- and intra-crystalline penetration of the activating agent into cellulose, which disrupts the strong, water-mediated hydrogen bonding of the natural polymer chains [37]. The relevance of this step to the success of the reaction is demonstrated by the erratic results that may be obtained if it is not carried out properly. The following data on % acetyl content of cellulose, after one day of treatment with 50 wt % acetic anhydride in pyridine, at 30 °C, drive home the point: No activation, 8.8%; pre-treatment with chloroform/pyridine, 26.4%; same pre-treatment with ethanol/chloroform, 27.6% [31]. Poor yields also result from limited accessibility of the OH groups, due to polymer hornification that may occur, inadvertently, during activation. For example, microcrystalline cellulose was thermally activated under reduced pressure (2 h, 60 °C, 2 mm Hg), the pressure was brought to atmospheric, and the polymer was dissolved in LiCl in *N,N*-dimethylacetamide, DMAc, at 100 °C for 8 h. Reaction with acetic anhydride/pyridine for 18 h at room temperature produced a DS of only 0.3, instead of the expected DS of 3. Increasing the "activation" time (6 h, 60 °C, 2 mm Hg) resulted in a still lower DS of 0.1 [38]. Previous studies on thermal treatment of celluloses, under atmospheric and reduced pressures, have indicated that the extent of some derivatizations under *heterogenous* reaction conditions (e.g., carboxymethylation), but not others (e.g., acetylation) are favorably affected by thermal activation [39–41].

Of the activation procedures available, alkali treatment has been most extensively employed. Mercerization, followed by washing of the residual base has been employed as a pre-treatment in the HRC scheme. Alkali treatment results in the irreversible transformation of cellulose I (where the polymer chains have parallel conformation) into the less-ordered cellulose II (antiparallel conformation), i.e., Ic is reduced. Another consequence of mercerization is to increase the α-cellulose content, due to extraction of the hemicelluloses and other residual, non-cellulosic material, e.g., wax. Therefore, this activation procedure has been employed with cellulose samples with high Ic and DP, in particular cotton linters. However, it should be employed selectively, not routinely, because the reactivity of cellulose is profoundly affected by its supra-molecular structure, i.e., cannot be explained *solely* by its crystallinity. Examples are known where mercerization decreases the reactivity of cellulose toward acetylation or nitration [42]. Also, regenerated cellulose samples did not dissolve in certain organic solvent systems, e.g., SO_2-dietheylamine-DMSO, which easily dissolves their native counterparts [43]. These results have been attributed to differences in hydrogen-bonding patterns within native and regenerated celluloses [42–44].

Another activation treatment, suitable for most celluloses (although with great variation of the time required, 1 to 48 h) is polar solvent displacement at room temperature. The polymer is treated with a series of solvents, ending with the one that will be employed in the derivatization step. Thus, cellulose is treated with the following sequence of solvents, before it is dissolved in LiCl/DMAc: water, methanol, and DMAc [37, 45–48]. This method, however, is both laborious, needs ca. one day for microcrystalline cellulose, and expensive, since 25 mL of water; 64 mL of methanol, and 80 mL of DMAc are required to activate one gram of cellulose. Its use may be reserved for special cases, e.g., where cellulose dissolution with almost no degradation is relatively important [49].

Several procedures have been suggested for heat-mediated cellulose activation, e.g., by using the reaction solvent itself as the heating medium. This activation, first proposed by Ekmanis, is based upon the fact that the vapor pressure of DMAc, near, or at its boiling point is sufficiently high to induce efficient fiber penetration and swelling [50]. Heat activation is considered more advantageous than polar activation (by solvent exchange), not only because it requires less LiCl in subsequent dissolution, but also because it is a one-step procedure [49].

Water can be removed from cellulose by heating the polymer/DMAc slurry first at 150 °C for 1 h, then to the boiling point of the solvent, followed by distillation of 25% of the volume of the latter [51]. This activation scheme may be problematic, however, because it may not lead to complete drying of the system, probably because residual water is tightly bound to the polymer [52]. Indeed, differential scanning calorimetric (heating) curves of cotton linters have indicated that vaporization of bound water *terminates* at 150 °C [53]. Additionally, refluxing celluloses in DMAc/LiCl produce amber to brownish solutions, this was attributed to oxidative degradation of the polymer at high temperature [54]. A subsequent study has indicated that polymer degradation may take place by two processes: The first one occurs at temperatures above ca. 80 °C, and involves N,N-dimethylacetoacetamide, $CH_3CO - CH_2CON(CH_3)_2$, a primary auto-condensation product of DMAc. It leads to a slow, thermal degradation, due to the reaction of this intermediate with the reducing end of the polymer, to produce furan structures. The other (faster) reaction involves the N,N-dimethyl keteniminium ion $(CH_2 = C = N^+(Me)_2)$, whose precursor is the enol tautomer of DMAc, $CH_2 = C(OH)N(Me)_2$. Formation of the solvent-originated cation, and its precursor has been demonstrated by employing 1-decene, and allyl alcohol, respectively, as the trapping agents. Formation of the enol is accelerated if there is free rotation around the $C - N$ (amide) bond. ^{1}H NMR spectroscopy has been employed to determine the temperature of coalescence (i.e., of free rotation) of the peaks of (magnetically non-equivalent) N-methyl groups of DMAc. In a LiCl/DMAc solution, this coalescence occurs above ca. 90 °C. That is, the formation of the N,N-dimethylketeniminium ion is accelerated at a temperature well below that employed in solvent activation. This cation is an extremely reactive electrophile, capable of random chain cleavage, resulting in pronounced and rather fast changes in the molecular weight distribution of cellulose [55]. In order to avoid degradation, the solvent-activation scheme may be carried out under reduced pressure, hence at a lower temperature. However, this resulted in incomplete dissolution of fibrous celluloses, e.g., bagasse and sisal [38]. Therefore, it is safer to keep the temperature of the cellulose slurry as low as practical, and to suspect extensive degradation if the color of the solution becomes dark amber, or brownish.

Cellulose activation has been achieved by heating the polymer with dry LiCl, at 110 °C, under reduced pressure, 2 mm Hg, followed by addition of DMAc. It is important to introduce the solvent while the system is maintained under reduced pressure, in order to avoid hornification [56]. As expected, the activation conditions employed were found to be dependent on cellulose structure, samples with high DP and high Ic required pre-treatment, i.e., mercerization (cotton linters), and/or longer activation time. This solubilization scheme causes very little change in DP of the starting cellulose, $\leq 6\%$, for microcrystalline, bagasse and cotton linters [57].

4
Cellulose Dissolution

The basic requirement for cellulose dissolution is that the solvent is capable of interacting with the hydroxyl groups of the AGU, so as to eliminate, at least partially, the strong inter-molecular hydrogen-bonding between the polymer chains. There are two basic schemes for cellulose dissolution: (i) Where it results from physical interactions between cellulose and the solvent; (ii) where it is achieved via a chemical reaction, leading to covalent bond formation "derivatizing solvents". Both routes are addressed in details below.

4.1
Non-Derivatizing Solvents

First, we address cellulose dissolution schemes that do not lead to derivatization. Treatment of cellulose with highly polar solvents, capable of strongly interacting with the hydroxyl groups of the AGU may achieve activation/dissolution in a single step. Ideally, the solvent (either mono- or poly-component) should conform, as much as practical, to the principles of green chemistry. That is, it should be safe, inexpensive, and can be regenerated in a pure state, so that it can be recycled into the process. From the application point of view, it should be able to dissolve celluloses of varying DP and Ic, without excessive degradation. In addition to solution stability, the solvent should be strongly dipolar in order to stabilize the highly polar activated complexes of the acyl-transfer reactions involved, but should not compete with cellulose for the derivatizing agent.

Alkali hydroxide solutions possess some of these requirements, e.g., those of cost, ease of recycling, and high polarity. Indeed, a clear solution was obtained when microcrystalline cellulose was swollen by 8 to 9 wt % aqueous solution of NaOH, the latter was frozen, thawed, then diluted to 5 wt % alkali. The same result was obtained with other celluloses of higher DP, provided that the polymer is either amorphous, or is present as cellulose II. Native samples were only partially soluble (26 to 37 wt % of the original cellulose), because the alkali pre-treatment, before solution freezing, did not completely disrupt the polymer long-range order. Interestingly, addition of urea enhances the solubility of cellulose in sodium hydroxide solution. For example, bagasse and cotton linter celluloses may be dissolved in 6 wt % NaOH solution at 0 °C, but the solution turns into gel upon increasing the temperature. Addition of 2 to 4 wt % urea inhibits gel formation [58]. The effect of urea is to weaken inter-molecular hydrogen bonding between cellulose chains, this favors solvation of the OH groups of the polymer, enhancing its solubility in the alkaline solution [59]. There is no interest, however, in using alkali hydroxides as solvents in ester formation, because the base present will most certainly consume the drivatizing agent, e.g., acid anhydride, before the latter reacts with cellulose.

Ni(tren)(OH)$_2$ and Cd(tren)(OH)$_2$ [tren = tris(2-aminoethyl)amine] dissolve cellulose by coordinating C$_2$ – OH and C$_3$ – OH of the AGU [60]. These solutions have been employed as reaction media for homogenous etherification, e.g., in the synthesis of carboxymethyl cellulose, CMC [60–62]. Subsequent analysis of the products by chain degradation, HPLC, and ^{1}H NMR spectroscopy have shown the following order of substitution in the AGU: C$_2$ ≥ C$_6$ > C$_3$. This order is similar to that observed for the commercial product, obtained by the heterogenous process [63, 64]. Several molten salt hydrates, e.g., LiClO$_4$ · 3H$_2$O, LiX · nH$_2$O (X = I$^-$, NO$_3^-$, CH$_3$CO$_2^-$, ClO$_4^-$), Zn(NO$_3$)$_2$ · XH$_2$O; FeCl$_3$ · 6H$_2$O, and eutectic mixtures, e.g., LiClO$_4$ · 3H$_2$O/MgCl$_2$ · 6H$_2$O and LiClO$_4$ · 3H$_2$O/Mg(ClO$_4$)$_2$ · H$_2$O dissolve celluloses, even those with very high DP [65–68]. The interaction involved is probably of the hard acid-hard base type, e.g., between the Li$^+$/O(H) – Cell and/or Cl$^-$/HO – Cell. As can be seen from the structures of these solvent systems, however, they contain water of hydration and will compete with cellulose for the derivatizing agent. Additionally, the question of their recycling needs to be addressed.

From the synthetic point of view, however, more flexibility is achieved by employing cellulose solutions in non-aqueous, non-derivatizing solvents. There are few single solvents that dissolve cellulose, including N-alkylpyridinium halides, e.g., N-ethylpyridinium chloride (m.p. 118–120 °C); tertiary amine oxides, e.g., N-methylmorpholine-N-oxide · H$_2$O (employed in the production of Lyocell fibers [69, 70]), and more recently ILs, that may, or may not be liquids at room temperature. Examples of ILs are 1-methyl-3-butylimidazolium chloride (m.p. 66 °C [71]), tetrafluoborate, hexafluorophophate and/or thiocyanide [72], and 1-allyl-3-metheylimidazolium chloride (m.p. ca. 17 °C, decomposition temperature 273 °C [73]). On heating, 1-methyl-3-butylimidazolium chloride dissolves cellulose. The amount of polymer solubilized depends on the temperature, being 3 wt % and 10 wt %, at 70 °C and 100 °C, respectively. Microwave heating resulted in the formation of viscous pastes containing up to 25 wt % cellulose. At high polymer content (> 10 wt %) the solutions are optically anisotropic, a property that may be useful in the formation of high-strength fibers. The formation of efficient Cl$^-$ – HO – Cell hydrogen bonding has been invoked to explain the solubilizing power of LIs. This explanation agrees with the deleterious effect of small concentrations of water, e.g., 1%, on cellulose solubilization by the ionic liquid. These solutions can be extruded; the cellulose fibers regenerated did not show significant degradation [72]. More recently, the enzyme laccase has been encapsulated by cellulose, the latter obtained from a (super-cooled) solution in 1-butyl-3-methylimidazolium chloride. Pre-coating by cellulose resulted in increased enzyme reactivity, presumably by providing a stabilizing microenvironment for the enzyme [74]. The advantage of 1-allyl-3-methylimidazolium chloride is that it is liquid at room temperature, unlike the structurally related 1-propyl-3-methylimidazolium chloride [75]. A cellulose solution of DP = 650 (5 wt %) can be readily obtained in this room temperature IL in 15 min, without pre-treatment, by heating at $100\,^{\circ}C$, thus preventing excessive degradation of the polymer. A transparent, highly viscous solution, containing 10 wt % cellulose can be also obtained in this medium [73, 76].

At present, the prices of ILs are much higher than those of conventional solvents. On the other hand, ILs conform to the principles of green chemistry because they are inert, highly polar, and have no measurable vapor pressure. In principle, their use offers more flexibility, because their physico-chemical properties may be adjusted by changing their structures, in particular the nature of the counter ion, the heterocyclic ring, and the length of-, and presence of unsaturation in the alkyl chain [77]. Simple schemes for product separation and solvent recycling have to be developed, if they are to be employed on a large scale.

More extensive work has been carried out on binary, or ternary mixtures, which will be designated as solvent "systems". The most investigated ones include inorganic or organic electrolytes in strongly dipolar aprotic solvents. Examples are LiCl in DMAc, in N-methyl-2-pyrrolidinone, in 1,3-dimethyl-2-imidazolidinone, and tetra-n-butyl ammonium fluoride trihydrate in DMSO, TBAF/DMSO. The solvent system LiCl/DMAc has been most extensively employed because it is capable of dissolving different celluloses including samples of high DP and Ic, e.g., cotton linters and bacterial cellulose, the last two after mercerization. The cellulose dissolved has been examined with ^{13}C NMR spectroscopy [78, 79], size exclusion chromatography, SEC, [80–85] and light scattering, LS [84, 86].

Electron microscopy has been employed in order to probe the mechanism of dissolution of different celluloses (DP from 458 to 1776, Ic from 0.42 to 0.58) in LiCl/DMAc, and in an alkaline solution of iron sodium tartarate (EWNN). Both systems have very different solubilization mechanisms. EWNN attacks the surface of fibers intensively and causes a large swelling. LiCl/DMAc, on the other hand, follows the naturally preformed structures after penetration into the cell walls, and dissolves cellulose without noticeable swelling. Both solvent systems affect the ends of the fibers as well as sites where the wall structure has been damaged. During dissolution, fragments of fibrillar bundles are formed, these show increasing cross-stripes with progressive dissolving times, indicating that the solvent acts preferentially on the less-ordered regions of the fibrils [87].

The structure of cellulose/solvent system complexes has been described by several schemes, differing essentially in the role played by the Li^+ and Cl^-

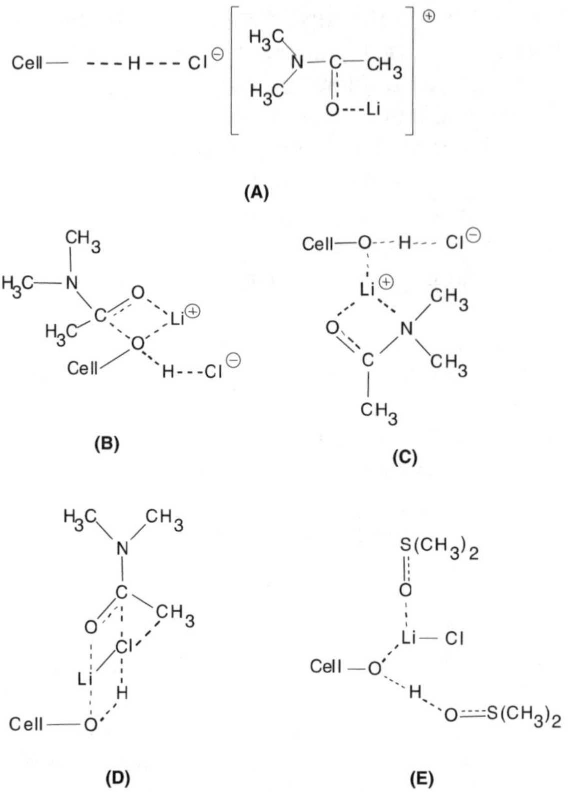

Fig. 1 Structures suggested for explaining the interactions of cellulose with LiCl/aprotic solvent systems. Redrawn from [45, 88–91]; for simplicity, partial charges are not shown

ions (Fig. 1). In structure (A), the complex between Li$^+$ and the oxygen of the solvent CO group results in formation of a macro-cation [Li(DMAC)]$^+$, leaving the Cl$^-$ free to form hydrogen bonding with the OH group of the AGU. The repulsion among macro-cations formed allows solvent penetration within the natural polymer chains [45]. Alternatively, the lithium ion binds simultaneously to the OH group of cellulose and the solvent, the latter binding can occur either with the CO group of DMAc, structure (B) [88], or with the CO group and the amide nitrogen, structure (C) [89]. LiCl may also be present in an un-dissociated form, as shown in structures (D), and (E). In the former, the electrolyte is bound to DMAc and the cellulose OH groups, forming a sandwich-type structure [90]. In the latter (suggested for dissolution in LiCl/DMSO), only Li$^+$ is bound to the S = O dipole of the solvent and the oxygen atom of the OH groups, whereas the chloride ion is not involved in bonding [91].

A comment on these structures is relevant to understanding the mechanism of cellulose dissolution by these solvent systems. Support for association of the Li^+ ion, a hard acid, with the oxygen atom of the amide CO group comes from studies on the protonation of amides, which occurs largely on the same atom, i.e., very little N-protonation has been detected [92, 93]. Therefore, structures A and C are not very different because the Li^+ – N association, if it occurs, is probably weak. There are problems associated with structures (D) and (E) in which the chloride ion plays a minor role (D), or no role (E) in cellulose dissolution. In view of the enhanced nucleophilicity of halide ions in dipolar aprotic solvents [94], and the fact that cellulose dissolution depends on the anion, being more efficient for LiCl than for LiBr, i.e., more efficient for the harder anion [95], it is possible to conclude that the chloride ion is the strongest general base present in this solvent system. This conclusion does not agree with (E), which shows that the cellulose OH groups interact with the solvent, not with the chloride ion.

The relative importance of the halide anion – HO – Cell interactions can be inferred from application of the Taft–Kamlet–Abboud equation to the UV-Vis absorbance data of solvatochromic probes, dissolved in cellulose solutions in different solvent systems, including LiCl/DMAc and LiCl/N-methyl-2-pyrrolidinone [96]. According to this equation, the microscopic polarity measured by the indicator, E_T (indicator), in kcal mol^{-1}, is correlated with the properties of the solvents by Eq. 1:

$$E_T(\text{indicator}) = \text{Constant} + s(\pi^* + d\delta) + a\alpha + b\beta + h(\delta_H^2) \qquad (1)$$

That is, E_T(indicator) is modelled as a linear combination of a dipolarity/ polarizability term $[s(\pi^* + d\delta)]$, two hydrogen-bonding terms, in which the solvent is the hydrogen-bond donor ($a\alpha$), or the hydrogen-bond acceptor ($b\beta$), and a cavity term $[h(\delta_H^2)]$. The later is redundant when the Frank–Condon principle is obeyed. The parameters π^*, α, and β, are known as solvatochromic parameters, because they are determined by the use of solvatochromic indicators [97, 98]. The results obtained have revealed that the most important solvatochromic parameters are $\alpha_{\text{cellulose}}$ (the H-bond donation ability of cellulose) and $\beta_{\text{LiCl/DMAc}}$ (the basicity of the solvent system). Therefore, the most dominant interaction is the Cl^- – HO – Cell, a conclusion in variance with structures (D) and (E). Solvatochromic probes have been also employed to probe the surfaces of different celluloses, as well as those of cellulose derivatives, e.g., CMC and cellulose tosylate, Cell – Tos. It has been shown that the "acidity" (H-bond donating power) of native cellulose is significantly larger in the crystalline than in the amorphous regions. For cellulose derivatives, this acidity decreases as a function of increasing DS, due the concomitant decrease in the number of free Cell – OH [99]. Therefore, cellulose dissolution in LiCl/DMAc may be pictured as a chloride ion-driven exchange of cellulose for DMAc in the coordination sphere of the lithium ion [100].

In any solvent system, the essential factors required for dissolution of cellulose include: adequate stability of the electrolyte/solvent complex; co-operative action of the solvated ion-pair on hydrogen bonding of cellulose; sufficient basicity (hardness) of the anion; adequate volume of the electrolyte/solvent complex [95]. The relative importance of these requirements has been investigated by several techniques, as shown by the following examples:

(i) Obtaining 3 wt % cellulose solution requires ca. 4 wt % and 10 wt % LiCl, in DMAc and N,N-dimethylformamide, DMF, respectively. This agrees with the fact that LiCl forms a stronger complex with the former solvent [101];

(ii) The strength of cation-solvent association has been directly determined by electro-spray ionization mass spectroscopy of alkali metal chlorides in DMAc and/or DMF. In both cases, the order found was $Li^+ > Na^+ > K^+ > Cs^+$, i.e., parallel to the order of hardness of the cation. For the same electrolyte, LiCl, in different solvents, the same technique indicated the following order of strength of cation-solvent association: N,N-dimethylpropionamide $>$ DMAc $\gg$ DMF. That is, the association increases as a function of increasing the negative charge on the oxygen atom of the solvent $C = O$ group, as well as its polarizability [80, 85];

(iii) For the same solvent, DMAc, LiCl is more efficient than LiBr in dissolving cellulose, since the later halide ion in less basic than the former one. Use of LiBr may be required, however, in special cases. For example, attempted bromination of microcrystalline cellulose by a mixture of N-bromosuccinimide and triphenylphosphine in LiCl/DMAc yielded chlorodeoxycellulose almost exclusively; most certainly due to the *preferential* nucleophilic attack of the chloride ion on the intermediate cellulose phosphonium salt $(Cell - O - P^+Ph_3X^-)$ [102].

Finally, dissolution of non-activated cellulose in LiCl/DMAc, and in ionic liquids has been accelerated by microwave irradiation [72, 103, 104], although the effect of microwave heating on the DP of the polymer has not been investigated. This last point is relevant in view of the fact that ILs are heated with exceptional efficiency by microwaves [105], so that care must be taken to avoid excessive localized heating that can induce chain degradation of the polymer during its dissolution.

Another important aspects of solubilization are the physical state of the dissolved polymer as well as the thermo-chemistry and kinetics of the dissolution reaction. It is known that a clear cellulose solution is a *necessary, but not sufficient* condition for the success of derivatization. The reason is that the polymer may be present as an aggregate, as will be discussed below. Additionally, dissolution of activated cellulose requires less time at low temperature, e.g., 2 h at 40 °C, and more than 8 h at 70 °C [106]. These aspects will be commented on below.

As indicated earlier, clear cellulose solutions are not necessarily molecularly dispersed; they may contain aggregates of still ordered cellulose molecules [107]. The structure of these aggregates has been described in terms of a "fringed" micellar structure. Figure 2a shows a schematic possi-

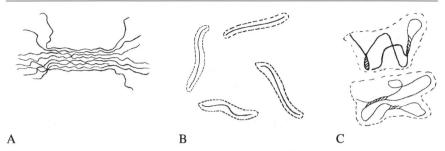

A   B   C

Fig. 2 Schematic representation of cellulose structures in solution: Part **A** shows the "fringed" micellar structure. Parts **B** and **C** show possible chain conformations of celluloses of different DP. For high molecular weight cellulose, **C**, intra-molecular hydrogen bonding is possible

bility for such an aggregate, composed of laterally aligned chains, forming a rather compact, and probably geometrically anisotropic core, immiscible with the solvent. The "coronas" at both ends of the particle consist of solvated amorphous cellulose chains. Formation of a fringed micellar structure is backed by experimental evidence. For example, increasing cellulose concentration results in a pronounced increase in molar mass of the particle, although its dimensions increase only slightly. The geometric anisotropy of the central part of the micelle is expected to be associated with optical anisotropy. Additionally, it may be visualized by an appropriate experimental technique. Both expectations have been confirmed by use of shear-induced birefringence, and electron microscopy [108]. The number of chain molecules, forming the aggregate, and the thickness of the coronas increase as a function of increasing both cellulose concentration, and the interfacial tension between the particle core and the solvent system [109]. Parts (B) and (C) of Fig. 2 refer to mono-disperse solutions of a small DP (B), and large DP (C) cellulose molecules. The former part shows that the length of the short cellulosic chain is practically equal to its persistent length, i.e., there is neither coiling, nor interactions between the chains. In Fig. 2c, the flexibility of the long chain polymer permits the formation of strong intra-molecular hydrogen bonds, provided that the OH groups reside for some time within a "critical distance" from each other, sufficient for the van der Waals interactions to operate [110]. Consequently, the properties of cellulose, in particular its DP, Ic, and concentration affect its state of solution, hence its derivatization. For the same cellulose, the accessibility of the hydroxyl groups increases as a function of decreasing solution concentration. For different celluloses, at the same concentration, only the outer surface of the fringed micellar core is accessible, the area of this part decreases as a function of increasing DP and Ic.

LS data have indicated the presence of aggregates in the LiCl/DMAc solvent system, whose size depends on the pre-treatment employed, DP, concentration of LiCl, and presence of water. Molecularly dispersed cellulose solutions are obtained at low polymer- and high LiCl concentrations [111]. Non-cellulosic material may lead to further aggregation. Whereas hardwood Kraft pulps were found to be completely soluble in this solvent system, softwood Kraft pulps were not, due to relatively higher contents of mannan, lignin, and nitrogen-containing compounds (originated from degraded proteins) [112]. Therefore, one of the reasons that mercerization *may lead* to better derivatization results, e.g. higher DS, is the effect of (sodium hydroxide-mediated) removal of non-cellulosic material on the physical state of cellulose in solution. LS was employed to study the structure of cellulose dissolved in *N*-methylmorpholine-*N*-oxide monohydrate. The results were indicative of aggregates that comprise several hundred cellulose chains, whose radius of gyration exceeded 160 nm. Additionally, aggregate heterogeneity is indicated by the bimodal distribution of the scattering intensity function, due to the presence of large and small aggregates. Activation of cellulose prior to dissolution leads to particles consisting of significantly fewer cellulose molecules, compared to un-activated polymer [113]. A microcrystalline cellulose sample was activated by heating under reduced pressure [56], and then dissolved in 6 wt % LiCl in DMAc. LS measurements of the clear solution (Malvern 4700 MW system, He/Ne laser light, 20 °C) have indicated that cellulose is present as a decamer, with an average radius of gyration of 62 nm [114]. Larger aggregation numbers (determined by SEC) have been reported for several celluloses in LiCl/DMAc [79, 82, 84, 115]. The (much more expensive) solvent system LiCl/1,3-dimethyl-2-imidazolidinone seems to be superior than LiCl/DMAc for cellulose dissolution, although the polymer concentration employed, 0.4 wt %, is smaller than that usually employed in synthesis of esters [116].

Some comments are worth mentioning with regard to determination of the molecular weight of celluloses dissolved in these solvent systems:

(i)   The protocol of solution preparation may affect the results obtained: A cellulose solution in 8 wt % LiCl/DMAc was prepared, then diluted for carrying out SEC analysis. The molecular weight calculated in the presence of 1 wt % LiCl was higher than that obtained for a 3 wt % electrolyte solution. This was attributed to a decrease in the intermolecular interactions and extent of aggregation during preparation of the former sample [117];

(ii) Other factors that may affect the results include final concentrations of the components, and presence of impurities: It has been shown that aggregates of cellulose from diverse sources (microcrystalline, softwood Kraft pulp, hardwood sulfite pulp) that form in 6 and 9 wt % LiCl in DMAc could be disintegrated on dilution to 2.6 wt % LiCl, provided that certain ratios of electrolyte/cellulose are maintained [118]. DMAc is unable to completely disrupt the inter-molecular hydrogen-bonding between the polymer chains in solutions of low LiCl, or high cellulose concentration. Additionally, traces of water present in solution affect the results [119, 120];

(iii) Unfavorable intrinsic properties of the cellulose solution may lead to errors: The basic equation of LS is given by [121]:

$$\frac{Kc}{R_\theta} = \frac{1}{M_w P_\theta} + 2Bc,$$ (2)

where:

$$K = \frac{2\pi^2 n_0^2 (dn/dc)^2}{N_{Av}\lambda^4},$$ (3)

$$R_\theta = \frac{I_s r^2}{I_0},$$ (4)

$$P(\theta) = 1 + \frac{16\pi^2 \overline{R}_g^2}{3\lambda^2} \mathrm{sen}^2 \left(\frac{\theta}{2}\right),$$ (5)

and c = solute concentration (in g/L); M_w = weight-averaged molecular weight; B = second virial coefficient; R_g = particle radius of gyration; n_0 = refractive index of the solvent at the wavelength of incident light, λ; dn/dc = solute specific refractive index increment; N_{Av} = Avogadro's number; I_0 and I_s = intensity of the incident, and scattered light, respectively, where the latter is measured at an angle of θ, and distance r between the sample and the detector of scattered-light. In applying this equation to cellulose solutions in LiCl/DMAc, one is faced with the problem of the effect of dn/dc employed on M_w. First, careful manipulation of the highly hygroscopic solvent is a must, because higher dn/dc will be obtained in the presence of adventitious water [122]. Second, dn/dc is small, being 0.061 for cellulose in 8.33 wt % LiCl in DMAc, compared with 0.324 for pure LiCl/DMAc solution [120]. The former refractive index increment is smaller than those of solutions of typical synthetic polymers, e.g., polystyrene in toluene (0.11) and poly(acrylamide) in water (0.183) [123]. It is also smaller than dn/dc of cellulose and its derivatives in a variety of solvents. Examples are cellulose in cuoxam (0.198), in Cd-tren (0.233) and in Ni-tren (0.199) [62], cellulose tricarbanilates in THF, 0.165 [124], ethylhydroxyethylcellulose in water (0.145), hydroxyethylcellulose, HEC, in water (0.141) and hydroxypropylcellulose in water, 0.143 [125]. As shown by Eq. 3, the term dn/dc is quadratic, i.e., an error of x % in dn/dc leads to an error of $2x$ % in M_w [123]. For example, LS has been employed in order to calculate M_w of microcrystalline cellulose in 8.33 wt % LiCl/DMAC. Based on $dn/dc = 0.061$, the M_w calculated was 246 078 g/mol. This would have been 250 961 or 241 279 g/mol if the dn/dc employed were 0.060 or 0.062, respectively [114];

(iv) A different strategy for the determination of the M_w (or DP) of cellulose involves its transformation into a completely substituted derivative

(DS $\approx$ 3). This reaction should be carried out under mild conditions, so that chain degradation is kept at minimum, and the group introduced should be able to disrupt the hydrogen bonding in cellulose. An example is cellulose tricarbanilates (reaction with phenyl isocianate in LiCl/DMAc, DS = 3), where introduction of the bulky substituent eliminates hydrogen bonding due to steric crowding, and increases dn/dc, vide supra [126];

(v) It is also worth mentioning that the concentrations employed in the determination of the state of aggregation of cellulose, e.g., by LS, are lower than those typically employed in the synthesis. Since M_w is cellulose concentration dependent, it is probable that cellulose is not present as mono-disperse during its derivatization.

In summary, although clear, light-colored cellulose solutions are required to start the synthesis, there is no guarantee, a priori, that the targeted DS will be obtained. The reasons are that the state of aggregation of cellulose is dependent on the structural characteristics of the starting material, is sensitive to the pre-treatment employed, and the impurities present. This may result in non-reproducible aggregation states, and may lead to oscillation in cellulose reactivity. Typically, effects of these oscillations may not be readily apparent, because:

(i) The derivatizing agent is employed in excess of the targeted DS;

(ii) The reaction time is usually long enough (typically overnight) for sluggish reactions to go to completion.

Attention is now focused on the thermochemistry of cellulose dissolution. Consider a typical scheme, where cellulose is activated at high temperature, then the slurry is cooled to a much lower temperature. During cooling, the crystallinity of *undissolved* cellulose decreases as a function of temperature and time. For example, a bagasse cellulose sample of Ic = 0.81 was suspended in 8.3 wt % LiCl in DMAc, at 155 °C. The slurry was cooled, and the Ic of undissolved cellulose was determined. It was found to be 0.70, at 112 °C, after 80 min of contact time with the solvent system, and 0.58, at 78 °C, after 100 min contact time [106]. This temperature effect can be explained by considering the three stages of dissolution of semi-crystalline polysaccharides. The first involves transition of the crystalline part to a (hypothetical) liquid amorphous state, the second is associated with dissolution and solvation of the macromolecule, whereas the third involves mixing of the solvated polymer chains with the solvent to give an infinitely dilute solution. The corresponding overall enthalpy of solution, $\Delta H_{solution}$, is given by [127, 128]:

$$\Delta H_{solution} = \Delta H_{fusion} + \Delta H_{transition} + \Delta H_{interaction} + \Delta H_{mixing}, \tag{6}$$

where ΔH_{fusion}, $\Delta H_{transition}$, $\Delta H_{interaction}$, and ΔH_{mixing} refer to the enthalpies of: (hypothetical) fusion of the cellulose crystalline regions; solvent-mediated transition of the amorphous component from vitreous to a highly elastic state; interaction between cellulose and the solvent system; and thermo-chemical effects of the mixing process, respectively. Analysis of the right-hand side of Eq. 6 reveals that the only positive term is the enthalpy of fusion. Note that $\Delta H_{interaction}$ includes enthalpies of interaction of DMAc with cellulose and with LiCl, both processes are exothermic [129, 130]. That is, LiCl-mediated dissolution of activated cellulose is expected to be exothermic, being more favorable at lower temperatures.

Solvent penetration into cellulose structure leads to its decrystallization. It is instructive, therefore, to examine the dependence of the rate of this process on DP and Ic of cellulose. This can be carried out by determining the dependence of Ic on time and temperature; use of non-isothermal conditions permits calculation of the rate constants of decrystallization, and the corresponding activation parameters from the same kinetic run [131, 132]. This has been carried out for celluloses whose DP and Ic cover the ranges 150 to 780 and 0.65 to 0.83, respectively [106, 133]. The results obtained are surprising because the rate constants and activation parameters calculated are only slightly dependent on the physico-chemical properties (DP and Ic) of the starting celluloses. Additionally, decrystallization is associated with a small $\Delta^{\ddagger}H$ and a large, negative $\Delta^{\ddagger}S$, whereas one expects the inverse, at least for the entropy. In other words, $|\Delta^{\ddagger}S|$ is expected to be small, its sign may be positive, since decrystallization is associated with an increase in the degrees of freedom of the polymer chains.

It should be born in mind, however, that the activation parameters calculated refer to the sum of several reactions, whose enthalpy and/or entropy changes may have different signs from those of the decrystallization proper. Specifically, the contribution to the activation parameters of the interactions that occur in the solvent system should be taken into account. Consider the energetics of association of the solvated ions with the AGU. We may employ the extra-thermodynamic quantities of transfer of single ions from aprotic to protic solvents as a model for the reaction under consideration. This use is appropriate because recent measurements (using solvatochromic indicators) have indicated that the polarity at the surface of cellulose is akin to that of aliphatic alcohols [99]. Single-ion enthalpies of transfer indicate that Li^+ is more efficiently solvated by DMAc than by alcohols, hence by cellulose. That is, the equilibrium shown in Eq. 7 is endothermic:

$$Li - DMAC + cellulose \leftrightarrows DMAc + Li - cellulose \qquad (7)$$

The inverse holds for Cl^-, i.e., the equilibrium depicted by Eq. 8 is exothermic [134].

$$Cl^- - DMAC + cellulose \leftrightarrows DMAc + Cl^- - cellulose \qquad (8)$$

As discussed above, solvatochromic data for cellulose solutions in LiCl-DMAc indicate that $Cl^- - HO - Cell$ hydrogen bonding is more important for dissolution than Li-cellulose interactions. If decrystallization is rate limiting, and considering that the equilibria shown in Eqs. 7 and 8 occur *prior to* decrystallization, then the activation parameters calculated represent the sum of the three reactions. That is, the endothermicity associated with breaking of the intra-molecular hydrogen bonding (between the cellulose chains) is partially offset by the exothermicity of the hydrogen bonds formed between cellulose and the chloride ion. This partial cancellation may explain the small $\Delta^{\ddagger}H$ calculated. In addition to an increase in the degrees of freedom of the chain, ion association with the polymer most certainly contributes to the overall $\Delta^{\ddagger}S$. That is, the change from crystalline cellulose to polymer-LiCl complex may be associated with: entropy increase, due to decrystallization; entropy decrease, due to ion complexation by the polymer [135]. The mobility may be further reduced due to electrostatic (dipole-dipole) interactions, and polymer-chain aggregation, vide supra. It would be very interesting to assess the relative contribution of the above-mentioned interactions, for example by measuring the thermodynamic parameters (enthalpy, entropy, and Gibbs free energy) of each step. This task is made relatively easy with the advent of power compensating titration calorimeters, running in the quasi-isothermal mode.

4.2
Derivatizing Solvents

Certain solvents react with cellulose. Dissolution occurs because the (solvent-based) group introduced disrupts hydrogen bonding within the polymer, by a combination of steric interactions and decrease of the number of OH groups available for bonding. If this dissolution scheme is employed, then the functional group introduced should be readily removable, e.g., by hydrolysis, after further derivatization. Examples are (solvent system, cellulose derivative formed): N_2O_4/DMF, nitrite; HCO_2H/H_2SO_4, formate; F_3CCO_2H, trifluoroacetate; Cl_2CHCO_2H, dichloroacetate; paraformaldehyde/DMSO, hydroxymethyl; $ClSi(CH_3)_3$, trimethysilyl. One problem of this approach to cellulose dissolution is poor reproducibility, due to side reactions, and formation of undefined structures. Nevertheless, N_2O_4/DMF has been employed in the preparation of cellulose inorganic esters, e.g., sulfate [136–139]. The mechanism involves the heterolytic cleavage of N_2O_4, followed by formation of cellulose nitrite. The presence of water in the medium leads to preferential nitrite formation at the C_6 position [137]. Other possibilities for cellulose nitrite formation include substitution of DMSO for DMF, and of NOCl; nitrosyl sulfuric acid; nitrosyl hexachloroantimonate or nitrosyl hexafluoroborate for N_2O_4.

Paraformaldehyde/DMSO dissolves cellulose rapidly, with negligible degradation, and forms the hydoxymethyl (methylol) derivative at C_6 [140–142]. Therefore, cellulose derivatives at the secondary carbon atoms are easily obtained after (ready) hydrolysis of the methylol residue. Additionally, fresh formaldehyde may add to the methylol group, resulting in longer methylene oxide chains, that can be functionalized at the terminal OH group, akin to non-ionic, ethylene oxide-based surfactants [143, 144].

There are several schemes for the synthesis of cellulose formates: (slow) reaction of the polymer with formic acid; faster reaction in the presence of a mineral acid catalyst, e.g., sulfuric or phosphoric acid. The latter route is usually associated with extensive degradation of the polymer chain. Reaction of $SOCl_2$ with DMF produces the Vilsmeier–Haack adduct $(HC(Cl) = N^+(CH_3)_2Cl^-)$ [145]. In the presence of base, cellulose reacts with this adduct to form the unstable intermediate $(Cell - O - CH = N^+(CH_3)_2Cl^-)$ from which cellulose formate is obtained by hydrolysis. The DS ranges from 1.2 to 2.5 and the order of reactivity is $C_6 > C_2 > C_3$ [140–143, 146].

A comment on the properties of the base employed in reactions that involve the formation of the Vilsmeier–Haack adduct is in order, because several derivatives of cellulose are obtained by this route. Preparation of Cell – Tos has been attempted in LiCl/DMAc, by reacting the polymer with TosCl/base. Whereas the desired product was obtained by employing triethylamine, use of pyridine (Py) resulted in the formation of chlorodeoxycellulose. In order to explain these results, the following reaction pathways have been suggested [147]:

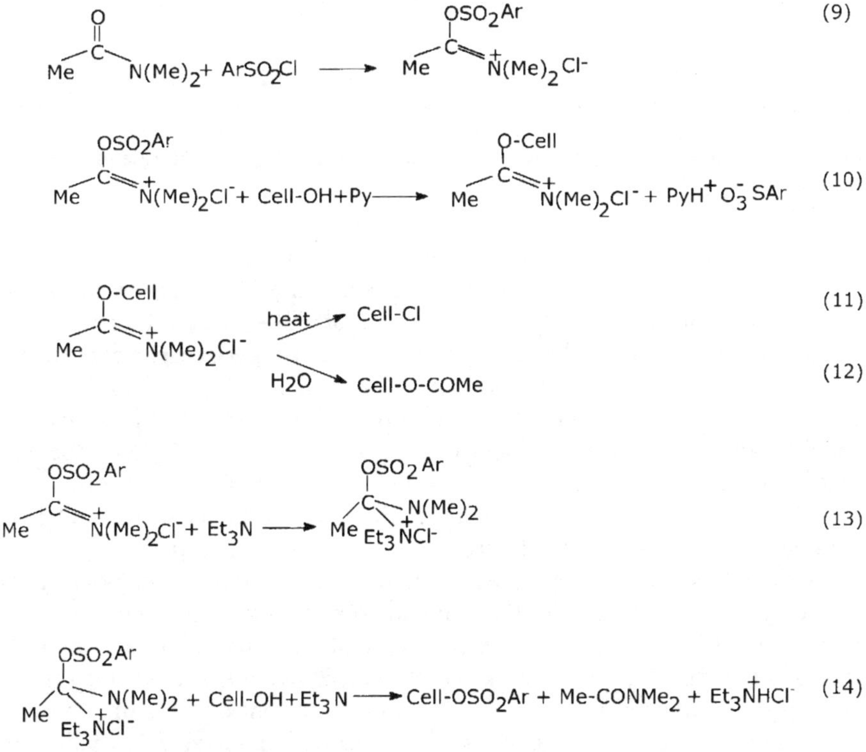

Fig. 3 Suggested mechanisms for the reaction of cellulose with tosyl chloride in the presence of pyridine (Py) and/or triethylamine (Et$_3$N) base catalyst. Redrawn from [147]

That is, the difference between the mechanisms of action of the two bases lies in the ability of Et$_3$N to add to the Vilsmeier–Haack adduct to form a tetrahedral intermediate, susceptible to S_N2 attack by (at least partially deprotonated) cellulose. This leads to formation of the desired Cell – Tos.

An alternative, *single* mechanism for both bases may be formulated, however, by taking into account the differences in basicity (pK_a = 5.25 and 11.01, for Py and Et$_3$N, respectively [148]), and hydrophobicity between the two bases. A quantitative measure of the latter property is given by log P, the partition coefficient of the solute between *n*-octanol and water (log P = log([solute]$_{n\text{-octanol}}$/[solute]$_{water}$), 0.65 and 1.45 for Py and Et$_3$N, respectively [149]. This unified mechanism is shown in Fig. 4, where B refers to the base employed.

$$ArSO_2Cl + Cell\text{-}OH + B \rightleftharpoons Cell\text{-}OSO_2Ar + BH^+Cl^- \tag{15}$$

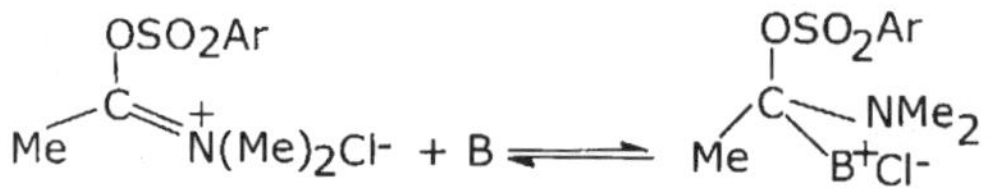

$$\underset{Me}{\overset{OSO_2Ar}{\underset{}{\overset{|}{C}}}}\!\!=\!\!\overset{+}{N}(Me)_2Cl^- + B \rightleftharpoons \underset{Me}{\overset{OSO_2Ar}{\underset{}{\overset{|}{C}}}}\!\!\underset{B^+Cl^-}{\overset{NMe_2}{\diagdown}} \tag{16}$$

$$Cell\text{-}OSO_2Ar + BH^+Cl^- \rightleftharpoons Cell\text{-}Cl + BH^+O_3SAr \tag{17}$$

Fig. 4 A unified mechanism for the reaction of cellulose with tosyl chloride in the presence of base (pyridine and/or triethylamine)

Note that the synthesis is carried out by adding TosCl and the base to a solution of cellulose in LiCl/DMAc. Consequently, the first reaction shown in Fig. 4 depicts direct, base-catalyzed formation of Cell – Tos, Eq. 15. Formation of the Vilsmeier–Haack adduct occurs as shown in Eq. 9 of Fig. 3 and, therefore is not repeated in Fig. 4. Equation 16 is similar to Eq. 13, except that it occurs for both nucleophiles. The formation of this tetrahedral intermediate should lead to Cell – Tos, by analogy to Eq. 14. Finally, Eq. 17 shows the formation of chlorodeoxycellulose, from *cellulose – Tos*, by an S_N2 reaction. The equilibria shown in Eqs. 15 and 16 lie much further to the right when Et$_3$N is employed, because it is a much stronger base, with a consequent better yield of Cell – Tos. We now address the reason that the reaction shown by Eq. 17 is operative for Py, but not for Et$_3$N. The position of this equilibrium depends on the strength of association of the ion pair BH$^+$Cl$^-$; weak interactions shift the equilibrium to the right hand side. A quantitative measure of this association may be inferred from the distribution coefficient of ion pairs between water and an immiscible organic solvent. Strongly associated ion pairs are hydrophobic and are, consequently, more soluble in the organic phase. For a series of alkylammonium picrates, log (distribution coefficient) increases linearly as a function of increasing the hydrophobicity of alkyl group, from methyl to n-pentyl [150]. By analogy, the association between the much more hydrophobic triethylammonium ion and the chloride ion is expected to be stronger than the corresponding association in PyH^+Cl^-. That is, the anion of the latter electrolyte is relatively free to react with Cell – Tos, by an S_N2 mechanism, leading to the formation of chlorodeoxycellulose.

As derivatives, cellulose trifluoroacetates have several interesting properties: they are thermally stable, up to 250 °C; easily hydrolyzable (few minutes) and the primary OH groups are almost completely functionalized [151, 152]. Provided that extensive polymer degradation does not take place during dissolution [153], they serve as an attractive starting material for functionalization at the secondary OH groups. Additionally, they form mesophases at a relatively low polymer concentration (4 wt %), therefore can be employed for regeneration of strong cellulose fibers. These esters can be obtained by reacting cellulose with trifluoroacetic acid, TFA, in organic solvents (e.g., chlorinated solvents [153]), or with a mixture of the acid and its anhydride [154]. TFA can be employed as a reaction medium, e.g., for the synthesis of esters (acetate to decanoate) and/or mixed esters of cellulose, of high DS, by the reaction with carboxylic acid anhydrides [151, 152]. The relatively strong dichloroacetic acid may be employed as a solvent for cellulose. Thus a mixture of this acid and the corresponding anhydride reacts slowly with cellulose to generate dichloroacetate esters with a DS from 1.6 to 1.9, in which position 6 is almost completely functionalized. These products are soluble in DMF, DMSO, pyridine, and THF, and are thermally stable up to 280 °C, but become insoluble at temperatures > 150 °C [155].

In summary, cellulose dissolution can be also achieved by derivatizing solvents, provided that severe degradation is avoided. The importance of this scheme is that it may lead to an *inverse pattern of functionalization*, provided that special reaction conditions are employed. Namely, the secondary OH groups may be specifically functionalized. This is attributed to the fact that the primary, i.e., more reactive, hydroxyl group is protected by a blocking moiety (nitrite, methylol, reactive acyl group), introduced during cellulose dissolution. Note that the above-discussed question of decrystallization is probably of less importance in this scheme. The reason is that the initial derivatization by the solvent contributes to disruption of intra-molecular hydrogen bonding, hence leads to decreased polymer crystallinity.

5
Cellulose Functionalization

This area has witnessed an explosive growth as a result of the introduction of novel synthetic schemes, designed to prepare a variety of products, or intermediates from which further products may be obtained. The reason is that many of the problems associated with the two-phase (industrial) reactions are avoided if the HRC scheme is employed, since, in principle, the whole cellulosic chain is accessible for derivatization (see, however, the discussion on polymer aggregation in solution). Some of the advantages of the HRC scheme include: Reproducible control of DS and regioselectivity; production of derivatives which are regularly substituted along the natural polymer backbone; negligible degradation of the natural polymer during the reaction; synthesis of products that are either difficult or impossible to obtain by the heterogenous reaction, e.g., esters of long-chain carboxylic acids; synthesis of intermediates, e.g., nitrites, sulfonates and trifluoroacetates that may be transformed into other functional groups, e.g., by nucleophilic substitution reactions. In the following account, we concentrate on the solvent systems inorganic and/or organic electrolytes in dipolar aprotic solvents, since these were most extensively employed.

5.1
Ester Preparation in Non-Derivatizing Solvents

The phase diagram of cellulose solutions in LiCl/DMAc has shown that unstable mesophases are formed at relatively high cellulose content, 10 to 15 wt % [156, 157]. Due to the above-mentioned cellulose aggregation in these media, use of such high concentrations is not of interest because the reaction will be only *apparently homogenous*. Thus low cellulose concentrations are employed, typically 2–3 wt %. Polymer solubilization may be easily visualized by employing a pair of polaroid glasses, in order to guarantee that the solution is isotropic [42, 50].

Functionalization by acyl chlorides and carboxylic anhydrides were among the first experiments carried out under the HRC scheme [158–160]. In the acyl chloride route, a tertiary base is routinely employed as a catalyst. The base has a two-fold role: to activate the acyl compound further (by forming the corresponding acylammonium ion), and to scavenge the HCl formed, in order to avoid excessive degradation of the precursor polymer, or the product during the reaction. The tertiary bases employed include pyridine, 4-N,N-dimethylaminopyridine, triethylamine, and N,N-dimethylaniline. The last base is ca. 23 times more basic than pyridine [148], and can be precipitated as perchlorate salt, and recycled [161]. Triethylamine is preferred over pyridine because it is a stronger base, i.e., the corresponding ammonium ion is a weaker acid, leading to less polymer degradation [162].

The acyl groups introduced included 4-phenylbenzoyl, phenylacetyl, 4-methoxybenzoyl, acetyl, 2,4-dichlorophenoxyacetyl, and 2,2-dichloropropionyl. Introduction of the last pair of acyl groups is important because they are bioactive (insecticides), i.e., the product can be employed in controlled-release formulations [159]. The structures of all these esters were determined by FTIR and NMR spectroscopy, whereas their solution properties, including the formation of mesophases, were studied by viscosimetric and LS techniques [126, 159, 163–165]. Poor solubility, combined with complex substituent signals precludes ^{1}H NMR analysis when the DS of alkanoates (RCO between C_6 and C_{18}) falls below 1. In this case, the cellulose esters were submitted to aminolysis by pyrrolidine. The (stable) amides were analyzed by gas chromatography using authentic acylpyrrolidine standards [166]. A recent work on the acetylation of microcrystalline cellulose by acetyl chloride has shown that the DS, obtained under comparable reaction conditions is lower in the presence of pyridine catalyst, than in its absence, this difference increases as a function of increasing the pyridine/AGU molar ratio. This may indicate pyridine-mediated ester hydrolysis [167]. Cellulose methacrylate (DS up to 1.3) and hemicellulose stearate (DS between 0.32 and 1.51) were also obtained by this procedure [168, 169]. The former esters undergo Uv-mediated cross-linking to form gels.

Acid anhydrides have been employed with, and without the use of a base catalyst. For example, acetates, propionates, butyrates, and their mixed esters, DS of 1 to ca. 3, have been obtained by reaction of activated cellulose with the corresponding anhydride, or two anhydrides, starting with the one with the smaller volume. In all cases, the distribution of both ester groups was almost statistic. Activation has been carried out by partial solvent distillation, and later by heat activation, under reduced pressure, of the native cellulose (bagasse, sisal), or the mercerized one (cotton linters). No catalyst has been employed; the anhydride/AGU ratio was stoichiometric for microcrystalline cellulose. Alternatively, 50% excess of anhydride (relative to targeted DS) has been employed for fibrous celluloses. In all cases, polymer degradation was minimum, and functionalization occurs preferentially at C_6 (^{13}C NMR spectroscopic analysis [52, 56, 57]).

Diketene, the dimer of ketene, reacts readily (30 min) with cellulose in LiCl/DMAc and LiCl/N-methyl-2-pyrrolidinone to produce cellulose acetoacetate. This ester is important because there exist several routes for further functionalization of the substituent introduced. Examples are through its methylene group, by condensation with bis(aldehydes) via aldol chemistry, or with bis(acrylates) via Michael addition chemistry. On the other hand, the CO group undergoes condensation with diamines or dihydrazines. The T_g of the products are relatively low, reaching 136 °C at DS = 2.7, this temperature showed no correlation with DP of the starting cellulose, but was highly dependent on its DS. The reaction with a mixture of diketene and an anhydride produces mixed esters, e.g., acetoacetate and acetate, propionate and/or butyrate. Depending on the final DS, and the ratio of both substituent groups, esters were obtained whose solubility ranges from water to THF [51, 170, 171]. However, the substituent distribution was not statistic, the acetoacetyl group is favored [52]. This may be explained as follows: at the reaction temperature, 110 °C, diketene decomposes rapidly to produce acetylketene, a very reactive intermediate, isolated and characterized in solid matrices [172, 173]. No kinetic data are available on the reaction of acetylketene with cellulose, and indeed with other nucleophiles. Therefore, n-butylketene may be taken as a convenient, albeit less reactive model for acetylketene, and water as a model for nucleophiles, including cellulose. The relative rate constants for the reactions with water are: 81 475 : 2.2 : 1.2 : 1, for n-butylketene, acetic anhydride, propionic anhydride, and butyric anhydride, respectively [174–176]. That is, the reaction of cellulose with acetylketene (to produce the corresponding acetoacetate ester) is expected to be much faster than its reaction with any aliphatic anhydride, even when the differences between volumes of the acetoacetyl and carboxyl groups are taken into consideration (1 : 1.37 : 1.73 : 1.77 for acetyl, propionyl, butyryl and acetoacetyl group, respectively; volumes calculated by Abraham's additive method [177]).

Recently, use of LiCl/DMAc and LiCl/1,3-dimethyl-2-imidazolidinone as solvent systems for acetylation of cellulose by acetic anhydride/pyridine has been compared. A DS of 1.4 was obtained; the substituent distribution in the products synthesized in both solvents was found to be the same, with reactivity order $C_6 > C_2 > C_3$. Therefore, the latter solvent system does not appear to be better than the much less expensive LiCl/DMAc, at least for this reaction. It appears, however, to be especially efficient for etherification reactions [178]. It is possible, however, that the effect of cellulose aggregation is more important for its reaction with the (less reactive) halides than with acid anhydrides; this being the reason for the better performance of the latter solvent system in ether formation, since it is more efficient in cellulose dissolution.

Another, recently introduced solvent system is TBAF/DMSO, preliminary experiments employed 2.9 wt % cellulose and 16.5 wt % TBAF and excess acetic anhydride. The relatively low DS obtained (0.83) is most probably due to water-mediated hydrolysis of the anhydride [179]. Later, the cellulose solution was partially dehydrated by distilling off ca. 30% of the solvent, before addition of the anhydride. This procedure resulted in a higher DS, 1.4 for mercerized sisal [180]. This solvent system is interesting because of the above-mentioned importance to the solubilization of the anion – HO – Cell hydrogen bonding, the fluoride ion being a harder base than the chloride ion (of LiCl/DMAc). Furthermore, the voluminous cation acts as a "spacer" between individual cellulose chains, preventing their re-attachment [100]. Both factors may explain the ease of solubilization of different celluloses in this medium. For instance, clear solutions of microcrystalline cellulose were obtained in 15 min, at room temperature, whereas fibrous sisal required 30 min at room temperature, plus 60 min at 60 °C [180]. This electrolyte is very hygroscopic, i.e., absorbs water *well beyond the trihydrate*. Periodic determination of water in the DMSO solution is, therefore, advisable in order to employ the appropriate amount of derivatizing agent, vide supra on the effect of water present on the product DS. The problem, however, is that the Karl-

Fischer titration may not be reliable for water concentration determination in the presence of highly hygroscopic electrolytes, e.g., LiCl/DMAc [119]. This conclusion has been also verified for TBAF/DMSO, by adding known amounts of water to the solvent system, followed by determination of the water content by Karl-Fischer titration. Whereas the added water ranged from 0.23 to 1.19 mol l^{-1}, that determined by titration ranged from 0.21 to 0.48 mol l^{-1}. An attempt has also been made to determine the water concentration by the solvatochromic indicator 2,6-dichloro-4-(2,4,6,-triphenyl-1-pyridinio)phenolate, whose pK_a in water is conveniently low, 4.78, so that its solvatochromic band should not be suppressed by traces of acidic impurities [181]. The idea was to use a calibration curve, polarity versus [water] in DMSO (no electrolyte), in order to determine [water] in the presence of TBAF. This attempt was, however, unsuccessful for water concentrations $\leq$ 1.2 mol l^{-1} because the color of the dye solution changed from violet (characteristic color of the solvatochromic band) to dark yellow within 15 min. The reason for this side reaction has not been investigated. Since the nucleophilicity of the fluoride ion is greatly enhanced when it is desolvated [182], an ipso-substitution of the dye chlorine atom by the fluoride ion may not be ruled out [183].

Finally, acetylation of cellulose (DP of 650) has been carried out in the IL, 1-allyl-3-methylimidazolium chloride, under different conditions. The reaction progress has been studied as a function of reaction time, under the following conditions: 80 °C; 4 wt % cellulose; 5 : 1 acetic anhydride/AGU ratio. As expected, the DS increased with time, being 0.94, 1.61, 2.21, 2.49 and 2.74 after 0.25, 1.0, 4.0, 8.0, and 23 h, respectively. At 100 °C, a DS of 2.3 was obtained in 3 h, by using the same reagent/AGU molar ratio. Increasing the latter to 6.5 decreased the DS from 2.43 to 2.38, this has been attributed to the effect of the extra anhydride on the structure of IL. The order of reactivity (total DS of 0.94) was found to be $C_6 > C_3 > C_2$, i.e., similar to that observed for acetylation in LiCl/DMAc [52b], but not in LiCl/1,3-dimethyl-2-imidazolidinone, where the corresponding order is $C_6 > C_2 > C_3$ [178].

There are some problems associated with the use of functional derivatives of carboxylic acids. Long-chain acid anhydrides are not commercially available, and one half of the acylation reagent is not utilized. Acyl chlorides require the use of tertiary base catalysts, whose double role has been explained before. Some of the intermediate acyl ammonium compounds formed are, however, insoluble in the solvent system. Examples include RCO − N$^+$Et$_3$Cl$^-$ in LiCl/DMAc, where RCO refers to the propionyl, hexanoyl, and stearoyl moiety, respectively. Hexanoyl- and stearoyl-pyridinium chlorides are also insoluble in the same solvent system [185].

In principle, some of these problems may be avoided if the carboxylic acid *proper* is employed for cellulose esterification. This approach, however, is not attractive because low yield, and polymer degradation are expected. The reason is that the equilibrium for esterification of aliphatic alcohols (models for cellulose) is not in favor of the product, and requires (strong) acid catalysis in order to proceed at a reasonable rate[184]. Use of this catalyst will almost certainly result in polymer degradation. Direct acylation of cellulose by carboxylic acid has been recently attempted. Cellulose was homogenized in aqueous alcohol, the slurry was treated with a mixture of fatty acid and a base catalyst (several soaps; NaOH; K_2CO_3). The solvent was distilled off, the remaining mass was heated at 195 °C. The DS increased with time, and levelled off at 0.23 (for octanoate) after 5 h reaction time. In addition to the low DS obtained, the DP decreased to ca. 40% of its starting value, after the first hour [186].

Schemes are available, however, that start from the free carboxylic acid, plus an "activator". Dicyclohexylcarbodiimide, DCC, has been extensively employed as a promoter in esterification reactions, and in protein chemistry for peptide bond formation [187]. Although the reagent is toxic, and a stoichiometric concentration or more is necessary, this procedure is very useful, especially when a new derivative is targeted. The reaction usually proceeds at room temperature, is not subject to steric hindrance, and the conditions are mild, so that several types of functional groups can be employed, including acid-sensitive unsaturated acyl groups. In combination with 4-pyrrolidinonepyridine, this reagent has been employed for the preparation of long-chain fatty esters of cellulose from carboxylic acids, as depicted in Fig. 5 [166, 185, 188]:

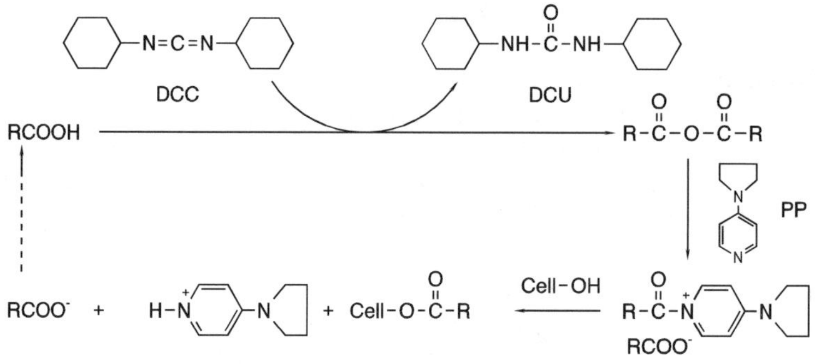

Fig. 5 DCC-mediated synthesis of cellulose esters, from [23]

First, the acid anhydride is produced by the reaction of the free acid with DCC. Nucleophilic attack by 4-pyrrolidinonepyridine on the anhydride results in the corresponding, highly reactive acylpyridinium carboxylate; this leads to the formation of cellulose ester, plus a carboxylate anion. The latter undergoes a DCC-mediated condensation with a fresh molecule of acid to produce another molecule of anhydride. N,N-carbonyldiimidazole (CDI) may substitute DCC for acid activation, the intermediate being N-acylimidazol, which readily reacts with cellulose to give the ester and imidazole, see Fig. 6. Adamantyl derivatives of cellulose are important because this ester moiety is a base-sensitive protecting group in deoxyribonucleoside chemistry [189], and its incorporation may lead to polymer derivatives with antitumor or antibacterial acitivity [190, 191]. These esters have been prepared with several reagents. For example, products with a DS of up to 1.4 have been prepared by CDI-activated acid in LiCl/DMAc [192]. Later, it has been prepared by the reaction of dissolved cellulose with the corresponding acyl chloride, and/or with acid activated by either CDI or Tos – Cl, vide infra. The DS range obtained was from 0.24 to 2.12, and some of the products showed anti-inflammatory effect, although this effect is not directly related to the DS of the ester [167].

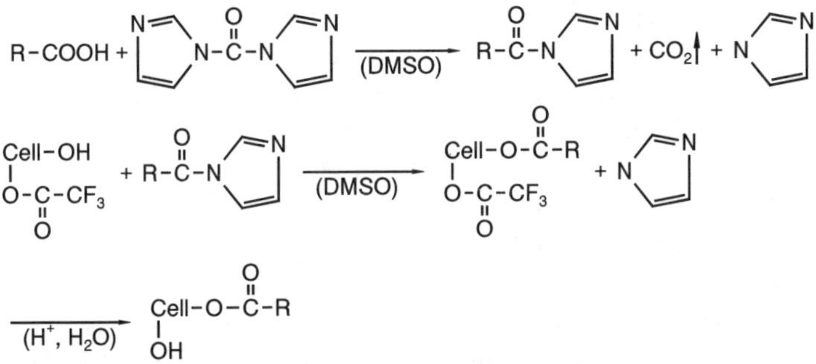

Fig. 6 CDI-catalyzed synthesis of cellulose trifluoroacetate esters, from reference [23]

In-situ activation of carboxylic acids can be also carried out by Tos – Cl/triethylamine, according to Fig. 7:

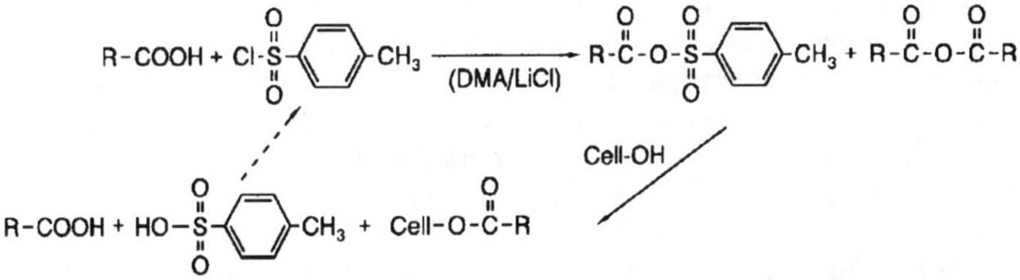

Fig. 7 Use of tosyl chloride as an acid "activator" for the synthesis of cellulose esters, from [23]

As shown, an asymmetric carboxylic-sulfonic acid anhydride is formed, but the cellulose attack occurs on the C = O group, since a nucleophilic attack on sulfur is slow, and the tosylate moiety is a much better leaving group than the carboxylate group [193]. Similar to other acylation reactions, there is a large preference for tosylation at the C_6 position, and cellulose tosylates with a DS from 0.4 to 1.4 are soluble in dipolar aprotic solvents [194]. Cellulose esters of long-chain fatty acids, e.g., dodecanoate to eicosanoate have been prepared in LiCl/DMAc with this activation method, with almost complete functionalization, DS 2.8–2.9 [195]. More recently, the effects of the molar ratio carboxylic acid (hexanoic to octadecanoic)/Tos – Cl/pyridine on the DS, and the solubility in organic solvents of the esters produced has been investigated. Under the same reaction conditions, the DS obtained increased as a function of increasing the chain length of the carboxylic acid, being 1.40 and 1.76 (no pyridine catalyst) and 1.76 and 1.92 (use of pyridine) for cellulose octanoate and octadecaonate, respectively [167]. This relationship between the chain-length of the reagent and DS appears to be general, vide infra.

The same procedure has been employed to increase the hydrophobic character of cellulose, by introducing fluorine-containing groups, e.g., 2,2-difluoroethoxy; 2,2,2-trifluoroethoxy; and octafluoropentoxy. Incorporation of hydrophobic moieties into cellulose is expected to increase the polymer compatibility with other materials, e.g., synthetic polymers. Note that an important part of incompatibility is due to the highly hydrophilic character of cellulose. Decreasing this character is expected to affect T_g of the derivative, as shown by cellulose propionate/octafluoropentoxy acetate (total DS = 3.0, partial DS in each moiety = 1.5), whose T_g is only 53 °C. The products are more stable than their fluorine-free counterparts, and the terminal $CF_2 - H$ group affects T_g much less than OH-substituted trifluoroethoxy cellulose derivatives [196, 197].

The Tos – Cl/carboxylic acid activation scheme has been employed in order to introduce a fluorescent probe, namely antharcene-9-carboxylate, into the structure of cellulose. Interestingly, the fluorescence spectra of the cellulose-bound ester, and that of the free acid were found to be the same [198]. Water soluble cellulose esters, with a DS range of 0.4 to 3.0 were similarly prepared, by reacting cellulose tosylate with 3,6-dioxahexanoic acid and/or 3,6,9-trioxadecanoic acid, without a base catalyst, to yield: Cell – O – $CO - CH_2(OCH_2CH_2)_2OCH_3$, and Cell – O – CO – $CH_2(OCH_2CH_2)_3OCH_3$, respectively. Although these esters start to dissolve in water at a DS as low as 0.4, they are also soluble in common organic solvents, e.g., acetone and ethanol, and are thermally stable up to 325 °C [199]. These products are interesting because their structure is akin to that of non-ionic surfactants, produced, e.g., by the reaction of fatty alcohols with ethylene oxide. The difference is that the cellulose derivatives carry a single number of oxyethyelene units, and not the usual bell-shaped distribution that is present in commercial non-ionic surfactants [144].

The Vilsmeier–Haack type adduct, formed by the reaction of oxalyl chloride with DMF can be also be employed for the activation of carboxylic acids, as shown in Fig. 8 [200].

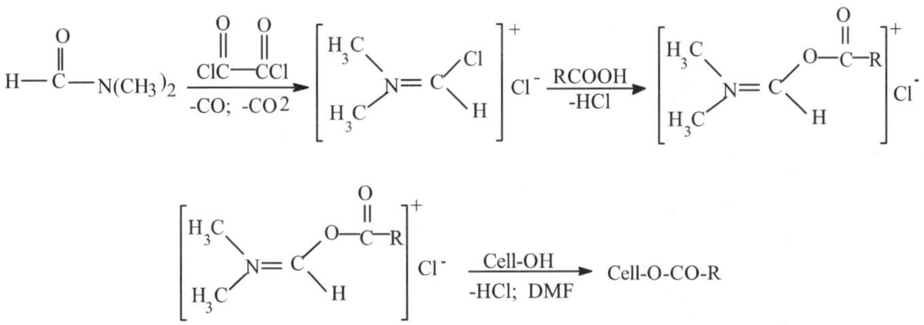

Fig. 8 Use of oxalyl chloride in the synthesis of cellulose esters in LiCl/DMAc, from [200]

The imminium chloride formed was transformed, in-situ, into the corresponding carboxylic acid derivative, this was added to a solution of cellulose in LiCl/DMAc. Palmitic, stearic, adamantane-1-carboxylic, and 4-nitrobenzoic acids were employed. The DS of the corresponding esters increased as a function of increasing the ratio oxalyl chloride/AGU. The solubility of the products obtained in aprotic solvents was tested; GPC results have indicated negligible degradation of the polymer [200].

Another possibility to obtain esters is by trans-esterification, an equilibrium reaction where the use of special procedures is required, in order to shift it to the product. For cellulose, advantage has been taken of the availability of vinyl carboxylates, since one of the products, namely vinyl alcohol, tautomerizes to volatile acetaldehyde (b.p. 21 °C). Thus it was possible to obtain cellulose acetate, DS = 2.7, by reacting 10 mol vinyl acetate per AGU, at 40 °C, for 70 h. The same reaction conditions produced cellulose dodecanoate, with a DS of 2.6, a remarkable result when the large volume of the acyl group is considered. It was found that the DS could be controlled by adjusting the ratio of vinyl ester/AGU. For example, at a ratio of 2.3, the following DS were obtained: 1.04, 0.86, and 0.95 for cellulose acetate, butyrate and benzoate, respectively [179]. The same strategy has been applied to the reaction of native, and mercerized sisal cellulose solutions in TBAF/DMSO. Interestingly, under comparable experimental conditions, the DS of the latter cellulose derivative was higher for dodecanoate than acetate [180]. As shown above, the same result has been obtained for the reaction of TBAF/DMSO-solubilized microcrystalline cellulose with carboxylic acids, activated with Tos – Cl [167]. This result has been attributed to long-chain mediated separation of aggregated cellulose chains by the voluminous acyl moiety, as shown in Fig. 9 [180].

The DS increase as a function of increasing the chain-length of the acyl group may be also attributed to hydrophobic interactions between the cellulosic surface, whose hydrophobic character increases as a function of increasing DS, and the acylating species. This cooperative interaction, Fig. 10, may contribute to the activation enthalpy, as a result of desolvation of the entering species. Since association between the chains attached to cellulose and those

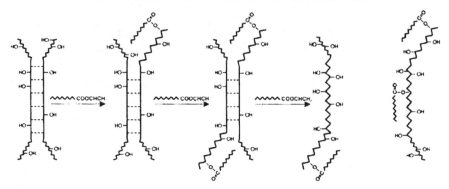

Fig. 9 Scheme for explaining the relationship between chain-length of the derivatizing agent and DS obtained, from [180]

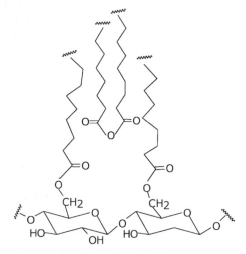

Fig. 10 Scheme for explaining the effect of hydrophobic interactions on the DS obtained. For simplicity, a part of the fatty chain is only shown

of the entering reagent is involved in a pre-equilibrium, before the rate limiting step, it may also affect the activation entropy. The occurrence of such hydrophobic interactions has been employed in order to explain:

(i)   The effects of ionic and/or non-ionic micellar aggregates on the reactivity of substrates of increasing degree of hydrophobicity [201, 202];

(ii) The phase behavior of mixtures of non-ionic surfactants (ethoxylated alcohols) and hydrophobically modified ethyl(hydroxyethyl)cellulose (by introduction of 1.7 mole % of nonylphenyl groups) [203];

(iii) Solution properties (viscosity, surface tension and LS) of mixtures of anionic and cationic surfactants with HEC, and HEC modified by introduction of (neutral) perfluoroalkyl and/or (cationic) trimethylammonium or alkyldimethylammonium groups [204, 205].

The HRC scheme makes regioselectivity control feasible, by the following reaction sequence: protection of a targeted Cell-OH group/functionalization of the remaining Cell-OH groups/removal of the protective group introduced. Esters and mixed esters where substitution has been regioselectively controlled are important reference compounds, e.g., for structure elucidation by spectroscopic techniques; for comparison of products prepared by using the homogenous and heterogenous processes; for DSC analysis; and for the direct imaging of a single crystal by atomic-force microscopy [206–209]. For example, cellulose was partially acetylated by acetic anhydride/pyridine in LiCl/DMAc to give a preferential reaction at O-6. Controlled deacetylation by treatment with hydrazine monohydrate in DMSO gave polymer functionalized at this position only, DS = 0.6, as shown by ^{13}C NMR spectroscopy [210]. Other amines that deacetylate the secondary positions include dimethylamine and hexamethylene diamine. Figure 11 shows several pathways to an important inorganic cellulose derivative, namely sulfate. The relevant point of the synthetic schemes shown is that the appropriate regioselectivities of the intermediate esters and/or ethers are central to obtaining sulfates with the desired structures.

The primary OH group can be selectively blocked by the bulky triphenylmethyl (trityl) moiety, followed by esterification at the secondary OH groups and removal of the protecting trityl group. Thus 2,3-di-O-acetyl cellulose has been obtained by this procedure. Moreover, regioselectively substituted mixed cellulose esters, acetate/propionate, were prepared by subsequent acy-

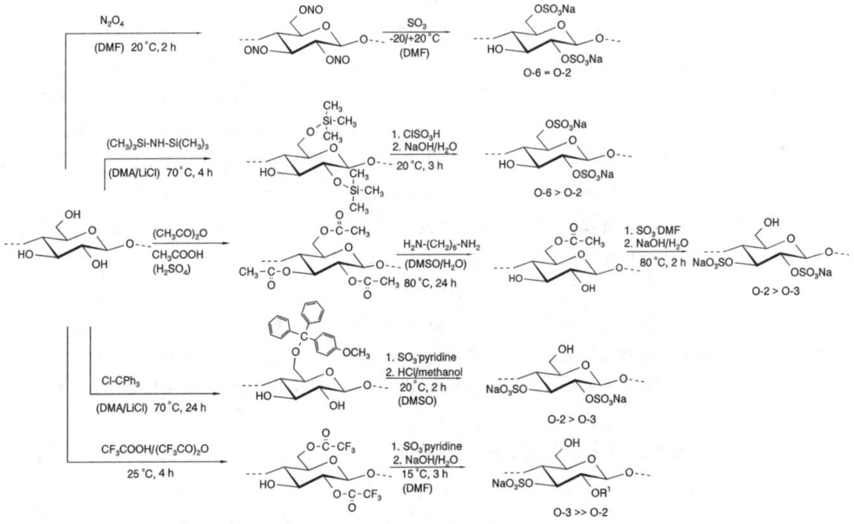

Fig. 11 Important synthetic pathways to regioselectivity in cellulose derivatives (R^1 is H or sulfuric half ester), from [23]

lation of the generated free hydroxyl groups [206–210]. Another protecting group for the primary OH is the thexyldimethylsilyl moiety [211]. This group has been employed for the synthesis of cellulose bearing the fluorescent moiety antharcene-9-carboxylic acid. The protecting group was introduced in the 6-position, the secondary OH groups were transformed in the corresponding tosylates, the latter group was substituted by the fluorescent moiety [198]. Synthesis of 6-O-acetyl-2,3-di-O-methyl cellulose, and 3-O-functionalized cellulose derivatives has been carried out by using this approach [212, 213]. Thus, cellulose was dissolved in LiCl/DMAc, then reacted with thexyldimethylsilyl chloride, in the presence of imidazole to yield the corresponding 2- and 6-O-silylated cellulose diether. Treatment of the latter with (reactive) alkyl- and allyl halides produced a cellulose derivative, completely functionalized at O-3. The silicon-bearing moiety was cleaved in THF to yield 3-O-allyl- and 3-O-methyl-ethers of cellulose (DS of 1.06), these can be further derivatized at the remaining positions, e.g., to give the corresponding 3-O-substituted-2,6-di-O-acetyl cellulose [214].

5.2
Esterification in Derivatizing Solvents

As previously discussed, solvents that dissolve cellulose by derivatization may be employed for further functionalization, e.g., esterification. Thus, cellulose has been dissolved in paraformaldehyde/DMSO and esterified, e.g., by acetic, butyric, and phthalic anhydride, as well as by unsaturated methacrylic and maleic anhydride, in the presence of pyridine, or an acetate catalyst. DS values from 0.2 to 2.0 were obtained, being higher, 2.5 for cellulose acetate. ^{1}H and ^{13}C NMR spectroscopy have indicated that the hydroxyl group of the methylol chains are preferably esterified with the anhydrides. Treatment of cellulose with this solvent system, at 90 °C, with methylene diacetate or ethylene diacetate, in the presence of potassium acetate, led to cellulose acetate with a DS of 1.5. Interestingly, the reaction with acetyl chloride or activated acid is less convenient; DMAc or DMF can be substituted for DMSO [215–219]. In another set of experiments, polymer with high α-cellulose content was esterified with trimethylacetic anhydride, 1,2,4-benzenetricarboxylic anhydride, trimellitic anhydride, phthalic anhydride, and a pyridine catalyst. The esters were isolated after 8 h of reaction at 80–100 °C, or 1 h at room temperature (trimellitic anhydride). These are versatile compounds with interesting elastomeric and thermoplastic properties, and can be cast as films and membranes [220].

The solvent system N_2O_4/DMF has been employed for the preparation of inorganic esters, e.g., phosphates and sulfates [221] as well as organic esters. The latter products were obtained by reacting the polymer with acyl chlorides, or acid anhydrides in the presence of a pyridine base. The nitrite ester formed has been successfully trans-esterified by the reaction with RCOCl producing C_6, C_8, C_{12}, C_{16} and C_{18} acyl chains. The reaction with acetic an-hydride produced a cellulose acetate with a DS of 2.0. ^{13}C NMR spectroscopy has indicated that esterification at O-2 occurs when the DS is kept low, around 0.5 [222, 223]. Similar results have been obtained for acetylation with acetyl chloride and/or anhydride in chloral/DMF/pyridine, a DS up to 2.5 was obtained [218].

An interesting (single component) derivatizing solvent is TFA. During dissolution, cellulose undergoes partial trifluoroacetylation, with a DS of 1.0–1.5. The trifluoroacetate ester produced reacts with the desired acyl compound. For example, cellulose was reacted with TFA for 24 h, the solution was treated with a mixture of trifluoracetic anhydride and the desired carboxylic anhydride, namely acetic, propionic, and 3-nitrophthalic, respectively. In another set of experiments, the solution after cellulose dissolution was treated with acyl chlorides, e.g., acetyl, acryl, cinnamoyl, benzoyl, and 4-nitrobenzoyl. The reaction progress was followed by IR spectroscopy; the spectra indicated that partial trans-esterification has occurred. That is, some of the CF_3CO groups introduced have been substituted by the (more basic) RCO moiety. Isolation of the products after several hours of reaction indicated that the DS of trifluoroacetate is ca. 1.4, whereas that of the other group ranged from 0.5 to 1.6 [224]. Esters with a total DS of between 2.9 and 3.0 were prepared by reacting cellulose in TFA with acid anhydrides ($C_2 - C_{10}$), at 60 °C. The viscosity and dependence of T_g on the chain-length of the acyl moiety were investigated [151]. Mixed aliphatic esters were also prepared in this solvent by reacting dissolved polymer with a mixture of acetic anhydride plus other aliphatic acids [152].

Another, better controlled approach, is to isolate the intermediate, then submit it to further reaction, usually in an inert (with regard to derivatization) organic solvent. This decreases the side reactions that result, unavoidably, from the long contact of the polymer with the derivatizing solvent, namely chain degradation, and methylol group condensation. Additionally, this approach allows a better control of regioselectivity, vide supra for the selectivity of the trifluoroacetyl group for the O-6 position. As expected, the important intermediates are those discussed in the preceding paragraph, including formate [141–143], trifluoroacetate [154], dichloroacetate [155], and the trialkylsilyl derivatives of cellulose [225–227]. The group introduced first (from the solvent) usually remains attached to the polymer backbone during subsequent functionalization, and is usually removed during the work-up, e.g., by acid or base hydrolysis, or by nucleophilic displacement with the fluoride ion (for the trialkylsilyl derivatives [189]). Some reactions where this scheme has been employed are (name of the final product, first group introduced; reference): cellulose sulfate, trifluoroacetate; cellulose sulfate, formate; cellulose sulfate, dichloroacetate [228, 229]; CMC, trifluoroacetate; CMC, formate; CMC, trimethylsilyl [143, 230]; cellulose 4-nitrobenzoate, trifluoroacetate; cellulose 4-nitrobenzoate, trimethylsilyl [229, 231].

Reaction progress, DS of product, distribution of substituent in the esters and mixed esters obtained, and regioselectivity of the reaction are most readily determined by a combination of spectroscopic and chromatographic techniques. These will be briefly commented on; in-depth covering of this subject for CA has been recently published [232]. FTIR is useful for following the reaction progress because the $\nu_{C=O}$ band of the reagent, e.g., acetyl chloride or acetic anhydride is reasonably well separated from $\nu_{C=O}$ of the produced ester, and the solvent amide I band (for DMF and/or DMAc). By running the experiment in the compensation mode, it is possible to assign the OH and C = O groups present in CA [233, 234]. ^{1}H and ^{13}C NMR spectroscopy have been extensively applied for determination of DS and the distribution of the acyl groups among C_2, C_3, and C_6, respectively. Whereas cellulose triacetate can be easily studied by ^{1}H NMR spectroscopy, partially substituted samples show complex spectra, resolved in some cases, due to the presence of un-substituted AGU, plus mono-, di-, and tri-substituted units [235]. Reaction of partially substituted CA with acetyl-$d3$-chloride or acetic anhydride-$d6$, and comparison of the integrations of the ^{1}H NMR spectra, before and after acetylation with the per-deutero acyl compound allows determination of DS [236, 237]. Alternatively, the remaining OH groups can be functionalized with a group (e.g., propionate) different from the one present (e.g., acetate), the DS can be calculated from the integration of the ^{1}H NMR spectrum [180, 199]. ^{13}C NMR chemical shifts of the carbonyl group, $\delta_{C=O}$, of CA are separated (in DMSO-$d6$) at 168.71 to 168.93, 169.22 to 169.60, and 169.83 to 170.04 ppm respectively, and can be employed for substituent position assignment [210]. The use of an efficient ^{13}C nuclei-relaxation additive (e.g., Cr(III) acetylacetonate) and the inverse-gated decoupling method [238] allows integration of the peaks of the CO groups present, hence the determination of their distribution among the different positions within the AGU [51, 56, 57, 239]. A more complex structural problem is to determine the distribution of the functional groups along the polymer chain. This has been solved by ^{13}C NMR spectroscopy of cellulose samples esterified with a ^{13}C-labelled acetyl group [240].

An alternative technique to NMR spectroscopy is chromatography. The partially functionalized sample is completely functionalized with a group different from the one present, the product carefully de-polymerized, its structure examined with a chromatographic technique. For example, partially substituted CA was further derivatized with methyl vinyl ether, the product hydrolyzed, the monomers produced examined with gas chromatography [241]. HPLC has been advantageously applied for the determination of substitution pattern for CAs with DS 0.8 to 3.0, by employing the same approach, i.e., further derivatization of the partially derivatized polymer with methyl trifluoroacetate, followed by de-polymerization. The results obtained by this technique compared favorably with those obtained by NMR [242].

6
The Challenge Ahead: Meeting the Requirements of Green Chemistry

The impetus for interest in the HRC scheme is, and will be, the new opportunities that this approach opens to cellulose chemistry. New, elegant synthetic schemes have been devised in order to control several aspects of the reaction, namely DS, regularity of substitution along the polymer backbone, and regioselectivity of substitution. New products, with properties of practical interest, e.g., low T_m, high thermal stability, bioactivity, film- and membrane forming ability, solubility in a wide range of solvents have been obtained. Effort has been made in order to solve the complex analytical problems involved, consequent of the different substitution patterns within the different AGU's of the polymer. The possibilities are still being explored, so that a great deal of emphasis is being placed on new reagents and synthetic strategies. On the basis of recent knowledge, also highlighted in this review, it may be expected that new cellulose solvent systems may be designed, which are consistent with the principles of green chemistry.

While this works progresses, a part of our attention should be focused on potential industrial applications. In this regard, the path is set, because important principles of green chemistry are inherent to cellulose derivatives, namely the raw material is renewable, and the products are biodegradable. With regard to these principles, consider the following:

(i)   In many procedures, after dissolution and addition of the derivatizing agent, the reaction is run "overnight", as a matter of convenience. More work should be done, however, on the kinetics of the different steps, so that the rate constants, and activation parameters of this (second order) reaction are determined under different conditions. This should result in economy of time, power consumption, and a decrease of side reactions, in particular polymer chain degradation and product deacylation. The few kinetic data available on the reaction of Rayon grade cellulose with anhydride/Tos – Cl/pyridine in LiCl/DMAc (acetic and/or pentanoic anhydride) indicate an optimum reaction temperature between 50–70 °C, and reaction time from 8–10 h [243, 244];

(ii) Practical considerations should be always considered in polymer activation. E.g., the convenient solvent-exchange scheme is time, and material consuming, whereas water consumption should be taken into account when deciding to use mercerization as a pre-treatment. Several solvent systems dissolve cellulose directly, and microwave radiation appears to accelerate this process. Effects of this combined treatment on the DP of both cellulose and the final product should be assessed;

(iii) Economy of reagents and solvents should always be in focus. Where possible, stoichiometric reagent/AGU ratios should be the target. Use of catalysts should be linked to their recovery. Recycling of the solvent and any excess reagent, e.g., acid anhydride has been successively carried out.

Although this recovery has not been optimized, reaction-grade quality DMAc and acetic anhydride were regenerated by distillation [52, 56, 57]. Even if they are employed on a large scale, room temperature ILs will probably continue to be more expensive than the classical solvents employed today, e.g., DMF or DMAc. Therefore, industrial use of ionic liquids rests on developing efficient schemes for solvent recuperation and recycling. A scheme for water separation from the IL should be worked out, in view of the deleterious effect of small concentrations of water on the cellulose-dissolution ability of these solvents. In this respect, the fact that high purity 1-allyl-3-methylimidazolium chloride can be regenerated from the acetylation reaction mixture by distillation, and that the same DS was obtained by acetylating cellulose in fresh and/or in recycled solvent is very encouraging [73];

(iv) Another problem that should be either suppressed, or minimized is reagent consumption by side reactions, e.g., hydrolysis of the anhydride by adventitious water from the system. Drying by solvent distillation is a good step in this direction [180].

In conclusion, the perspective for the HRC scheme is bright because it may be employed for obtaining specialty products, where the particularities of polymer structure, and the consistency of its properties are central to performance. Examples include new nano-composites; new "smart" polymers that are capable of responding reversibly to slight changes in the properties of the medium (pH, temperature, ionic strength, the presence of certain substances, light, electric field) [245]; and bio-compatible polymeric materials. Having the principles of green chemistry in mind helps us get there.

Acknowledgements O.A. El Seoud thanks the State of São Paulo Research Foundation (FAPESP) and the National Council for Scientific and Technological Research (CNPq) for research and travel funds, and Prof. T. Heinze and his research group for their kind help during a short stay at Friedrich-Schiller-Universität in Jena.

References

1. www.investor.bayer.de, visited August 15, 2004
2. Amass W, Amass A, Tighe B (1998) Polym Intern 47:89
3. Anastas PT, Warner JC (1998) Green Chemistry: Theory and Practice. Oxford University Press, New York, p 30
4. Tundo P, Anastas P, Black DS, Breen J, Collins T, Memoli S, Miyamoto J, Polyakoff M, Tumas W (2000) Pure Appl Chem 72:1207
5. Anastas PT, Zimmerman JB (2003) Environ Sci Technol 95 A
6. Lörcks J (1998) Polym Degrad Stab 59:245
7. Yoshioka M, Shiraishi N (2000) Mol Cryst Liq Cryst Sci Technol A 353:59
8. Braunegg G, Lefebvre G, Genser K (1998) J Biotechnol 65:127
9. Riedel U, Nickel J (1999) Angew Makromol Chem 272:34

10. Kaplan DL (ed) (1998) Biopolymers from Renewable Resources: Macromolecular Systems-Materials Approach. Springer, Berlin Heidelberg New York
11. Chiellini E, Gil H, Braunegg G, Buchert J, Gatenholm P, van de Zee M (2001) Biorelated Polymers: Sustainable Polymer Science and Tecnology. Kluwer, New York
12. Chiellini E, Solaro R (eds) (2003) Recent Advances in Biodegradable Polymers and Plastics. Wiley-VCH, Weinheim
13. Carollo P (2004) In: Rustemeyer P (ed) Cellulose Acetates: Properties and Applications. Wiley-VCH, Weinheim, p 335
14. Rustemeyer P (2004) In: Rustemeyer P (ed) Cellulose Acetates: Properties and Applications. Wiley-VCH, Weinheim, p 267
15. Buchanan CM, Edgar KJ, Wilson AK (1991) Macromolcules 24:3060
16. Puls J, Altaner C, Saake B (2004) In: Rustemeyer P (ed) Cellulose Acetates: Properties and Applications. Wiley-VCH, Weinheim, p 239
17. Samios E, Dart RK, Dawkins JV (1997) Polym 12:3045
18. O'Sullivan AC (1997) Cellulose 4:173
19. Klemm D, Phillip B, Heinze T, Heinze U, Wagenknecht W (1998) Comprehensive Cellulose Chemistry, Vol 1. Wiley-VCH, Weinheim p 9
20. Klemm D, Phillip B, Heinze T, Heinze U, Wagenknecht W (1998) Comprehensive Cellulose Chemistry, Vol 1. Wiley-VCH, Weinheim p 147
21. Klemm D, Phillip B, Heinze T, Heinze U, Wagenknecht W (1998) Comprehensive Cellulose Chemistry, Vol 1. Wiley-VCH, Weinheim p 43
22. Klemm D, Phillip B, Heinze T, Heinze U, Wagenknecht W (1998) Comprehensive Cellulose Chemistry, Vol 1. Wiley-VCH, Weinheim p 25
23. Heinze T, Liebert T (2001) Prog Polym Sci 26:1689
24. Gardner KH, Blackwell J (1974) Biopolymers 13:1975
25. Kroon-Batenburg LMJ, Kroon J, Nordholt MG (1986) Polym Commun 27:290
26. Kondo T, Sawatari C (1996) Polym 37:393
27. Hinterstoisser B, Salmén L (2000) Vibrational Spectroscop 22:111
28. Jeffries R, Jones DM, Roberts JG, Selby K, Simmons SC, Warwicker JO (1969) Cellul Chem Technol 3:255
29. Glasser WG, Samaranayake G, Dumay M, Davé V (1995) J Polym Sci B 33:2045
30. Samios E, Dart RK, Dawkins JV (1997) Polym 12:3045
31. Krässig H (1986) Ullman's Encyclopedia of Industrial Chemistry, Campbell FT, Pfefferkorn R, Rousaville JF (eds) 5th ed. VCH, Weinheim p 375
32. Toyoshima I (1993) Cellulosics: Chemical, Biochemical and Material Aspects, Kennedy JF, Phillips GO, Williams PA (eds). Ellis Horwood, Chichester, p 125
33. Hermans PH, Heicken D, Weidinger A (1959) J Polym Sci 35:145
34. Tang L-G, Hon DN-S, Zhu Y-Q (1997) J Appl Polym Sci 64:1953
35. Rowland SP (1977) In: Arthur JC Jr (ed) Textile and Paper Chemistry and Technology, ACS Smp ser No. 49. Am Chem Soc, Washington, DC p 20
36. Krässig HA (ed) (1993) In: Cellulose: Structure, Accessibility, and Reactivity. Gordon and Breach, Yverdon
37. Callais PA (1986) Derivatization and Characterization of Cellulose in Lithium Chloride and N,N-Dimethylacetamide Solutions. Ph.D. Thesis, University of Southern Mississippi, USA
38. Marson GA (1999) Acylation of Cellulose in Homogeneous Medium. MSc Thesis, University of São Paulo, Brazil
39. Sefain MZ, Nada AMA, El-kalyoubi SF (1980) Cellul Chem Technol 14:139
40. Sefain MZ, Nada AMA (1985) Ibid 19:257
41. Youssef MAM, Nada AMA, Ibrahim AA (1989) Ibid 23:505

42. Shimizu Y, Kimura K, Masuda S, Hayashi J (1993) In: Kennedy JF, Phillips GO, Williams PA (eds) Cellulosics: Chemical, Biochemical and Material Aspects. Ellis Horwood, Chichester, p 67

43. Kamide K, Okajima K, Matsui T, Kowasaka K (1984) Polym J 16:857

44. Isogai A, Ishizu A, Nakano J, Atalla RH (1985) In: Atalla RH (ed) The Structure of Cellulose 1: Characterization of the Solid State, ACS Symposium Ser No 340. Am Chem Soc, Washington, DC, p 292

45. McCormick CL, Callais PA, Huchinson BH (1985) Macromolecules 18:2394

46. McCormick CL, Callais PA (1986) Polym Prepr 27:91

47. McCormick CL, Callais PA (1987) Polym 28:2317

48. Kurata S, Suzuki I, Ikeda I (1992) Polym Internat 29:1

49. Dupont A-L (2003) Polym 44:4117

50. Ekmanis JL (1987) Am Lab News Jan-Feb:10

51. Edgar KJ, Arnold KM, Blount WW, Lawniczak JE, Lowman DW (1995) Macromolecules 28:4122

52. Regiani AM, Frollini E, Marson GA, Arantes GM, El Seoud OA (1999) J Polym Sci A 37:1357

53. Hatakeyama T, Nakamura K, Hatakeyama H (2000) Thermochim Acta 352–353:233

54. Dawsey TR, McCormick CL (1990) JMS-Rev Macromol Chem Phys C30:405

55. Potthast A, Rosenau ST, Sixta JH, Kosma P (2003) Polym 44:7

56. Marson GA, El Seoud OA (1999) J Appl Polym Sci 74:1355

57. El Seoud OA, Marson GA, Ciacco, GT, Frollini E (2000) Macromol Chem Phys 201:882

58. Zhou J, Zhang L (2000) Polym J 32:866

59. Petropavlovskii GA, Zimina TR (1994) Russ J Appl Chem 67:629

60. Heinze T, Liebert T, Klüfers P, Meister F (1999) Cellulose 6:153

61. Liebert T, Heinze T (1998) Macromol Symp 130:271

62. Saalwächter K, Burchard W, Klüfers P, Kettenbach G, Mayer P, Klemm D, Dugarmaa S (2000) Macromolecules 33:4094

63. Feddersen RL, Thorp SN (1993) In: Whistler RL, BeMiller JN (eds) Polysaccharides and their Derivatives, 3rd ed. Academic Press, New York, p 537

64. Heinze T, Pfeiffer K (1999) Angew Makromol Chem 266:37

65. Fischer S, Voigt W, Fischer K (1999) Cellulose 6:213

66. Fischer S, Leipner H, Brendler E, Voigt W, Fischer K (2000) In: El-Nokaly MA, Soini HA (eds) Polysaccharide Applications, Cosmetics and Pharmaceuticals. ACS Symposium Series, Washington, DC p 143

67. Leipner H, Fischer S, Brendler E, Voigt W (2000) Macromol Chem Phys 201:2041

68. Striegel AM, Timpa JD, Piotrowiak P, Cole RB (1997) Int J Mass Spectrom Ion Proc 162:45

69. Woodings CR (1995) Int J Macromol 17:305

70. O'Driscoll C (1996) Chemistry in Britain 32:27

71. Holbrey JD, Reichert WM, Nieuwenhuyzen M, Johnson S, Seddon KR, Rogers RD (2003) Chem Commun 1636

72. Swatloski RP, Spear SK, Holbrey JD, Rogers RB (2002) J Am Chem Soc 124:4974

73. Wu J, Zhang J, Zhang H, He J, Ren Q, Guo M (2004) Biomacromolecules 5:266

74. Turner MB, Spear SK, Holbrey JD, Rogers RD (2004) Macromolecules 5:1379

75. Ngo HL, LeCompte K, Hargens L, McEwen AB (2000) Thermochim Acta 357-358:97

76. Ren Q, Wu J, Zhang J, Zhang J, He JS (2003) Acta Polym Sin 3:448

77. Wasserscheid P, Welton T (eds) (2002) Ionic liquids in synthesis. Wiley-VCH, Weinheim

78. Dawsey TR (1994) In: Gilbert RD (ed) Cellulosic Polymers, Blends and Composites. Carl Hanser Verlag, New York, p 157
79. Striegel AM, Timpa JD (1995) Carbohydr Res 267:271
80. Silva AA, Laver ML (1997) Tappi J 80:173
81. Kvernheim AL, Lystad E (1989) Acta Chem Scand 43:209
82. Hasegawa M, Isogai A, Onabe F (1993) J Chromatogr 635:334
83. Striegel AM, Timpa JD (1996) In: Potschka M, Dubin PL (eds) Strategies in Size Exclusion Chromatography. ACS Symposium Series, Washington, DC, 635, p 366
84. Sjöholm E, Gustafsson K, Eriksson B, Brown W, Colmsjö A (2000) Carbohydr Polym 41:153
85. Striegel A (1997) Carbohydr Polym 34:267
86. El Seoud OA, Regiani AM, Frollini E (2000) In: Frollini E, Leao AL, Mattoso LHC (eds) Natural Polymers and Agrofibers Composites. São Carlos, Brazil, p 73
87. Pionteck H, Berger W, Morgenstern B, Fengel D (1996) Cellulose 3:127
88. El-Kafrawi A (1982) J Appl Polym Sci 27:2435
89. Turbak AS (1994) Tappi J 64:94
90. Vincendon M (1985) Macromol Chem 186:1787
91. Petrus L, Gray DG, BeMiller JN (1995) Carbohydr Res 268:319
92. Fersht AR (1971) J Am Chem Soc 93:3504
93. Kresge JA, Fitzgerald PH, Chiang Y (1974) J Am Chem Soc 96:4698
94. Cary FA, Sundberg RJ (1990) Advanced Organic Chemistry, 3rd ed., Part A. Plenum Press, New York, p 257
95. Morgenstern B, Kammer H (1996) Trip J 4:87
96. Spange S, Reuter A, Vilsmeier E, Heinze T, Keutel D, Linert W (1998) J Polym Sci A 36:1945
97. Kamlet MJ, Abboud J-LM, Taft RW (1981) Prog Phys Org Chem 13:485
98. Reichardt C (2003) Solvents and Solvent Effects in Organic Chemistry, 3rd edn. VCH, Weinheim, p 389
99. Spange S, Fischer K, Prause S, Heinze T (2003) Cellulose 10:201
100. Berger W, Philipp B (1988) Cellul Chem Technol 22:387
101. Morgenstern B, Berger W (1993) Acta Polym 44:100
102. Furuhata K, Koganei K, Chang H-S, Aoki N, Sakamoto M (1992) Carbohydr Res 230:165
103. Satgé C, Verneuil B, Branland P, Granet R, Krausz P, Rozier J, Petit C (2002) Carbohydr Polym 49:373
104. Satgé C, Granet R, Verneuil B, Branland P, Krausz P (2004) Comptes Rend Chimie 7:135
105. Varma RS, Namboordiri VV (2001) Chem Commun 643
106. Marson G, El Seoud OA (1999) J Polym Sci A 37:3738
107. Buchard W (1993) Trip J 1:192
108. Schulz L, Burchard W, Dönges R (1998) In: Heinze T, Glasser WG (eds) Cellulose Derivatives: Modification, Characterization, and Nanostructures. ACS, Symposium Series 688, Washington, DC, p 218
109. Morgenstern B, Kammer H-W (1999) Polym 40:1299
110. Menger FM (1993) Acc Chem Res 26:206
111. Röder T, Morgenstern B, Glatter O (2000) Lenzinger Berichte 79:97
112. Sjöholm E, Gustafsson K, Pettersson B, Colmsjö A (1997) Carbohydr Polym 32:57
113. Röder T, Morgenstern B (1999) Polym 40:4143
114. Ruiz N (2004) Acylation of Cellulose Under Homogeneous Reaction Conditions. MSc Thesis, University of São Paulo, Brazil

115. Westermark U, Gustafsson K (1994) Holzforschung 48:146
116. Yanagisawa M, Shibata I, Isogai A (2004) Cellulose 11:169
117. Strlic M, Kolenc J, Kolar J, Pihlar B (2002) J Chromatogr A 964:47
118. Röder T, Morgenstern B, Glatter O (2000) Macromol Symp 162:87
119. Potthast A, Rosenau T, Buchner R, Röder T, Ebner G, Bruglachner H, Sixta H, Kosma P (2002) Cellulose 9:41
120. Röder T, Morgenstern B, Schelosky N, Glatter O (2001) Polym 42:6765
121. Zimm BH (1948) J Chem Phys 16:1093
122. Kamide K, Miyazaki Y, Abe T (1979) Polym J 11:523
123. Brandrup J, Immergut EH (eds) (1989) Polymer Handbook, 3rd ed. Wiley, New York
124. Evans R, Wearne RH, Wallis AFA (1989) J Appl Polym Sci 37:3291
125. Nilsson S, Sundelöf L-O, Porsch B (1995) Carbohydr Res 28:265
126. Terbojevich M, Cosani A, Camilot M, Focher B (1995) J Appl Polym Sci 55:1663
127. Basedow M, Ebert KH, Feigenbutz W (1980) Makromol Chem 181:1071
128. Myasoedova VV, Pokrovskii SA, Zav'yalov NA, Krestov GA (1991) Russ Chem Rev 60:954
129. Paul RC, Banait JS, Narula SP (1975) J Electroanal Chem 66:111
130. Taniewska-Osinska S, Wozincka J (1981) Thermochim Acta, 47:57
131. Mason TJ, Lorimer JP (1983) Computers Chem 7:159
132. Maskill H (1990) Educ Chem 27:111
133. Ramos LA, Ciacco GT, Assaf JM, El Seoud OA, Frollini E (2002) Fourth International Symposium on Natural Polymers and Composites—ISNaPol, p 42
134. Hefter G, Marcus Y, Waghorne WE (2002) Chem Rev 102:2773
135. Mark JE, Eisenberg A, Graessley WW, Mandelkern L, Samulski ET, König JL, Wignall GD (1992) Physical Properties of Polymers, 2nd ed. ACS, Washington, DC, p 150
136. Klemm D, Heinze T, Philipp B, Wagenknecht W (1997) Acta Polymerica 48:277
137. Golova LK, Kulichikhin VS, Papkov SP (1986) Vysokomol Soedin A28:1795
138. Wagenknecht W, Philipp B, Schleicher H, Beierlein I (1977) Faserforsch Textiltech 28:421
139. Johnson DC, Nicholson MD, Haigh FG (1976) J Appl Polym Sci Appl Polym Symp 28:931
140. Fujimoto T, Takahashi S, Tsuji M, Miyamoto M, Inagaki H (1986) J Polym Sci Polym Lett 24:495
141. Schnabelrauch M, Vogt S, Klemm D, Nehls I, Philipp B (1992) Angew Makromol Chem 198:155
142. Philipp B, Wagenknecht W, Nehls I, Ludwig J, Schnabelrauch M, Kim HR, Klemm D (1990) Cellul Technol Chem 24:667
143. Liebert T, Klemm D, Heinze T (1996) JMS-Pure Appl Chem A 33:613
144. Hiemenz, PC, Rajagopalan (1997) Principles of Colloid and Surface Chemistry, 3rd ed. Marcel Dekker, New York, pp 297–355
145. Jutz C (1976) Adv Org Chem 9:225
146. Vigo TL, Daighly BJ, Welch CM (1972) J Polym Sci B Polym Phys 10:397
147. McCormick CL, Dawsey TR, Newman JK (1990) Carbohydr Res 208:183
148. CRC Handbook of Chemistry and Physics (1993) 73rd ed. CRC Press, Boca Raton, FL, Sects 8–36
149. Leo AJ, Hansch C (1999) Perspect Drug Discov 17:1
150. Demhlow EV, Demhlow SS (1983) Phase Transfer Catalysis. Verlag-Chemie, Weinheim, p 12
151. Salin BN, Cemeris M, Mironov DP, Zatsepin AG (1991) Khim Drev 3:65
152. Salin BN, Cemeris M, Malikova OL (1993) Khim Drev 5:3

153. Hawkinson DE, Kohout E, Fornes RE, Gilbert RD (1991) J Polym Sci B Polym Phys 29:1599
154. Liebert T, Schnabelrauch M, Klemm D, Erler U (1994) Cellulose 1:249
155. Liebert T, Klemm D (1998) Acta Polymerica 49:124
156. Conio G, Corazza P, Bianchi E, Tealdi A, Ciferri A (1981) Mol Cryst Liq Cryst 69:273
157. Guo J-X, Gray DG (1994) In: Gilbert RD (ed) Cellulosic Polymers, Blends and Composites. Hanser Publ, New York, p 25
158. McCormick CL, Dawsey TR (1990) Macromolecules 23:3606
159. McCormick CL, Lichatowich DK (1979) J Polym Sci Polym Lett Ed 17:479
160. McCormick CL, Chen TS (1982) In: Symor RB, Stahl GA (eds) Macromolecular Solutions, Solvent-Property Relationships in Polymers. Pergamon Press, New York, p 101
161. Ibrahim AA, Nada AMA, Hagemann U, El Seoud OA (1996) Holzforschung 50:221
162. Diamantoglou M, Kundinger EF (1995) In: Kennedy JF, Phillips GO, Williams PA, Picullel L (eds) Cellulose and Cellulose Derivatives, Physicochemical Aspects and Industrial Applications. Woodhead Publishing, Cambridge, p 141
163. Terbojevich M, Cosani A, Focher B, Gastaldi G, Wu W, Marsano E, Conio G (1999) Cellulose 6:71
164. Pawlowski WP, Sanakar SS, Gilbert RD (1987) J Polym Sci A Polym Chem 25:3355
165. Pawlowski WP, Gilbert RD, Fornes R, Purrington S (1988) J Polym Sci B Polym Phys 26:1101
166. Samaranayake G, Glasser WG (1993) Carbohydr Res 22:79
167. Heinze T, Liebert T, Pfeiffer K, Hussain MA (2003) Cellulose 10:283
168. Marsano E, De Paz L, Tambuscio E, Bianchi E (1998) Polym 39:4289
169. Sun RC, Fang JM, Tomkinson J, Hill CAS (1999) J Wood Chem Technol 19:287
170. Witzemann JS, Nottingham WD, Rector FD (1990) J Coating Technol 62:101
171. Witzemann JS, Nottingham WD (1991) J Org Chem 56:1713
172. Clemens RC (1986) Chem Rev 86:241
173. Wentrup C, Heilmayer W, Kollenz G (1994) Synthesis 1219
174. Larsson L, Hansen B (1956) Svensk Kem Tidskr 68:521, CA, 51, 7062e
175. Robertson RE, Rossall B, Redmond WA (1971) Can J Chem 49:3665
176. Allen AD, Kresge AJ, Schepp NP, Tidwell TT (1987) Ibid 65:1719
177. Abraham MH (1990) Chem Soc Rev 73
178. Takaragi A, Minoda M, Miyamoto T, Liu HQ, Zhang LN (1999) Cellulose 6:93
179. Heinze T, Dicke R, Koschella A, Kull AH, Klohr E-A, Koch W (2000) Macromol Chem Phys 201:627
180. Ciacco GT, Liebert TF, Frollini E, Heinze T (2003) Cellulose 10:125
181. Tada EB, El Seoud OA (2002) J Phys Org Chem 15:403
182. Hefter GT (1991) Pure Appl Chem 63:1749
183. March J (1992) Advanced Organic Chemistry, 4th edn. Wiley, New York, p 659
184. March J (1992) Advanced Organic Chemistry, 4th edn. Wiley, New York, p 393
185. Samaranayake G, Glasser WG (1993) Carbohydr Res 22:1
186. Gerardeau S, Aburto J, Vaca-Garcia C, Alric I, Borredon E, Gaset A (2001) In: Biorelated Polymers: Sustainable Polymer Science and Tecnology. Kluwer, New York, p 53
187. Oera J (2003) Esterification. Wiley-VCH, Weinheim, p 21
188. Glasser WG, McCartney BK, Samaranayake G (1994) Biotechnol Prog 10:214
189. Greene TW, Wuts PGM (1991) Protective groups in organic synthesis. Wiley, New York, p 71
190. Gerzon K, Kau D (1967) J Med Chem 10:189
191. Orzeszko A, Gralewska R, Starosciak BJ, Kazimierczuk Z (2000) Acta Biochim Pol 47:87

192. Vaca-Garcia C, Thiebaud S, Borredon MW, Gozzelino G (1998) JAOCS 75:315
193. Shimizu Y, Hayashi J (1988) Sen-i Gakkaishi 44:451
194. Siegmund G, Klemm D (2002) Polym News 27:84
195. Sealey JE, Samaranayake G, Todd JG, Glasser WG (1996) J Polym Sci B Polym Phys 34:1613
196. Sealey JE, Frazier CE, Samaranayake G, Glasser WG (2000) J Polym Sci B Polym Phys 38:486
197. Glasser WG, Becker U, Todd JG (2000) Carbohydr Polym 42:393
198. Koschella A, Haucke G, Heinze T (1997) Polym Bull 39:597
199. Heinze T, Schaller J (2000) Macromol Chem Phys 201:1214
200. Hussain MA, Liebert T, Heinze T (2004) Polym News 29:14
201. Tascioglu S (1996) Tetrahedron 34:11113
202. Bunton CA (1997) J Mol Liq 72:231
203. Thuresson K, Lindman B (1997) J Phys Chem B 101:6460
204. Kastner U, Hoffmann H, Donges R, Ehrler R (1994) Colloids Surf A 82:279
205. Kastner U, Hoffmann H, Donges R, Ehrler R (1996) Ibid 112:209
206. Iwata T, Azuma J, Okamura K, Muramoto M, Chun B (1992) Carbohydr Res 244:277
207. Iwata T, Okamura K, Azuma J, Tanaka F (1996) Cellulose 3:107
208. Iwata T, Fukushima A, Okamura K, Azuma J (1997) J Appl Polym Sci 65:1511
209. Iwata T, Doi Y, Azuma J (1997) Macromolecules 30:6683
210. Kamide K, Saito M (1994) Macromol Symp 83:233
211. Koschella A (2000) Use of regioselective synthesis for obtaining new functionalized polymers, and use of NMR spectroscopy for the characterization of unconventional substitution patterns. PhD Thesis, University of Jena, Germany
212. Stein A, Klemm D (1995) Papier (Darmstadt) 49:723
213. Koschella A, Klemm D (1997) Macromol Symp 120:115
214. Koschella A, Heinze T, Klemm D (2001) Macromol Biosci 1:49
215. Arai K, Ogiwara Y (1980) Sen-i Gakkaishi 36:T82–T84
216. Morooka T, Norimoto M, Yamada T, Shiraishi N (1982) J Appl Polym Sci 127:4409
217. Miyagi Y, Shiraishi N, Yokota T, Yamashita S, Hayashi Y (1983) J Wood Chem Technol 3:59
218. Seymour RB, Johnson EL (1978) J Polym Sci Polym Chem Ed 16:1
219. Leoni R, Baldini A (1982) Carbohydr Polym 2:298
220. Saikia CN, Dutta NN, Borah M (1993) Thermochim Acta 219:191
221. Wagenknecht W, Nehls I, Philipp B (1992) Carbohydr Res 237:211
222. Mansson P, Westfeld L (1980) Cellul Chem Technol 14:13
223. Shimizu Y, Nakayama A, Hayashi J (1993) Sen-i Gakkaishi 49:352
224. Emel'yanov YG, Grinshpan DD, Kaputskii FN (1988) Khim Drev 1:23
225. Schempp W, Krause T, Seifried U, Koura A (1984) Papier (Darmstadt) 38:607
226. Klemm D, Schnabelrauch M, Stein A, Philipp B, Wagenknecht W, Nehls I (1990) Papier (Darmstadt) 44:624
227. Mormann W, Demeter J (1999) Macromolecules 32:1706
228. Klemm D, Heinze T, Stein A, Liebert T (1995) Macromol Symp 99:129
229. Liebert T (1995) Cellulose esters as hydrolytically instable intermediates and pH-sensitive carriers. Ph.D. Thesis, University of Jena, Germany
230. Liebert T, Heinze T (1998) ACS Symposium Series 688. American Chemical Society, Washington, DC, p 61
231. Stein A, Klemm D (1988) Macromol Rapid Commun 9:569
232. Heinze T, Liebert T (2004) In: Rustemeyer P (ed) Cellulose Acetates: Properties and Applications. Wiley-VCH, Weinheim, p 167

233. Krasovskii AN, Polyakov DN, Mnatsakanov SS (1993) Russ J Appl Chem 66:918
234. Krasovskii AN, Polyakov DN (1996) Ibid 69:1049
235. Hikichi K, Kakuta Y, Katoh T (1995) Polym J 7:659
236. Buchanan CM, Hyatt JA, Lowman DW (1989) J Am Chem Soc 111:7312
237. Deus C, Friebolin H, Siefert E (1991) Makromol Chem 192:75
238. Braun S, Kalinowski HO, Berger S (1996) 100 and More Basic NMR Experiments. VCH, Weinheim, p 99
239. Kowasaka K, Okajima K, Kamide K (1988) Polym J 20:827
240. Iijima H, Kowasaka K, Kamide K (1992) Polym J 24:1077
241. Bjorndal H, Lindberg B, Rosell KG (1971) J Polym Sci Polym Symp 36:523
242. Lee CK, Gray GR (1995) Carbohydr Res 269:167
243. Tosh B, Saikia CN (2000) Trends in Carbohydrate Chemistry 6:143
244. Tosh B, Saikia CN, Dass NN (2000) Carbohydr Res 327:345
245. Galaev IYu (1995) Russian Chem Rev 64:471

Adv Polym Sci (2005) 186: 151–209
DOI 10.1007/b136820
© Springer-Verlag Berlin Heidelberg 2005
Published online: 30 August 2005

Chitosan Chemistry: Relevance to the Biomedical Sciences

R. A. A. Muzzarelli (✉) · C. Muzzarelli

Institute of Biochemistry, Faculty of Medicine, Polytechnic University, Via Ranieri 67, 60100 Ancona, Italy
muzzarelli@univpm.it

Abstract Chitin is well characterized in terms of analytical chemistry, is purified from accompanying compounds, and derivatized in a variety of fashions. Its biochemical significance when applied to human tissues for a number of purposes such as immunostimulation, drug delivery, wound healing, and blood coagulation is currently appreciated in the context of biocompatibility and biodegradability. Physical forms (nanoparticles, nanofibrils, microspheres, composite gels, fibers, films) are as important as the chemical structures. Besides being safe to the human body, chitin and chitosan exert many favourable actions, and some chitin based products such as cosmetics, nutraceuticals, bandages, and textiles are presently commercially available. This chapter puts emphasis on the development of new drug carriers and on the interaction of chitosans with living tissues, two major topics of the most recent research activities.

Keywords Chitosan · Nanoparticles · Microspheres · Chemically modified chitosans · Polyelectrolyte complexes · Oral and nasal administration · Nerve, cartilage and bone regeneration · Wound dressing

1
Introduction

Because chitin is the most abundant compound of nitrogen, it represents the major source of nitrogen accessible to countless living terrestrial and marine organisms.

Moreover, the lipo-chitooligosaccharides, also known as nod factors, permit nitrogen fixation by which plants and symbiotic Rhizobia bacteria can reduce atmospheric nitrogen to the ammonia that is utilized by the plant, thus making available nitrogen compounds to other living organisms.

These two statements alone would suffice to highlight the outstanding biochemical significance of chitin, chitosan and their derivatives, which justifies their growing importance in a vast number of fields, mainly medicine, agriculture, food and non-food industries. They have emerged as a family of polysaccharides with highly sophisticated functions, and their versatility is still a challenge to the scientific community and industry. Chitins and chitosans are endowed with biochemical activity, excellent biocompatibility, and complete biodegradability, in combination with low toxicity. The large body of knowledge related to chitin extends beyond the borders of classic scientific fields, and includes ecology, microbiology, zoology, entomology, enzymology, just to mention a few of the disparate areas where chitin plays a role.

The present chapter, therefore, will present certain topics based on a selection of references, mainly in view of providing a perception of the current developments and great potential of chitin today. The reader is referred to books and reviews [1–17] where basic information and specific subjects are treated in a more systematic way. These polysaccharides are described not only in encyclopaedias, handbooks, monographs and articles, but also in the American Standard Testing Materials standard guides and in the Pharmacopoeias of various countries [10, 16, 17].

In the last quarter of the 20th century, certain scientific research topics have been prominent in various periods; they can be roughly related to:

(i)   technological advances (spinning, coloring, uptake of soluble species, cosmetic functional ingredients);
(ii)   biochemical significance (blood coagulation, wound healing, bone regeneration, immunoadjuvant activity);
(iii)   inhibition of the biosynthesis (insecticides);
(iv)   chitin enzymology (isolation and characterization of chitinases, their molecular biology, biosynthesis, and hydrolases with unspecific chitinolytic activity);
(v)   collection of metal ions and chemical derivatization;
(vi)   combinations of chitosan with natural and synthetic polymers (grafting, polyelectrolyte complexation, blends, coatings);
(vii)   use of chitosan as a dietary supplement and a food preservative (anticholesterolemic dietary products, antimicrobial coatings for grains and exotic fruits).

A deeper knowledge on chitosan was obtained, but also a fruitful integration of interdisciplinary interests.

The literature published in the period October 2003–June 2004 includes a number of articles dealing with associations of chitosan with polysaccharides (most often polyuronans), namely carrageenan [18], alginate [19–22], hydroxypropyl guar [23], hyaluronan [24–29], cellulose [30–34], gellan [35–37], arabinogalactan [38] and xanthan [39, 40], all of them more or less directly related to the preparation of drug vehicles. Chitosans were also modified with pendant lactosyl, maltosyl and galactosyl groups for better targeting to certain cells [41–45], and succinyl chitosan has been described as a long-lasting chitosan for systemic delivery [46].

A number of articles considered the association of chitosan with polylactic acid or similar compounds [47–49]; another group of articles presented new data on highly cationic chitosans [50–55]. More data have also been made available on the delivery of growth factors [56] and ophthalmic drugs [57, 58], on the activation of the complement, macrophages [59–61] and fibroblasts [62], on mucoadhesion [63] and functionalization of chitin [64]. The development of new carriers for the delivery of drugs, and the interactions of chitosans with living tissues seem therefore to be major topics in the current research on chitosan. Therefore, this chapter will place emphasis on these aspects.

2
Chitin and Chitosan

Most commonly, chitin means the skeletal material of invertebrates. At least 1.10^{13} kg of chitin are constantly present in the biosphere. α-Chitin occurs in the calyces of hydrozoa, the egg shells of nematodes and rotifers, the radulae

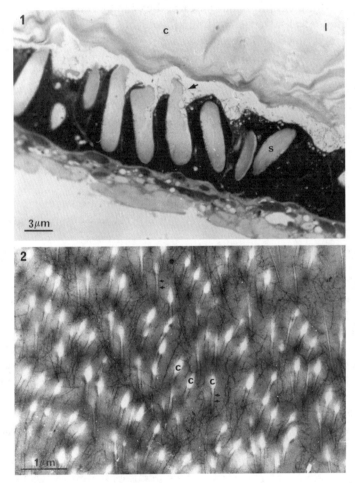

Fig. 1 The chitin-secreting gland of the marine worm *Riftia pachyptila*. Part 1: the central lumen (*I, upper-right corner*) contains the amorphous chitin secretion (c); the sub-lumen marked with an arrow is emptying its content in the central lumen. Part 2: chitin microfibrils sections (c); the filamentous network (*arrows*) that connects the edges of the crystallite sections, seems to contain protein (From Shillito et al., in: Chitin Enzymology, Vol. 1, R.A.A. Muzzarelli (ed.) pp. 129–136, Atec, Italy, 1993)

of mollusks and the cuticles of arthropods, while β-chitin is part of the shells of brachiopods and mollusks, the cuttlefish bone, the squid pen, and the pogonophora tubes. Chitin is found in exoskeletons, peritrophic membranes and cocoons of insects. In the fungal walls, chitin varies in crystallinity, degree of covalent bonding to other wall components, and degree of acetylation.

Chitin is synthesized according to a common pathway that ends with the polymerization of N-acetylglucosamine from the activated precursor uridine diphosphate-N-acetylglucosamine (UDP-GlcNAc). In this process, the nitrogen comes from glutamine. The pathway includes the action of chitin synthases that accept substrate UDP-GlcNAc and feed nascent chitin into the extracellular matrix. In crustacea, the Golgi apparatus is directly concerned with the synthesis and secretion of chitin [14]. The equation for the chitin synthesis reaction is:

$$\text{UDP-GlcNAc} + (\text{GlcNAc})_n \rightarrow (\text{GlcNAc})_{n+1} + \text{UDP}$$

Fungal chitin synthases are found as integral proteins of the plasma membrane and in chitosomes; a divalent cation, Mg(II), is necessary for enzyme activity but neither primers nor a lipid intermediate are required. The substrate and free GlcNAc activate the allosteric enzyme. UDP, the byproduct of the enzymatic activity, is strongly inhibitory to chitin synthase; however, it may be metabolized readily to UMP by a diphosphatase.

The chitin is modified to impart the structure required by the functions of each particular tissue, via crystallization, deacetylation, cross-linking to other biopolymers (Fig. 1), and, in certain cases, quinone tanning. The resulting complex structures are capable of exceptional performances [15].

In insects, for instance, chitin functions as scaffold material, supporting the cuticles of the epidermis and trachea as well as the peritrophic matrices lining the gut epithelium. Insect growth and morphogenesis are strictly dependent on the capability to remodel chitin-containing structures. For this purpose, insects repeatedly produce chitin synthases and chitinolytic enzymes in different tissues. Coordination of chitin synthesis and its degradation require strict control of the participating enzymes during development.

3
Isolated Chitins

Isolated chitins are highly ordered copolymers of 2-acetamido-2-deoxy-β-D-glucose and 2-amino-2-deoxy-β-D-glucose. The occurrence of the latter is explained by the fact that in vivo chitin is covalently linked to proteins via the nitrogen atom of approximately one repeating unit out of ten, therefore upon isolation a degree of deacetylation close to 0.10 is found. Chitobiose, O-(2-amino-2-deoxy-β-D-glucopyranosyl)-$(1 \rightarrow 4)$-2-amino-2-deoxy-

D-glucose, is the structural unit of native chitin. Bound water is also a part of the structure.

The molecular order becomes macroscopically evident when microfibrillar fragments of purified crustacean chitins are prepared in 3 M HCl at 104 °C: after removal of the acid, sonication yields colloidal suspensions that self-assemble spontaneously in a chiral nematic liquid crystalline phase, and reproduce the helicoidal organization that characterize the cuticles [65]. The polymorphic forms of chitin differ in the packing and polarities of adjacent chains in successive sheets; in the β-form all chains are aligned in parallel manner, whereas in α-chitin they are antiparallel. The molecular order of chitin depends on the physiological role, for instance, the grasping spines of *Sagitta* are made of pure α-chitin, whilst the centric diatom *Thalassiosira* contains pure α-chitin. According to Noishiki et al. [66] the β-chitin can be converted to α-chitin by swelling with 20% NaOH and then washing with water.

The solubility of chitin is remarkably poorer than that of cellulose, because of the high crystallinity of chitin, supported by hydrogen bonds mainly through the acetamido group. Dimethylacetamide containing 5–9% LiCl (DMAc/LiCl), and N-methyl-2-pyrrolidinone/LiCl are systems where chitin can be dissolved up to 5%. The main chain of chitin is rigid at room temperature, so that mesomorphic properties may be expected at a sufficiently high concentration [67, 68].

It is interesting to note that a partially deacetylated chitin, called water-soluble chitin, i.e., a polysaccharide with degree of acetylation close to 0.50, has been found particularly effective as a wound healing accelerator. This chitin can be prepared via alkaline treatment of chitin and ultrasonication: chitin is suspended in 40% NaOH aqueous solution and the resulting alkali chitin is dissolved by stirring with ice, the solution is further stirred at 25 °C for 60 h (so called homogeneous deacetylation) and then neutralized with HCl. Insolubilization of the product can be promoted with acetone. The product molecular weight drops from 1.64 MDa to 795 kDa after ultrasonication at 225 W for 1 h, and the degree of deacetylation is 0.50. It is therefore highly susceptible to lysozyme, and soluble in slightly acidic solutions. This chitin was used as a wound dressing material ([69], see also below).

Reacetylation of chitosan under proper conditions leads to products having the same solubility. Experiments showed that the amount of acetic anhydride was the most important factor affecting the N-acetylation degree of the chitosan. The effect of the means of adding materials and the amount of solvent on the reaction could not be ignored [70].

By enzymatic means, chitosan can be easily depolymerized by a variety of hydrolases including lysozyme, pectinase, cellulases, hemicellulases, lipases and amylases, among others, thus showing a peculiar vulnerability to enzymes other than chitosanases [71–76]. While pectinase is of particular importance, recent work indicates that pectinase would not only be active on chitosan but also on chitin [77].

4
Chitosans

Chitosans are those chitins that have nitrogen content higher than 7% and degree of acetylation lower than 0.40. The removal of acetyl groups from chitin is a harsh treatment usually performed with concentrated NaOH solution (either aqueous or alcoholic). Protection from oxygen, with a nitrogen purge or by addition of sodium borohydride to the alkali solution, is necessary to avoid undesirable reactions such as depolymerization and generation of reactive species. The amount of NaOH represents however an economic and ecological worry, therefore alternatives are being sought to keep the NaOH to a minimum: for instance, chitin is mixed with NaOH powder (weight ratio 1 : 5) by extrusion at 180 °C, and highly deacetylated and soluble chitosan is obtained with just one half of the NaOH needed for aqueous systems [78].

The presence of a prevailing number of 2-amino-2-deoxyglucose units in a chitosan allows the polymer to be brought into solution by salt formation. Chitosan is a primary aliphatic amine that can be protonated by selected acids, the pK of the chitosan amine being 6.3. The following salts, among others, are water soluble: formate, acetate, lactate, malate, citrate, glyoxylate, pyruvate, glycolate and ascorbate.

Therefore, chitosan is peculiar for its cationicity and the consequent capacity to form polyelectrolyte complexes and nitrogen derivatives, according to the chemistry of the primary amino group. The film-forming ability of chitosan is another important aspect that cannot be found with cellulose. This shows that chitosan is not intractable. For instance, chitosans are soluble in water when interchain hydrogen bond formation is prevented by partial random reacetylation of the amino groups, or by insertion of bulky substituents and side chains, or by glycosylation at C6 via oxazoline derivatives; recent examples of chitosan-bearing saccharide or betaine side chains are available [41, 42, 79, 80].

The crab tendon (consisting mainly of chitin) has strong mechanical properties due to its aligned molecular structure. Proteins and calcium phosphate were removed during deacetylation by using 50 wt % NaOH aqueous solution at 100 °C, and a subsequent ethanol treatment. The aligned molecular structure of the chitosan remained intact, and had a high tensile strength (67.9. 11.4 MPa). The tensile strength was further enhanced to 235.30 MPa by a thermal treatment at 120 °C, corresponding to the formation of the intermolecular hydrogen bonds [81].

In spite of the alteration due to deacetylation, chitosan from crab tendon possesses a crystal structure showing an orthorhombic unit cell with dimensions $a = 0.828$, $b = 0.862$ and $c = 1.043$ nm (fiber axis). The unit cell comprises four glucosamine units; two chains pass through the unit cell with an antiparallel packing arrangement. The main hydrogen bonds are O3 $\cdots$ O5 (intramolecular) and N2 $\cdots$ O6 (intermolecular) [82]. This material has also found medical uses (below).

The quality of chitosan can be assessed according to various methods [83–86]. In fact, chitosan comes from a harsh treatment that affects the characteristic properties, namely average molecular weight, degree of deacetylation, viscosity of solutions, presence of reactive terminal groups. Milling and sieving introduce mechanical and thermal stresses that further alter the quality of chitosans; therefore alternative routes are being explored for the production of chitosan pellets by extrusion and spheronization, spray-drying and supercritical CO_2 drying, which moreover makes the access to the polymer functional groups easy [87, 88].

Recent progress of basic and application studies in chitin chemistry was reviewed by Kurita (2001) with emphasis on the controlled modification reactions for the preparation of chitin derivatives. The reactions discussed include hydrolysis of main chain, deacetylation, acylation, N-phthaloylation, tosylation, alkylation, Schiff base formation, reductive alkylation, O-carboxymethylation, N-carboxyalkylation, silylation, and graft copolymerization. For conducting modification reactions in a facile and controlled manner, some soluble chitin derivatives are convenient. Among soluble precursors, N-phthaloyl chitosan is particularly useful and made possible a series of regioselective and quantitative substitutions that was otherwise difficult. One of the important achievements based on this organosoluble precursor is the synthesis of nonnatural branched polysaccharides that have sugar branches at a specific site of the linear chitin or chitosan backbone [89].

5
Chitin and Chitosan Derivatives of Major Importance

5.1
Polyelectrolyte Complexes

As a polycation, chitosan spontaneously forms macromolecular complexes upon reaction with anionic polyelectrolytes. These complexes are generally water-insoluble and form hydrogels [90, 91]. A variety of polyelectrolytes can be obtained by changing the chemical structure of component polymers, such as molecular weight, flexibility, functional group structure, charge density, hydrophilicity and hydrophobicity, stereoregularity, and compatibility, as well as by changing reaction conditions, such as pH, ionic strength, polymer concentration, mixing ratio, and temperature. This, therefore, may lead to a diversity of physical and chemical properties of the complexes. Nevertheless, chitosan polyelectrolyte complexes are inherently hydrophilic and have a high tendency to swell.

5.1.1
Complexes with Hyaluronic Acid

Polyelectrolyte complexes composed of various weight ratios of chitosan and hyaluronic acid were found to swell rapidly, reaching equilibrium within 30 min, and exhibited relatively high swelling ratios of 250–325% at room temperature. The swelling ratio increased when the pH of the buffer was below pH 6, as a result of the dissociation of the ionic bonds, and with increments of temperature. Therefore, the swelling ratios of the films were pH- and temperature-dependent. The amount of free water in the complex films increased with increasing chitosan content up to 64% free water, with an additional bound-water content of over 12% [29].

The optimum conditions for polyion complex formation between chitosan and hyaluronate were identified; the compression exerted to manufacture the implant had no role to play in the release kinetics [28, 92]. Various authors published data confirming that the combination of chitosan and hyaluronic acid is always susceptible to swelling, even in the presence of cross-linking.

5.1.2
Complexes with DNA

Currently, the major drawback of gene therapy is the gene transfection rate. The two main types of vectors that are used in gene therapy are based on viral or non-viral gene delivery systems. The viral gene delivery system shows a high transfection yield but it has many disadvantages, such as oncogenic effects and immunogenicity.

Many new polymers have moved from in vitro characterization to preclinical validation in animal models of cancer, diabetes, and cardiovascular disorders. Although the transfection efficiency of most polymeric carriers is still significantly lower than that of viral vectors, their structural flexibility allows for continued improvement in polymer activity. Also, simple manufacturing and scale-up schemes and the low cost of manufacturing are eventually likely to compensate for the performance gap between viral and polymeric vectors. Non-viral delivery systems for gene therapy have been increasingly proposed as safer alternatives to viral vectors. Chitosan is considered to be suitable for the gene delivery system since it is a biocompatible, biodegradable, and nontoxic cationic material. Chitosan protects DNA against DNase degradation and leads to its condensation. However, the use of chitosan for gene delivery might be limited by low transfection efficiency [93, 94].

Chemical modifications of chitosan assayed to enhance cell specificity and transfection efficiency were reviewed. Also, chemical modifications of chitosan were performed to increase the stability of chitosan/DNA complexes [95].

The urocanic-acid-modified chitosan showed good DNA binding ability, high protection of DNA from nuclease attack, and low cytotoxicity. The transfection efficiency of chitosan into 293T cells was much enhanced after coupling with urocanic acid [96].

The transfection mechanism of plasmid–chitosan complexes as well as the relationship between transfection activity and cell uptake was analyzed by using fluorescein isothiocyanate-labeled plasmid and Texas-Red-labeled chitosan. Several factors affect transfection activity and cell uptake, for example: the molecular mass of chitosan, stoichiometry of complex, serum concentration and the pH of the transfection medium. The level of transfection with plasmid–chitosan complexes was found to be highest when the molecular mass of chitosan was 40 or 84 kDa, the ratio of chitosan nitrogen to DNA phosphate was 5, and serum at pH 7.0 was 10%. Plasmid–chitosan complexes most likely condense to form large aggregates (5–8 μm), which absorb to the cell surface. After this, plasmid–chitosan complexes are endocytosed, and accumulate in the nucleus [97].

5.1.3
Complexes with Tripolyphosphate

A simple example of gel formation is provided by chitosan tripolyphosphate and chitosan polyphosphate gel beads; the pH-responsive swelling ability, drug-release characteristics, and morphology of the gel bead depend on polyelectrolyte complexation mechanism and the molecular weight. The chitosan beads gelled in pentasodium tripolyphosphate or polyphosphoric acid solution by ionotropic cross-linking or interpolymer complexation, respectively.

Chitosan microparticles were prepared with tripolyphosphate by ionic cross-linking, starting from chitosan acetate 1% and oil as an emulsion in the presence of the surfactant Tween-80 2%: the o/w 1 : 10 emulsion was introduced into tripolyphosphate solution by a spray gun. The microparticles were then washed; their sizes were in the 500–710 μm range. As the pH of tripolyphosphate solution decreased and the molecular weight of chitosan increased, the microparticles had a more spherical shape and smoother surface [98].

However, it is not mandatory to prepare an emulsion; in fact, Pan et al. [99] reported the identification of the formation conditions of the chitosan-tripolyphosphate nanoparticles, in terms of concentrations of chitosan and tripolyphosphate. They simply used a chitosan solution at pH 4 (4 ml, concentration 1–5 mg/ml) to which they added variable amounts of tripolyphosphate solution (0.5–5 mg/ml) through a needle under stirring at room temperature. Solution, suspension and aggregates were observed, depending on the concentrations used. The zone of existence of nanoparticles in suspension is 0.9–3.0 mg/ml for chitosan and 0.3–0.8 mg/ml for tripolyphosphate.

For the preparation of nanoparticles based on two aqueous phases at room temperature one phase contains chitosan and poly(ethylene oxide) and the other contains sodium tripolyphosphate. The particle size (200–1000 nm) and zeta potential (between + 20 mV and + 60 mV) could be modulated by varying the ratio chitosan/PEO-PPO. These nanoparticles have great protein-loading capacity and provide continuous release of the entrapped protein (particularly insulin) for up to one week [100, 101].

A freeze-drying procedure for improving the shelf life of the chitosan nanoparticles using various cryoprotective agents was also investigated and negligible differences between the freeze-dried and fresh particles were found [102]. These particles were used to address the difficulties in the nasal absorption of insulin, and for the entrapment and release of the hydrophilic anthracycline drug, doxorubicin [103, 104]. These nanoparticles were also used for the improved delivery of the drugs to the ocular surface, and cyclosporin A was used as a model drug. A review discussed various possibilities for forming particles [105].

5.1.4
Other Complexes

A hydrogel with high sensitivity was prepared with chitosan (DA = 0.18) and dextran sulfate; the maximum volume of the complex gel was observed in a dilute NaOH solution at pH 10.5, and was about 300 times as large as the volume at pH values below 9 [106, 107].

Chitosan samples with degrees of deacetylation of 65, 73, 85, and 92% were almost completely adsorbed onto the surfaces of cellulosic fibers, especially onto the surfaces of fines in a variety of cellulosic systems used in industrial operations. Adsorption increased as the degree of deacetylation of chitosan increased. The aggregation of the fine cellulosic particles was maximum at a dosage of about 10 mg/kg. The interactions between chitosan and the cellulosic substrates were dominated by a bridging mechanism at about pH 7 [32].

Microemulsions based on a heparin–chitosan complex suitable for oral administration based on ingredients acceptable to humans were studied with or without biologically active ingredients. Appropriate mixing and modifications of these microemulsions lead to nanometer-sized heparin–chitosan complexes [108].

The chitosan–heparin polyelectrolyte complex was covalently immobilized onto the surface of polyacrylonitrile membrane. The immobilization caused the water contact angle to decrease, thereby indicating an increase in hydrophilicity. Protein adsorption, platelet adhesion, and thrombus formation were all reduced but antithrombogenicity was improved [109].

For the chitosan–xanthan polyelectrolyte complex, the degree of swelling has been found to be influenced by the time of coacervation, the pH of the solution of chitosan used to form the hydrogel and the pH of the swelling solution. The molecular weight and the degree of acetylation of the chitosan also influence the swelling degree. The kinetics has shown that (a) the coacervate is formed in two distinct steps and (b) the storage modulus of the hydrogel reaches a stable plateau [40].

A review article was devoted to chitosan–poly(acrylic) acid based systems for gastric antibiotic delivery, based on different mixtures of amoxicillin, chitosan and poly(acrylic) acid. The extent of swelling was greater in the polyionic complexes than in the single-chitosan formulations. The amoxicillin diffusion from the hydrogels was controlled by the polymer–drug interaction. The property of these complexes to control the solute diffusion depends on the network mesh size, which is a significant factor in the overall behavior of the hydrogels. The gastric half-emptying time of the polyionic complex was significantly delayed compared to the reference formulation, showing mean values of 164.32 ± 26.72 and 65.06 ± 11.50 min, respectively. These polyionic complexes seem to be good for specific gastric drug delivery [110].

5.1.5
Metal Chelates

The subject of the separation and purification of metals with the aid of chitosan has been reviewed by Inoue (1998) who collected data relevant to chitosans modified with chelating functional groups as well [111].

In fact, one of the major applications of chitosan and some of its many derivatives is based on its ability to bind precious, heavy and toxic metal ions. Another article reviews the various classes of chitosan derivatives and compares their ion-binding abilities under varying conditions, as well as the analytical methods to analyze them, the sorption mechanism, and structural analysis of the metal complexes. Data are also presented exhaustively in tabular form with reference to each individual metal ion and the types of compounds that complex with it under various conditions, to help reach conclusions regarding the comparative efficacy of various classes of compounds [112].

Flakes and powders cause too large a pressure drop in chromatographic columns. Coating chitosan beads with a high porosity and large surface area together with cross-linking to impart insolubility solves the problem, as exemplified in early works by Muzzarelli et al., who combined chemical derivatization and cross-linking to produce rigid gels with high chelating capacity in column operations [113].

A cobalt(II)–chitosan chelate has been prepared by soaking a chitosan film in CoCl$_2$ aqueous solution. The chitosan chelated Co(II) through both oxygen and nitrogen atoms in the chitosan chain. The tetracoordinated, high-spin Co(II)–chitosan chelate could be used as a catalyst, and the polymerization of vinyl acetate was carried out in the presence of Na$_2$SO$_3$ and water at pH 7 and normal temperature. The polyvinyl acetate possessed a random structure [114, 115].

Chitosan (> 75% deacetylation, 800–2000 cps) was mixed with stock solutions of Cu(II), Fe(II), Cd(II) and Zn(II), prepared in 0.1 M HNO$_3$, and of Ca(II) and Mn(II), in 0.1 M HCl. It was found that, in the chelation of most metal ions by chitosan, 1 : 1 binding of chitosan is more dominant than 2 : 1 cooperative binding, but vice versa for Zn(II) and Cd(II). The chelation of Cu(II) by chitosan showed much higher reactivity when compared to other divalent metal ions. Cu(II), Fe(II), Cd(II) and Zn(II) showed strong reactivity and stability of their chelates. In contrast, the interactions between Ca(II) or Mn(II) and chitosan were almost negligible. These data confirm brilliantly previous data by Muzzarelli et al. [116].

A variety of approaches to the removal of metal ions have been the subject of recent articles; for example Gotoh et al. found that a water-soluble chitosan could remain in solution in the presence of sodium alginate, and the homogeneous solution of chitosan and alginate dispensed into a CuCl$_2$ solution gave gel bead particles that were then reinforced by a cross-linking reaction with glutaraldehyde to make them durable under acidic conditions. The adsorption of Cu(II), Co(II), and Cd(II) on the beads was significantly rapid and reached equilibrium within 10 min at 25 °C. Adsorption isotherms of the metal ions on the beads exhibited Freundlich and/or Langmuir behavior, in contrast to gel beads either of alginate or chitosan, which show a step-wise shape of adsorption isotherm [117].

5.2
Chitin Ethers and Esters

In concentrated NaOH, chitin becomes alkali chitin which reacts with 2-chloroethanol to yield O-(2-hydroxyethyl) chitin, known as glycol chitin: this compound was probably the first derivative to find practical use (as the recommended substrate for lysozyme). Alkali chitin with sodium monochloroacetate yields the widely used water-soluble O-carboxymethyl chitin sodium salt [118]. The latter is also particularly susceptible to lysozyme, and its oligomers are degraded by N-acetylglucosaminidase, thus it is convenient for medical applications, including bone regeneration.

The studies on the chemical synthesis of O-acyl chitins were followed by studies on their biocompatibility, and hence their potential use as materials for blood contacting surfaces has been investigated by measuring, inter alia, their critical surface tensions, clotting times, and plasma protein absorption [119].

The accessibility of chitin, mono-O-acetylchitin, and di-O-acetylchitin to lysozyme, as determined by the weight loss as a function of time, has been found to increase in the order: chitin < mono-O-acetylchitin < di-O-acetylchitin [120]. The molecular motion and dielectric relaxation behavior of chitin and O-acetyl-, O-butyryl-, O-hexanoyl and O-decanoylchitin have been studied [121, 122]. Chitin and O-acetylchitin showed only one peak in the plot of the temperature dependence of the loss permittivity, whereas those derivatives having longer O-acyl groups showed two peaks.

Among the O-acyl chitins, dibutyryl chitin, a diester of chitin at 3 and 6 positions, having the prerogative of being soluble in various solvents, such as methanol, ethanol, ethylene chloride and acetone was obtained using methanesulphonic acid as both catalyst and solvent; the dibutyryl chitin filaments were manufactured as follows: dry-spun fibres were obtained from a 23% solution of dibutyryl chitin in acetone into air (elongation at break 34–47%); the wet spinning was performed from a 16% solution in dimethylformamide into a water coagulating bath (elongation at break 8.3%). The tensile strength was found to be small, as justified by the low crystallinity and low overall internal orientation of the filaments [123].

The synthesis of dibutyryl chitin (DBC) using perchloric acid as a catalyst and butyric anhydride as acylation agent has been worked out under heterogeneous conditions on krill, shrimp, crab and insect chitins. The preferred krill chitin had degree of acetylation 0.98 and intrinsic viscosity 13.33 dL/g (determined in DMAc + 5% LiCl solutions), which corresponded to a viscosity average molecular weight of chitin of $M_v = 286.7$ kDa. The acylation mixture was prepared by pouring perchloric acid into butyric anhydride at about $-12\,°C$, and was added to the chitin powder placed in the reactor (ca. $0\,°C$ for 30 minutes, ca. $20\,°C$ later). Weight average molecular weight values were usually in the range 120–200 kDa. DBC fibres were spun using Pt–Au spinneret with a hole diameter of 80 μm; 14.5% solution of polymer in dimethyl formamide (DMF) was used as a dope [124].

Because O-acyl chitins appear to be scarcely susceptible to lysozyme, the susceptibility of DBC to lipases has been studied to obtain insight into its biodegradability in vivo. The changes in infrared and X-ray diffraction spectra of the fibers support the slow degradation of DBC by lipases [125, 126]. The chemical hydrolysis of DBC to chitin is the most recent way to produce regenerated chitin.

5.3
Oxychitin

Crustacean chitins were submitted to regiospecific oxidation at C-6 with NaOCl in the presence of the stable nitroxyl radical 2,2,6,6-tetramethyl-1-

piperidinyloxy (Tempo®) and NaBr at 25 °C in aqueous solution. The resulting oxychitins have anionic character and are fully soluble over the pH range 3–12; they lend themselves to metal chelation, polyelectrolyte complex formation with a number of biopolymers including chitosan, and to microsphere and bead formation. Oxychitin sodium salt coagulates a number of proteins, including papain, lysozyme and other hydrolases [127].

The similarly treated biomasses of *Aspergillus niger, Trichoderma reesei* and *Saprolegnia* yielded polyuronans in the sodium-salt form, fully soluble in water over the pH range 3–12. Yields were much higher than for the chitosan extraction. The polyuronans characterized by ^{1}H-NMR and Fourier-transform infrared (FTIR) spectrometry contain 20% and > 75% oxychitin, for *A. niger* and *T. reesei*, respectively. Since the fungi examined are representative of the three major types of cell walls, and are of industrial importance, it is concluded that the process is of wide applicability. The process allows upgrading the spent biomasses and the exploitation of their polysaccharides.

Oxychitin keeps the regenerative properties of chitin and chitosan; in a model study, surgical lesions in rat condylus were treated with *N,N*-dicarboxymethyl chitosan and 6-oxychitin sodium salt. Morphological data indicated that the best osteoarchitectural reconstruction was promoted by 6-oxychitin, even though healing was slower compared to *N,N*-dicarboxymethyl chitosan. Plates of Ti – 6Al – 4V alloy were plasma-sprayed with hydroxyapatite or with bioactive glass. Chitosan acetate solution was then used to deposit a chitosan film upon the plasma-sprayed layers, which was further reacted with 6-oxychitin to form a polyelectrolyte complex. The latter was optionally contacted with 1-ethyl-3-(3-dimethylaminopropyl)carbodiimide at 4 °C for 2 hr to form amide bonds between the two polysaccharides. Uniform flat surfaces exempt from fractures were observed at the electron microscope. The results are useful for the preparation of prosthetic articles possessing an external organic coating capable to promote colonization by cells, osteogenesis and osteointegration [128].

In a review by Bragdt et al. (2004) results and perspectives are given to change the salt-based oxidative systems for cleaner oxygen or hydrogen peroxide enzyme-based Tempo® systems. Moreover, several immobilized Tempo® systems have been developed [129].

5.4
Modified Chitosans

The Schiff reaction between chitosan and aldehydes or ketones yields the corresponding aldimines and ketimines, which are converted to *N*-alkyl derivatives upon hydrogenation with borohydride. Chitosan acetate salt can be converted into chitin upon heating [130]. The following are important examples of modified chitosans that currently have niche markets or prominent places in advanced research.

5.4.1
Thiolated Chitosans

The derivatization of the primary amino groups of chitosan with coupling reagents bearing thiol functions leads to the formation of thiolated chitosans. So far, three types of thiolated chitosans have been generated: chitosan–cysteine conjugates, chitosan–thioglycolic acid conjugates and chitosan–4-thio-butyl-amidine conjugates. Various properties of chitosan are improved by the immobilization of thiol groups. Due to the formation of disulfide bonds with mucus glycoproteins, the mucoadhesiveness is 6–100-fold augmented. The permeation of paracellular markers through intestinal mucosa can be enhanced 1.6–3-fold utilizing thiolated instead of plain chitosan. Moreover, thiolated chitosans display in situ gelling features, due to the pH-dependent formation of inter- and intramolecular disulfide bonds, with consequent cohesion and stability of carrier matrices based on thiolated chitosans, that provide prolonged controlled release of drugs [131].

5.4.2
N-Carboxymethyl Chitosan

By using glyoxylic acid, water-soluble N-carboxymethyl chitosan is obtained: the product is a glucan carrying pendant glycine groups [132]. N-Carboxymethylchitosan from crab and shrimp chitosans is obtained in water-soluble form by proper selection of the reactant ratio, i.e., with equimolar quantities of glyoxylic acid and amino groups. The product is in part N-mono-carboxymethylated (0.3), N,N-dicarboxymethylated (0.3) and N-acetylated depending on the starting chitosan (0.08–0.15) [133].

N-Carboxymethylchitosan as a 1.0% solution at pH 4.80 is a valuable functional ingredient of cosmetic hydrating creams in view of its durable moisturizing effect on the skin [134]. The film-forming ability of N-carboxymethylchitosan assists in imparting a pleasant feeling of smoothness to the skin and in protecting it from adverse environmental conditions and consequences of the use of detergents. N-Carboxymethyl chitosan was found to be superior to hyaluronic acid as far as hydrating effects are concerned.

In general these derivatives are safe, their chemical functions being the glycine moiety; the same holds for N,O-carboxymethyl chitosan, as demonstrated for instance by studies intended to assess the efficacy of N,O-carboxymethyl chitosan to limit adhesion formation in a rabbit abdominal surgery model. The inability of fibroblasts to adhere to N,O-carboxymethyl chitosan-coated surfaces suggests that it may act as a biophysical barrier [135].

5.4.3
Highly Cationic Chitosans

Trimethyl chitosan is a soluble derivative that shows effective enhancing properties for peptide and protein drug transport across mucosal membranes. Trimethyl chitosan (TMC) was synthesized by reductive methylation of chitosan in an alkaline environment at elevated temperature. The number of methylation steps affects the degree of quaternization of the primary amino group and methylation of 3- and 6-hydroxyl groups. The degree of quaternization was higher when using sodium hydroxide as the base compared to using dimethyl amino pyridine. O-Methylation resulted in decreased solubility of trimethyl chitosan [136].

Based on the fact that the transport of desmopressin across the intestinal mucosa in vitro was enhanced by applying trimethyl chitosan chloride, minitablets and granules were developed as solid oral dosage forms for the delivery of peptide drugs; they were suitable as a dosage form due to their ability, as components of multiple unit dosage forms, to disperse from each other, before disintegration, effectively increasing the area in which the polymer can assert its absorption-enhancing effect. Both the optimized minitablet formulation and the granule formulation showed suitable release profiles for the delivery of peptide drugs with TMC as an absorption enhancer in solid oral dosage forms [137, 138].

As an alternative, functionalized compounds such as choline dichloride, carrying the preformed trimethylammonium group, can react with chitosan to yield highly cationic chitosans; the other new cationic derivative being N-[(2-hydroxy-3-trimethylammonium) propyl] chitosan chloride as reported by Xu et al. [52] and by Lim and Hudson [139]. The Chitopearl® products (Fuji Spinning Co., Japan) belong to this class of chitosans, where the cross-linking compound contains two quaternary nitrogens.

5.4.4
Polyurethane-type Chitosans

Some other types of Chitopearl® spherical chitosan particles are produced from diisocyanates and are suitable for chromatographic purposes and as enzyme supports [140] (Fig. 2). Chitins of various origins in DMAC-LiCl solution react with excess 1,6-diisocyanatohexane. Upon exposure to water vapor for 2 days, flexible and opaque materials are produced; whose main characteristics are insolubility in aqueous and organic solvents, remarkable crystallinity, typical infrared spectrum, high N/C ratio (0.287) and relatively high degree of substitution (0.29), but no thermoplasticity. Chitosan similarly treated under heterogeneous conditions in anhydrous pyridine, yields reaction products with a lower degree of substitution (0.17). Microencap-

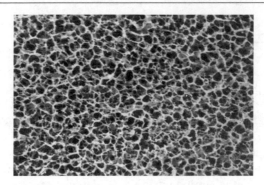

Fig. 2 The Chitopearl© surface aspect. Chitopearl© is manufactured by Fuji Spinning Co. Ltd., Tokyo, Japan

sulation of lactic acid bacteria based on the cross-linking of chitosan by 1,6-diisocyanatohexane has been performed [141].

5.4.5
Sugar-Modified Chitosans

Hall and Yalpani [142, 143] synthesized sugar-bound chitosan by reductive *N*-alkylation using sodium cyanoborohydride (NaCNBH₃) and unmodified sugar or sugar–aldehyde derivatives. In view of the specific recognition of sugars by cells, viruses and bacteria, Morimoto et al. [144–147] reported the synthesis of fucose-bound chitosans and their specific interaction with lectins or cells. Kato et al. [148] also prepared lactosaminated *N*-succinyl chitosan and its fluorescein derivative as a liver-specific drug carrier in mice through the asialoglycoprotein receptor.

N-Succinyl-chitosan is a good drug carrier for mitomycin-C in liver metastasis [149]; it has favorable properties such as biocompatibility, low toxicity and long-term retention in the systemic circulation after intravenous administration; the plasma half-lives of *N*-succinyl chitosan (Mw 3.4×10^5; succinylation degree 0.81 mol/sugar unit; deacetylation degree: 1.0 mol/sugar unit) were about 100 h in normal mice and 43 h in Sarcoma 180-bearing mice. The biodistribution of *N*-succinyl chitosan into other tissues was trace, apart from the prostate and lymph nodes. The maximum tolerable dose for the intraperitoneal injection of *N*-succinyl chitosan to mice was greater than 2 g/kg. The water-insoluble and water-soluble conjugates could be prepared using a water-soluble carbodiimide and mitomycin C. A review summarized the utilization of *N*-succinyl chitosan as a drug carrier for macromolecular conjugates of mitomycin-C and the therapeutic efficacy of the conjugates against various tumors [150, 151].

The difference between chitosan and succinyl chitosan can be appreciated if one compares the respective experimental results. In fact randomly

50% deacetylated chitin, otherwise called water-soluble chitosan, was examined, having been fluorescein isothiocyanate (FTC) labeled, in terms of biodegradability, body distribution and urinary excretion after the intraperitoneal administration to mice. The in vitro biodegradability was investigated by incubation with lysozyme and murine plasma and urine. FTC-Chitosan moved fast to the kidney and urine, and was scarcely distributed to the liver, spleen, and plasma. Most of the FTC-chitosan was excreted into the urine after 14 h, and the molecular weight of the excreted FTC-chitosan was as small as that of the product obtained by long in vitro incubation. Therefore, this chitosan is considered to be highly biodegradable and easily excreted in urine, and further it is suggested that it undergoes no accumulation in the body; however, plain water-soluble chitosan does not exhibit long retention in the body [152].

Reductive amination of N-succinyl chitosan and lactose using sodium cyanoborohydride in a phosphate buffer (pH 6.0) for 6 days was suitable for the preparation of lactosaminated N-succinyl chitosan (Fig. 3). Over 10% of dose/g-tissue was distributed to the prostate and lymph nodes at 48 h post-administration in both chitosan and lactosaminated N-succinyl chitosan. The labeled lactosaminated N-succinyl chitosan was easily distributed into not only the liver but also prostate, intestine, preputial gland and lymph nodes [153].

Another example of the versatility and usefulness of the fluorescein isothiocyanate conjugates is provided by the following study. Glycol chitosan was labeled with fluorescein isothiocyanate to investigate its biodistribution in tumor-bearing rats. Glycol-chitosan-doxorubicin formed micelle-like nanoaggregates spontaneously in aqueous media. A loading content of dox-

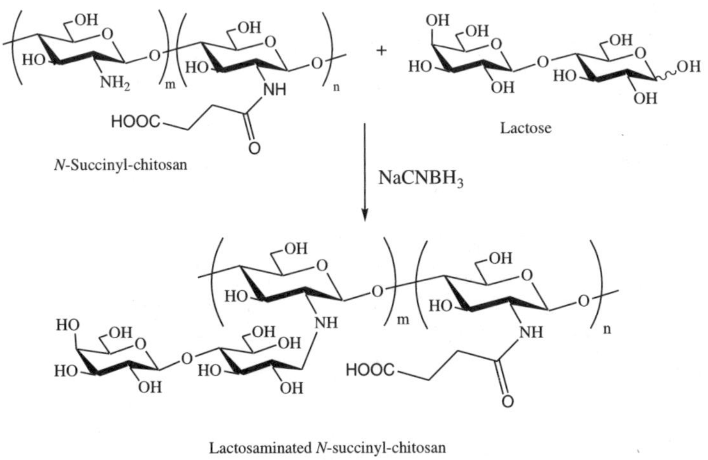

Fig. 3 Synthesis of lactosaminated N-succinyl chitosan

orubicin nanoaggregates as high as 38%, with 97% loading efficiency, could be obtained upon cross-linking with 1-ethyl-3-(3-dimethylaminopropyl) carbodiimide and N-hydroxysuccinimide at room temperature overnight. The FTC-glycol chitosan nanoaggregates were injected into the tail vein of tumor-bearing rats and were found to be distributed mainly in kidney, tumor and the liver. They were maintained at a high level for 8 days and their distribution in tumor tissues increased gradually. This suggests that chitosan nanoaggregates accumulate passively in the tumor tissue due to the enhanced permeability and retention effect. Tumor growth was suppressed over 10 days [154].

Galactosylated chitosan prepared from lactobionic acid and chitosan with 1-ethyl-3-(3-dimethylaminopropyl)-carbodiimide and N-hydroxysuccinimide was a good extracellular matrix for hepatocyte attachment [155] (Fig. 4). Furthermore, graft copolymers of galactosylated chitosan with poly(ethylene glycol) or poly(vinyl pyrrolidone) were useful for hepatocyte-targeting DNA carrier [156, 157].

Sialic acid, the ubiquitous sugar present in the mammalian-cell surface glycolipids and glycoproteins, is the essential epitope for many infections. Sialic-acid-containing polymers have been shown to be potent inhibitors of hemagglutination of human erythrocytes by influenza viruses. Sashiwa and Roy prepared sialic-acid-bound chitosan as a new family of sialic-acid-containing polymers using p-formylphenyl-α-sialoside [158] by reductive N-alkylation [159]. Since that derivative was insoluble in water, further N-succinylation was carried out and a water-soluble derivative was obtained exhibiting specific binding with wheat-germ agglutinin lectin.

Artificial glycopolymers having α-galactosyl epitope are of interest from the viewpoint of medical transplantation of pig liver since they can block im-

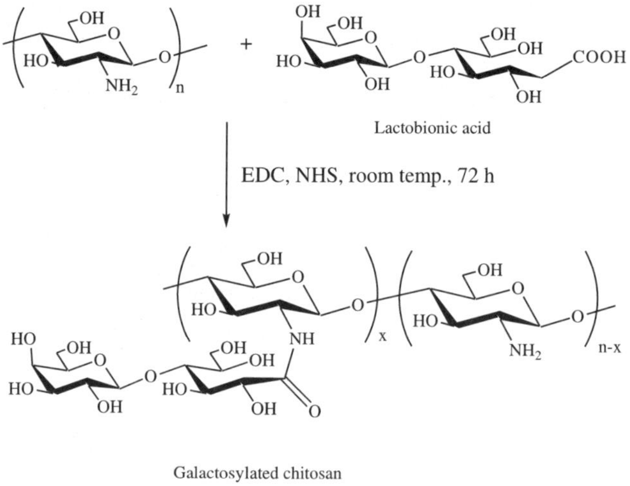

Fig. 4 Synthesis of galactosylated chitosan

mune rejection. Water-soluble α-galactosyl chitosan showed specific binding against α-galactosyl specific lectin (*Griffonia simplicifolia*) [160, 161].

5.4.6
Chitin–Inorganic Phosphate Composites

Besides the occurrence forms mentioned above, chitin is also found as a major component of the organic fraction of several biocomposites in which an organic matrix is associated with an inorganic compound. The relationship between the mineral phase and the organic phase implies molecular recognition. Chitin in mineralized biological systems has a crucial role in the hierarchical control of the biomineralization processes; the nacre of the mollusk shell is an example. The actual and future applications of mineral–chitin composites range from the medical field as bone repair (chitin–calcium-phosphate composites) to the industrial field as catalysts [162].

Chitin is utilized for tissue repair processes by acting as a temporary scaffold in a bone substitute, pending resorption of the implant and replacement by natural bone. Calcium phosphate cement is interesting for craniofacial and orthopedic repair because of its ability to self-harden in situ to form hydroxyapatite with excellent osteoconductivity. However, its poor strength, long hardening time, and lack of macroporosity limit its use. Xu et al. [163] developed fast-setting and anti-washout scaffolds with high strength. Chitosan, sodium phosphate, and hydroxypropyl methylcellulose were used to impart fast setting and resistance to washout. Absorbable fibers and mannitol porogen were incorporated into cement for strength and macropores. Flexural strength, work of fracture, and elastic modulus were measured against immersion time in a physiological solution. Hardening time was 69.5 ± 2.1 min for cement control, 9.3 ± 2.8 min for cement-HPMC-mannitol, 8.2 ± 1.5 min for cement–chitosan–mannitol, and 6.7 ± 1.6 min for cement–chitosan–mannitol-fiber. Immersion for 1 day dissolved mannitol and created macropores.

A composite material was produced from microporous coralline hydroxyapatite microgranules, chitosan fibers and chitosan membrane. Cylindrical microgranules were oriented along channel direction and aligned particles were supported with fibers and a chitosan membrane. The positive replica of mould channels was clasp fixed to produce thicker scaffolds. Light microphotographs of the developed complex structure showed good adhesion between the hydroxyapatite particles, the fibers and the supporting membrane. The composite material showed 88% (w/w) swelling in one hour and preserved the complex structure of the original material upon long-term incubation in physiological medium [164].

Human osteoblast-like MG63 cells were cultured on the macroporous chitosan scaffolds reinforced with hydroxyapatite or calcium phosphate invert glass were fabricated using a thermally induced phase separation technique.

The cell growth was much faster on the chitosan–hydroxyapatite scaffolds with the glass than on the chitosan-hydroxyapatite scaffold without the glass. The total protein content of cells increased over time on both composites. The cells on the chitosan-hydroxyapatite-glass also expressed significantly higher amount of alkaline phosphatase at days 7 and 11 and osteocalcin at day 7 than those on chitosan-hydroxyapatite [165].

A chitosan-bonded hydroxyapatite bone-filling paste was prepared as follows: chitosan (0.5 g) was dissolved in malic acid (0.5 g) solution made with saline, and a chitosan film was formed by mixing this sol with hydroxyapatite powder (2 g), followed by neutralization with 5% sodium polyphosphate. To help cells and blood vessels to penetrate this material, the tensile strength and elongation were optimized [166, 167]. Chitosan–hydroxyapatite porous microspheres were prepared using tripolyphosphate as coagulating agent. The size increased and the water sorption decreased with increasing hydroxyapatite contents. The ceramic particles were well embedded and homogeneously distributed within the polymer matrix [168].

Similarly, a composite of hydroxyapatite and a network formed via cross-linking of chitosan and gelatin with glutaraldehyde was developed by Yin et al. [169]. A porous material, with similar organic–inorganic constituents to that of natural bone, was made by the sol–gel method. The presence of hydroxyapatite did not retard the formation of the chitosan–gelatin network. On the other hand, the polymer matrix had hardly any influence on the high crystallinity of hydroxyapatite.

Results are different when solutions are used instead of suspensions. Chitosan–hydroxyapatite composites with a homogeneous nanostructure have been prepared by a coprecipitation method by Yamaguchi et al. [170].

Hydroxyapatite crystallites in the composites formed elliptic aggregations 230 nm in length and 50 nm in width. The typical length of the aggregations corresponded approximately to that of a chitosan molecule. The size of the constituent hydroxyapatite crystallites was found to be predominantly 30 nm in length and 10 nm in width, and the c-axes were well aligned in parallel with the chitosan molecules in the respective aggregations. The growth of the hydroxyapatite crystallites is considered to occur at nucleation sites. Yokogawa et al. used chitin fibres as supports on which to grow calcium phosphate [171].

Results were obtained on the calcium phosphate growth on phosphorylated chitin fibres using the urea/H_3PO_4 method and subsequently soaked in saturated $Ca(OH)_2$ solution and in simulated body fluid solution.

To obtain the phosphorylated chitin, fibres were soaked in saturated $Ca(OH)_2$ solution (pH 12.4) for 8 days. The $Ca(OH)_2$ solution was renewed every 4 days. After completion of the soaking period, the fibres were washed, filtered and dried under vacuum at 60 °C. This technique of phosphorylation and $Ca(OH)_2$ treatment has been found to be a useful method for creating favorable local conditions leading to the nucleation and growth of calcium phosphate. After $Ca(OH)_2$ treatment, the thin layer functioned as a nucleation layer for further calcium phosphate deposition.

Calcium phosphate cements have been developed during the last two decades. They are suitable for the repair and reconstruction of bone; they adapt immediately to the bone cavity and permit subsequent good osteointegration.

To make the cement injectable, several additives can be incorporated; however, the properties of the cement should be preserved: setting times suited to a convenient delay with surgical intervention, limited disintegration in aqueous medium, and sufficient mechanical resistance. Lactic acid, glycerol, glycerophosphate and chitosan were studied as adjuvants, in terms of injectability, setting time, disintegration and toughness [172–174].

Sodium carboxymethyl chitin and phosphoryl chitin had most evident influences on the crystallization of calcium phosphate from supersaturated solutions. They potently inhibited the growth of hydroxyapatite and retarded the rate of spontaneous calcium phosphate precipitation. These chitin derivatives were incorporated into the precipitate and influenced both the phase and morphology of the calcium phosphate formed (flaky precipitate resembling octacalcium phosphate instead of spherical clusters in the absence of polysaccharide) [175].

Muzzarelli et al. [176] studied the effects of N,N-dicarboxymethyl chitosan on the precipitation of a number of insoluble salts. The chelating ability of this modified chitosan interfered effectively with the physicochemical behavior of magnesium and calcium salts. N,N-Dicarboxymethyl chitosan mixed with calcium acetate and disodium hydrogen phosphate in suitable ratios yielded clear solutions from which an amorphous material was isolated containing an inorganic component about one half its weight. This compound was used for the treatment of bone lesions in experimental surgery and in dentistry. Bone tissue regeneration was promoted in sheep, leading to complete healing of otherwise non-healing surgical defects. Radiographic evidence of bone regeneration was observed in human patients undergoing apicoectomies and avulsions. The N,N-dicarboxymethyl chitosan–calcium phosphate chelate favored osteogenesis while promoting bone mineralization.

The in situ precipitation route towards obtaining composites of polymer and calcium phosphate is similar to the strategy employed in naturally occurring biocomposites and well may prove a viable method for the synthesis of bone substitutes.

Chitosan scaffolds were reinforced with beta-tricalcium phosphate and calcium phosphate invert glass [177]. Along the same line, composites of *Loligo* beta-chitin with octacalcium phosphate or hydroxyapatite were prepared by precipitation of the mineral into a chitin scaffold by means of a double diffusion system. The octacalcium phosphate crystals with the usual form of 001 blades grew inside chitin layers preferentially oriented with the 100 faces parallel to the surface of the squid pen and were more stable to hydrolysis to hydroxyapatite with respect to those precipitated in solution. In these in vitro experiments the compartmentalized space in the chitin governs the orientation of the crystals, even if epitaxial factors may play a role in the nucleation processes [178].

5.4.7
Enzymatic Modification of Chitosan

The early works by Muzzarelli et al. [179] showed that tyrosinase converts a wide range of phenolic substrates into electrophilic o-quinones [180]. Tyrosinase was used to convert phenols into reactive o-quinones which then underwent chemical reactions leading to grafting onto chitosan. A review article showed that in general the tyrosinase-catalyzed chitosan modifications resulted in dramatic changes in functional properties [181].

These derivatives were inspired by the chemistry of the cuticle tanning in vivo. Stable and self-sustaining gels are obtained from tyrosine glucan (a modified chitosan synthesized with 4-hydroxyphenylpyruvic acid) in the presence of tyrosinase. Similar gels are obtained from 3-hydroxybenzaldehyde, 4-hydroxybenzaldehyde and 3,4-dihydroxybenzaldehyde; all of them are hydrolyzed by lysozyme, lipase and papain. No cross-linking is observed for chitosan derivatives of vanillin, syringaldehyde and salicylaldehyde. With collagen + chitosan + tannin mixtures under the catalytic action of tyrosinase, partially crystalline, hard, mechanically resistant and scarcely wettable materials are obtained upon drying. In contrast, the products obtained from albumin, pseudocollagen and gelatin in the presence of a number of phenols and chitosan under comparable conditions are brittle. Phenoxyacetate is used in the production of penicillin and is often recycled; to remove p-hydroxylated derivatives of this precursor, tyrosinase is used followed by adsorption of the quinone species on chitosan. Volatile phenols (air pollutants) in the presence of tyrosinase are coupled (i.e. chemisorbed) onto chitosan films [182–185]. Results provided evidence that peroxidases can be used to graft phenolic moieties onto chitosan, such as dodecyl gallate [186].

In the case of gelatin and chitosan, the ability of two enzymes to catalyze the formation of gels from solutions was compared. A microbial transglutaminase catalyzed the formation of strong and permanent gels from gelatin solutions. Chitosan was not required for transglutaminase-catalyzed gel formation, although gel formation was faster, and the resulting gels were stronger if reactions were performed in the presence of this polysaccharide. Consistent with transglutaminase ability to covalently cross-link proteins, the transglutaminase-catalyzed gelatin–chitosan gels lost the ability to undergo thermally reversible sol–gel transitions characteristic of gelatin. Mushroom tyrosinase was also observed to catalyze gel formation for gelatin–chitosan blends. Tyrosinase-catalyzed gelatin–chitosan gels were weaker than transglutaminase-catalyzed gels [187].

Besides high-molecular-weight proteins, the functional groups studied include low-molecular-weight phenols of which arbutin, a natural phenol found in pears, is an example. Tyrosinase catalyses reactions that lead to the conversion of arbutin–chitosan solutions into gels. These gels can be rapidly broken by treatment with the chitosan-hydrolyzing enzyme chitosanase, demonstrating that the chitosan derivatives remain biodegradable [188].

Biologically significant quinones, such as menadione (vitamin K), plumbagin, ubiquinone (CoQ_{10}) and CoQ_3 were examined along with 1,2-naphthoquinone and 1,4-naphthoquinone for their capacity to react with five chitosans in freeze-dried or film form. The chitosans tested were: chitosan acetate salt, reacetylated chitosan, amorphous chitosan, 5-methylpyrrolidinone chitosan and N-carboxymethyl chitosan. CoQ_{10} and CoQ_3 did not react with the chitosans, whilst menadione and, even more, 1,4-naphthoquinone are reactive and yield deeply colored, modified chitosans. The reactivities of plumbagin and 1,2-naphthoquinone are modest or nil, depending on the chitosan tested. The maximum capacity of chitosans for 1,4-naphthoquinone corresponded to an amine/quinone molar ratio close to 2, indicative of saturation over a 12 h contact period: the relevant infrared spectra did not show the typical bands of 1,4-naphthoquinone. UV–vis measurements on methanol solutions indicated that chitosan acetate salt and reacetylated chitosan were most reactive with menadione. Menadione-treated chitosans gave infrared spectra containing typical quinone bands, and the films had altered surface properties, their contact angles with saline being much higher than for controls. The absence of reactivity between ubiquinone and N-carboxymethyl chitosan, both widely accepted functional cosmetic ingredients, could constitute the basis for the formulation of tooth pastes and gingival gels, possessing enhanced reparative properties due to the synergistic actions of intact ubiquinone and N-carboxymethyl chitosan [189].

6
Chitin and Chitosan in Various Forms

6.1
Nanoparticles

Ohya et al. reported poly(ethyleneglycol)-grafted chitosan nanoparticles as peptide drug carriers. The incorporation and release of insulin was dependent on the extent of the reaction of poly(ethyleneglycol) with chitosan [190].

Lee et al. reported a novel and simple method for delivery of adriamycin using self-aggregates of deoxycholic acid modified chitosan. Deoxycholic acid was covalently conjugated to chitosan via a carbodiimide-mediated reaction generating self-aggregated chitosan nanoparticles. Adriamycin was entrapped physically within the nanoparticles and the aggregates were spherical. They achieved about 49.6 wt % loading efficiency with slow release phenomena over time in phosphate buffer (pH 7.2) [191]. Kim et al. explored these self-aggregates of deoxycholic-acid-modified chitosan as DNA carriers. They have explained the critical aspects involved in the self-assembly formation of deoxycholic-acid-modified chitosan [192].

Nanoparticles of methotrexate were prepared using O-carboxymethyl chitosan. The amount of cross-linking agents on drug release in different media was evaluated [193].

6.2
Microspheres

A review of chitosan microspheres as carrier for drugs published recently by Sinha et al. provides insight into the exploitation of the various properties of chitosan to microencapsulate drugs. Various techniques used for preparing chitosan microspheres and evaluation protocols have also been reviewed, together with the factors that affect the entrapment efficiency and release kinetics of drugs [194].

Spray-drying of chitosan salt solutions provides chitosan microspheres having diameters close to 2–5 μm and improved binding functionality. The chitosan microsphere free-flowing powder is compressible and hence most suitable as a drug carrier [195–204]. The following are some examples.

By choosing the excipient type and concentration, and by varying the spray-drying parameters, control was achieved over the physical properties of the dry chitosan powders. The in vitro release of betamethasone showed a dose-dependent burst followed by a slower release phase that was proportional to the drug concentration in the range 14–44% w/w [200].

Chitosan microspheres containing chlorhexidine diacetate, an antiseptic, were prepared by spray-drying. Chlorhexidine from the chitosan microspheres dissolved more quickly in vitro than chlorhexidine powder. The minimum inhibitory concentration, minimum bacterial concentration and killing time showed that the loading of chlorhexidine into chitosan was able to maintain or improve the antimicrobial activity of the drug, the improvement being particularly high against *Candida albicans*. It should be noted that the drug did not decompose despite its thermal lability above 70 °C. Buccal tablets were prepared by direct compression of the microspheres with mannitol alone or with sodium alginate. After their in vivo administration the determination of chlorhexidine in saliva showed the capacity of these formulations to give a prolonged release of the drug in the buccal cavity [205].

The general chemical behavior of chitosan, however, should be considered in order to avoid certain difficulties stemming from its insolubility at pH higher than 6.5 and its reactivity under the thermal conditions of the sprayer. For example, spray-drying of 1–2% chitosan in acid solution at 168 °C seems to be easy, however, the release of a drug from the spray-dried chitosan depends on the acetic acid concentration due to the acetylation reaction occurring at the temperature to which the salt is exposed, certainly lower than 168 °C but high enough for side reactions to occur. In fact, the degree of acetylation of chitosan increased during spray-drying and affected its enzymatic degradability [203].

Chitosan can be spray-dried at neutral pH if a colloidal suspension is prepared with NaOH. Nevertheless, this preparation is prohibitively time-consuming due to the difficulties involved in washing and removal of excess alkali and salts.

Chitosan has been recently found to be soluble in alkaline media, viz. NH_4HCO_3 solutions, where it assumes the ammonium carbamate form Chit-$NHCO_2^-NH_4^+$, i.e., a transient anionic form that keeps it soluble at pH 9.6, while reversibly masking the polycationic nature of chitosan. Because ammonium carbamates and NH_4HCO_3 decompose thermally and liberate CO_2, NH_3 and water, this alkaline system is suitable for producing chitosan microspheres by spray-drying (Table 1) [206].

For the preparation of spray-dried polyelectrolyte complexes, the polyanion was dissolved in dilute NH_4HCO_3 solution and mixed with the chitosan carbamate solution just before spray-drying. The excess NH_4HCO_3 decomposed thermally between 60 and 107 °C; on the other hand, the carbamate function released carbon dioxide under the effect of the temperature at which the spray-drier was operated, thus regenerating chitosan at the moment of the polyelectrolyte microsphere formation (Fig. 5).

Fig. 5 Microspheres manufactured from the polyelectrolyte complex of chitosan carbamate and ammonium alginate in ammonium bicarbonate solution. Muzzarelli, original data, 2004

Table 1 Operating conditions adopted for spray-drying chitosan carbamate–polyanion mixtures in dilute NH_4HCO_3. From ref. [126]

Polyanion	Dry weight ratio chitosan/polyanion	NH₂/COOH Molar ratio	Total polysaccharide concentration, g/l	Solubility microsphere	Inlet air /°C
Alginic acid	1	0.9	4.2	Insoluble	155
Polygalacturonic acid	1	0.9	5.7	Insoluble	155
Carboxymethyl cellulose	1	n.a.	10.2	Insoluble	130
Carboxymethyl guaran	1	n.a.	8.5	Insoluble	135
Acacia gum	0.5	1.6	9.4	Insoluble	150
6-Oxychitin	1.25	1.0	7.5	Insoluble	160
Xanthan	2.5	6.0	3.7*	Soluble	155
Hyaluronic acid	1.66	3.9	4.0*	Swelling	145
Pectin	1	4.0	9.2	Soluble	145
κ-Carrageenan***	2	5.0	3.3*	Swelling	155
Guaran**	1	No	7.7	Soluble	140

* These low-polysaccharide concentrations were preferred to avoid excessive viscosity.
** Neutral polysaccharide.
*** Sulfated polysaccharide.
Flow rate 10 ml/min.

In most cases the microspheres were insoluble. The polysaccharides might be partially cross-linked via amido groups formed by the carboxyl groups of the polyanion and the restored free amino group of chitosan. The susceptibility to enzymatic hydrolysis by lysozyme was poor, mainly because lysozyme, a strongly cationic protein, can be inactivated by anionic polysaccharides [207].

Notwithstanding the chemical differences (alcohol groups in guaran, carboxyl groups in xanthan, and partially esterified carboxyl groups in pectin) these three polysaccharides in combination with chitosan in the microspheres appear to be able to bring chitosan into solution. This is particularly interesting if one considers the solubility of these three polysaccharides in water and their important applications in the food and pharmaceutical industries.

The multiple emulsion technique includes three steps: 1) preparation of a primary oil-in-water emulsion in which the oil dispersed phase is constituted of CH_2Cl_2 and the aqueous continuous phase is a mixture of 2% v/v acetic acid solution: methanol (4/1, v/v) containing chitosan (1.6%) and Tween (1.6, w/v); 2) multiple emulsion formation with mineral oil (oily outer phase) containing Span 20 (2%, w/v); 3) evaporation of aqueous solvents under reduced pressure. Details can be found in various publications [208, 209]. Chemical cross-linking is an option of this method; enzymatic cross-linking can also be performed [210]. Physical cross-linking may take place to a certain extent if chitosan is exposed to high temperature.

The emulsion technique is convenient when the drug is particularly sensitive to certain parameters connected to the spray-drying. The emulsion technique may be associated to cross-linking or other treatments of the microspheres. The following examples are self-explanatory.

Chitosan microspheres cross-linked with glutaraldehyde, sulphuric acid or heat treatment, have been prepared to encapsulate diclofenac sodium. Chitosan microspheres were produced in water in oil emulsion followed by cross-linking in the water phase. The cross-linking of chitosan took place at the free amino group in all cases, and lead to the formation of imine groups or ionic bonds. Polymer crystallinity increased after cross-linking. Microspheres had smooth surfaces, with sizes in the range 40–230 μm. Loading of diclofenac sodium was carried out by soaking the already swollen cross-linked microspheres in a saturated solution of diclofenac sodium [211].

Eudragit RS microspheres containing chitosan hydrochloride were prepared by the solvent evaporation method using an acetone/liquid paraffin solvent system, and their properties were compared with Eudragit RS microspheres without chitosan. The content of pipemidic acid, an antibacterial, increased in larger microspheres as a consequence of cumulation of undissolved pipemidic acid particles in larger droplets. Pipemidic acid release was faster from microspheres with chitosan [212].

Microspheres were prepared from carboxymethyl chitosan and alginate by emulsion phase separation. The encapsulated bovine serum albumin was quickly released in a Tris-HCl buffer (pH 7.2), whereas a small amount of bovine serum albumin was released under acid conditions (pH 1.0) because of the strong electrostatic interaction between NH_2 groups of carboxymethyl chitosan and COOH groups of alginic acid and a dense structure caused by a Ca^{2+} cross-linked bridge [213].

6.3
Hydrogels

Polymer scaffolds have many different functions in the field of tissue engineering. They are applied as space-filling agents, as delivery vehicles for bioactive molecules, and as three-dimensional structures that organize cells and present stimuli to direct the formation of a desired tissue. Much of the success of scaffolds in these roles hinges on finding an appropriate material to address the critical physical, mass transport, and biological design variables inherent to each application. Hydrogels are an appealing scaffold material because they are structurally similar to the extracellular matrix of many tissues, can often be processed under relatively mild conditions, and may be delivered in a minimally invasive manner. Consequently, hydrogels have been utilized as scaffold materials for drug and growth factor delivery, tissue replacements, bone formation and a variety of other applications. [27, 214].

Gel materials are utilized in a variety of technological applications and are currently investigated for advanced exploitations such as the formulation of intelligent gels and the synthesis of molecularly imprinted polymers.

One of the simplest ways to prepare a chitin gel is to treat chitosan acetate salt solution with carbodiimide to restore acetamido groups. Thermally not reversible gels are obtained by N-acylation of chitosans: N-acetyl-, N-propionyl- and N-butyryl-chitosan gels are prepared using 10% aqueous acetic, propionic and butyric acid as solvents for treatment with appropriate acyl anhydride. Both N- and O-acylation are found, but the gelation also occurs by selective N-acylation in the presence of organic solvents.

pH-Sensitive hydrogels were synthesized by grafting D,L-lactic acid onto the amino groups in chitosan without a catalyst; polyester substituents provide the basis for hydrophobic interactions that contribute to the formation of hydrogels [215, 216]. The crystallinity of chitosan gradually decreased after grafting, since the substituents are randomly distributed along the chain and destroy the regularity of packing between chitosan chains (Fig. 6).

A popular cross-linking agent for chitosan is glutaraldehyde, as proposed by Muzzarelli et al. [217]. Chitosan networks were obtained by reaction with glutaraldehyde in lactic acid solution (pH 4–5) at molar ratio amino groups/carbonyl functions about 10–20: reduction gave stable chemical gels.

Investigations on biocompatible hydrogels based exclusively on polysaccharide chains were reported; chitosan was linked with dialdehyde obtained from scleroglucan by controlled periodate oxidation [218]. The reaction took

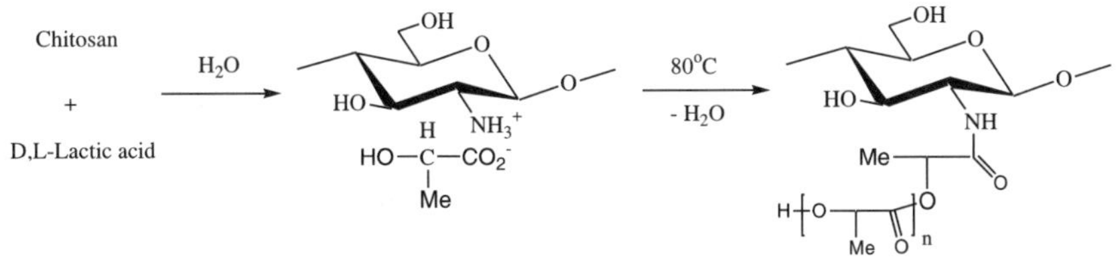

Fig. 6 Reaction of lactic acid with chitosan

place at pH 10 and the reduction of the resulting Schiff base was performed with NaCNBH$_3$. The swelling capacity of the hydrogel was remarkable, which is dependent on the highly hydrophilic character of both polysaccharides, and on the pH of the solutions.

Controlled release of growth factors from polymer scaffolds has been an attractive platform to regenerate tissues or organs in many tissue engineering applications. Growth factors and polymers can be adopted in tissue engineering as well as in cell transplantation. Development of polymer scaffolds that release growth factors in response to mechanical stimulation could provide a novel means to guide tissue formation in vivo [219]. Fast-setting calcium phosphate scaffolds with tailored macropore formation rates were developed for bone regeneration [219].

Microcapsules can be used for mammalian cell culture and the controlled release of drugs, vaccines, antibiotics and hormones. To prevent the loss of encapsulated materials, the microcapsules should be coated with another polymer that forms a membrane at the bead surface. The most well-known system is the encapsulation of the alginate beads with poly-L-lysine.

Chenite et al. reported on thermosensitive chitosan gels for encapsulating living cells and therapeutic proteins; they are liquid below room temperature but form monolithic gels at body temperature [220–223].

6.4
Films

Chitin films can be manufactured from DMAc solutions or by other approaches, for example, blend films of beta-chitin (derived from squid pens) and poly(vinyl alcohol) (PVA) were prepared by a solution casting technique from corresponding solutions of beta-chitin and PVA in concentrated formic acid. Upon evaporation of the solvent, the film having 50/50 composition was found to be cloudy [224].

Among polysaccharides, chitosan has peculiar filmogenic properties. For example chitosan films were prepared by wet casting followed by oven drying or infrared (IR) drying. While IR drying was found to be more efficient and uniform, oven drying showed a higher color index. The tensile strength of films dried by IR was less than that of other films, while no differences were observed in their elongation burst strength values. Water vapor and oxygen transmission rate values were slightly reduced in oven-dried and IR-dried films. The X-ray diffraction pattern of oven-dried films showed a different crystallinity nature [225].

The thermal treatment of chitosan salt solutions leads, however, to amide formation: the process of amidation in films consisting of chitosan formate, acetate and propionate proceeds rapidly in the air at 120 °C. The highest degree of amidation (up to 50%) was reached in chitosan formate. The amidation leads to significant strengthening of the films and reduces their solubility in aqueous media [226].

Other studies reached similar results for chitosan citrate and other salts. For instance, Yao et al. exposed to 65 °C a chitosan lactate solution for film formation and then heated it at 85 °C and 5–10 mm Hg for 3 hr, to obtain amide linkages [47, 227]. This is just an extension of existing technology for cotton fabrics to the area of chitosan chemistry: in fact, a number of polycarboxylic acids have been used as cross-linking agents by Yang and Andrews [228].

The surface of chitosan films was modified using acid chloride and acid anhydrides, thus forming amide linkages, and the modification proceeded to the depth at least 1 μm. The surface became more hydrophobic than that of non-modified film when a stearoyl group was attached to the films. The reaction of chitosan films with succinic anhydride or phthalic anhydride, however, produced more hydrophilic films. Selected modified films were subjected to protein adsorption study. The amount of protein adsorbed, determined by bicinchoninic acid assay, related to the types of attached molecules. The improved surface hydrophobicity affected by the stearoyl groups promoted protein adsorption. In contrast, selective adsorption behavior was observed in the case of the chitosan films modified with anhydride derivatives. Lysozyme adsorption was enhanced by hydrogen bonding and charge attraction with the hydrophilic surface, while the amount of albumin adsorbed was decreased possibly due to negative charges that gave rise to repulsion between the modified surface and albumin. It is therefore conceivable to fine-tune surface properties which influence its response to bio-macromolecules by heterogeneous chemical modification [229].

For improved mechanical and water-swelling properties of chitosan films, a series of transparent films were prepared with dialdehyde starch as a cross-linking agent. Fourier transform infrared and X-ray analysis results demonstrated that the formation of Schiff's base disturbed the crystallization of chitosan. The mechanical properties and water-swelling properties of the films were significantly improved. The best values of the tensile strength and breaking elongation were 113.1 MPa and 27.0%, respectively, when the dialdehyde starch content was 5%. All the cross-linked films still retained antimicrobial effects toward S. aureus and E. coli [230].

When the films were treated in either an oxygen plasma environment or under UV/ozone irradiation, the rates of oxidation were faster for the plasma process. Irradiation of chitosan solution showed that UV/ozone induces depolymerization. In both plasma and UV/ozone reactions, the main active component for surface modification was UV irradiation at a wavelength below 360 nm [231].

The intrinsic ionic conductivities of hydrated chitosan membranes investigated using impedance spectroscopy were as high as 10^{-4} S cm^{-1} [232].

For hydroxyethyl chitosan and hydroxypropyl chitosan prepared through the reaction of alkali–chitosan with 2-chloroethanol and propylene epoxide, respectively the values were of the same order, 10^{-4} S cm^{-1} in the hydrated state, while before hydration they were in the 10^{-10} S cm^{-1} range. Moreover, the crystallinity of hydroxyethyl and hydroxypropyl chitosan membranes was remarkably reduced, and their swelling indices increased significantly. However, these modified membranes did not exhibit significant changes in their tensile strength and breaking elongation [233].

Superficially phosphorylated chitosan membranes prepared from the reaction of orthophosphoric acid and urea in DMF, showed ionic conductivity about one order of magnitude larger compared to the unmodified chitosan membranes. The crystallinity of the phosphorylated chitosan membranes and the corresponding swelling indices changed pronouncedly, but these membranes did not lose either tensile strength or thermal stability [234].

The ductility of chitosan can be improved by blending and copolymerizing with poly(ethylene glycol), as manifested by modulus decrease and strain at break increase. For comparable poly(ethylene glycol) composition (ca. 30%), the properties of the solution-cast blend were better than those of the grafted copolymer. Therefore, blending may be a more efficient way to improve ductility of chitosan. Annealing of the blend leads to decreased intermolecular interactions, phase coarsening, and deterioration of its properties [235].

Being eatable, chitosan films are finding use in the preservation of exotic fruits. The application of chitosan coating to browning control and quality maintenance of fresh-cut Chinese water chestnut was investigated. Fresh-cut water chestnut were treated with 0.5, 1 or 2% chitosan solutions, placed into trays over-wrapped with plastic films and then stored at 4 °C. The chitosan coating delayed discoloration associated with reduced activities of phenylalanine ammonia lyase, polyphenol oxidase and peroxidase, as well as lower total phenolic content, and depressed the loss in eating quality associated with higher contents of total soluble solids, titratable acidity and ascorbic acid of water chestnut. The application of a chitosan coating extended the shelf life and maintained quality [236].

Mango fruits *(Mangifera indica)* were kept in carton boxes whose top surface was covered with either chitosan film or with low-density polyethylene (positive control) and stored at room temperature (27 $\pm$ 1 °C at 65% RH). The CO_2 and O_2 levels measured on day 3 were 23–26% and 3–6%, and at the end of the storage period they were 19–21% and 5–6%, respectively. Various quality parameters such as color, chlorophyll, acidity, vitamin C, carotenoid and sugar contents were studied. The fruits stored as such had a shelf-life of 9 ± 1 days, whereas those stored in low-density polyethylene showed off-flavor due to fermentation and fungal growth on the stalk and around the fruits, and were partially spoiled. On the other hand, fruits stored in chitosan-covered boxes showed an extension of shelf-life of up to 18 days and were exempt from any microbial growth and off-flavor. Being biodegradable and ecologically acceptable, chitosan films are a valid alternative to synthetic packaging films in the storage of freshly harvested exotic fruits [237].

Similarly, the effectiveness of chitosan and short hypobaric treatments, alone or in reciprocal combination, to control storage decay of sweet cherries, was investigated over 2 years. In combined treatments, sweet cherries were dipped in 1.0% chitosan and then exposed to 0.50 and 0.25 atm, or sprayed with chitosan (0.1, 0.5, and 1.0%) 7 days before harvest and exposed to 0.50 atm soon after harvest. A combined treatment with 1.0% chitosan and 0.50 atm was the best in controlling decay, showing in the first year, a synergistic effect in the reduction of brown rot and total rots [238].

Chitosan therefore finds applications in agriculture and agroindustries. Coating fruit and vegetables with chitosan has some advantages for the long-term storage of foods, because the film of chitosan acts as an active packaging that allows a gradual release of preservatives, thus inhibiting fungal growth and maintaining the appearance of the fruits for a longer time.

Fortune mandarins and Valencia oranges were coated with a mixture containing chitosan, to investigate its effect on maturity, decay and damage, and for the long term storage of fruit. The fruit was maintained in a damp storage room, ideal conditions for the growth of fungi, and the magnetic resonance imaging technique was used to monitor the process of ripening and decay of the chitosan-coated fruits. The chitosan on the mandarins and oranges produced excellent results in terms of percentage of weight loss and visual appearance [239].

The rate of in vivo biodegradation of subcutaneous implanted films was very high for chitin compared with that for deacetylated chitin. No tissue reaction was found with highly deacetylated chitosans, although they contained abundant primary amino groups [240].

Lopez et al. [241] carried out design and evaluation of chitosan/ethylcellulose mucoadhesive bilayered devices for buccal drug delivery. The mucoadhesive layer was composed of a mixture of drug and chitosan, with or without an anionic cross-linking polymer (polycarbophil, sodium alginate, gellan gum), and the backing layer was made of ethylcellulose. It was evident that hydrophilic chitosan was easily laminated onto hydrophobic ethylcellulose, and that a perfect binding between the mucoadhesive and the backing layers was achieved.

The permeability of the films to paracetamol as a model compound was dependent on film composition and was markedly increased after exposure to pectinolytic enzymes, used to mimic conditions in the colon. Similar formulations, applied as a film coat to tablets, were used with colonic conditions for an increased release rate [242].

The gastrointestinal transit and in vivo drug release behavior of a film-coated tablet formulation was investigated in five healthy human subjects using the technique of gamma scintigraphy. The film coating system consisted of a mixture of pectin, chitosan and hydroxypropylmethyl cellulose in a ratio of 6 : 1 : 0.37 applied to 750 mg cores at a coat weight gain of 9%. The amount of radioactive tracer released from the labeled tablets was minimal when the tablets were in the stomach and the small intestine. There was increased release of radioactivity when the tablets were in the colon due to increased degradation of the film coatings by pectinolytic enzymes resident in the colon. The pectin/chitosan/HPMC film coating system thus acted as a colonic delivery system [243].

Chitosan acetate and lactate salt films have been tested as wound-healing materials. Mechanical, bioadhesive and biological evaluation of the films were carried out. The results were compared to Omiderm®. Chitosan lactate exhibited a lower tensile strength, however, it was more flexible and bioadhesive than chitosan acetate. Chitosan lactate and Omiderm® did not cause any allergic reactions; in contrast, chitosan acetate produced skin irritation clearly due to the anion. Nevertheless, no sign of toxicity was encountered when the extracts of three preparations were administered parenterally [244].

Chitosan membranes were developed to provide an alternative means of evaluating transdermal drug delivery systems. They were prepared by cast-drying method. The effects of concentration of chitosan, sodium tripolyphosphate as well as cross-linking time on flux and lag time was studied using central composite design. A mathematical model was developed to assess the permeation of the drugs through different animal skins. The chitosan membrane at a particular composition simulated the permeation of diclofenac sodium through rat skin [245].

6.5
Fibers

The subject of fiber production from chitin and chitosan has been reviewed considering the interest in the exploitation of chitin as a textile material [246–251]. The production of fibers, however, is a challenging and difficult task.

A recent approach was based on the DMAc/LiCl and N-methyl-2-pyrrolidone/LiCl solvent systems [252, 253] that enabled better dry tensile strengths although they did not provide adequate wet tenacities, due to poor crystallinity and poor consolidation of the fibre. A further problem was the removal of Li from the fibre once it had formed, because the Li ion solvates the chitin amido groups. These fibres could not offer adequate knot resistance, and this jeopardized further studies for surgical applications, in spite of the efforts invested [254].

Nontoxic and noncorrosive solvent systems can be used when chitosan is considered instead of chitin for the manufacture of fibers. Improvements of the fiber quality rely mostly on technical shrewdness; for instance the use of potassium hydrogen phthalate at pH 4–5 imparted better mechanical properties [255].

Cross-linking agents have been proposed for the improvement of chitin fibres in the wet state. Epichlorohydrin is a convenient base-catalysed cross-linker to be used in 0.067 M NaOH (pH 10) at 40 °C. The wet strength of the fibres was considerably improved, whereas cross-linking had negligible effect on the dry fibre properties. Of course, the more extended the chemical modification, the more unpredictable the biochemical characteristics and effects in vivo. Every modified chitin or modified chitosan fibre should be studied in terms of biocompatibility, biodegradability and overall effects on the wounded tissues.

The present trend is to coat other polysaccharide fibers with a film of chitosan or modified chitosan, to impart novel characteristics to the textile. For example, a range of commercial chitosans were utilized for modification of sodium alginate/alginic acid fibres (prepared using a range of different fibre spinning conditions), and levels of chitosan incorporated onto/into base alginate fibres were estimated by elemental analysis. Tensile properties (% elongation and tenacity) of the resultant chitosan/alginate fibres were determined to assess their suitability for potential application in wound dressings. Modification of fibres with unhydrolyzed chitosans generally resulted in a significant reduction of tenacity, i.e., no increase in fibre strength was observed, implying that the unhydrolyzed chitosan is more like a coating rather than penetrating/reinforcing the alginate fibre.

Paradoxically, the reduction of chitosan molecular weight had a positive effect on its ability to penetrate the alginate fibres, not only increasing fibre chitosan content, but also reinforcing fibre structure and thus enhancing tensile properties. Hydrolyzed chitosan/alginate fibres demonstrated an antibacterial effect (in terms of bacterial reduction) with initial use, and had the ability to provide a slow release/leaching of antibacterially active components (presumably hydrolyzed chitosan fragments). The overall aims of maximizing chitosan content (to provide satisfactory antibacterial activity) whilst retaining desirable physical properties (i.e., satisfactory textile processing ability) were therefore achieved, with respect to the fibres having potential application as wound-dressing materials [256].

A novel fiber-reactive chitosan derivative was synthesized in two steps from a chitosan of low molecular weight and low degree of acetylation. First, a water-soluble chitosan derivative, N-[(2-hydroxy-3-trimethylammonium)-

propyl]chitosan chloride (HTCC), was prepared by introducing quaternary ammonium salt groups on the amino groups of chitosan. This derivative was further modified by introducing functional (acrylamidomethyl) groups, which can form covalent bonds with cellulose under alkaline conditions, on the primary alcohol groups of the chitosan backbone. The fiber-reactive chitosan derivative, O-acrylamidomethyl-HTCC showed complete bacterial reduction within 20 min at the concentration of 10 ppm, when contacted with *Staphylococcus aureus* and *Escherichia coli* ($1.5-2.5 \times 10^5$ colony forming units per ml [139]).

This approach to the use of chitosan seems to lead to satisfactory results, as testified by the Japanese production of socks, underwear and similar items.

6.6
Enteric Coatings

The pH of the stomach is acidic and increases in the small and large intestines. This pH variation in different segments of the gastrointestinal tract has been exploited for colon-specific delivery. Coating the drug core with pH-sensitive Eudragit® has been successfully used for colon drug delivery of Asacol®, Asamax® and Salofalc®. Those polymers are insoluble in acidic media, thereby providing protection to the drug core in the stomach and to some extent in the small intestine, but dissolve at a pH 6 or higher, releasing the drug in the colon. However, the pH of the gastrointestinal tract is subject to individual variations depending upon diet, disease, age, sex and the fed/fasted state. Due to the simplicity of the formulation of this device many marketed preparations utilize this approach.

The degradation of the drug carrier in the colon is mainly due to the bacterial flora. For this fermentation, the microflora produces a vast number of enzymes such as beta-glucuronidase, beta-xylosidase, alpha-arabinosidase, beta-galactosidase, nitroreductase, azoreductase, pectinases, deaminase, and urea dehydroxylase. Because of the presence of these enzymes only in the colon, the use of polymers degradable by bacteria for colon-specific drug delivery seems to be a more site-specific approach than other approaches. These polymers shield the drug from the environments of the stomach and the small intestine, and deliver the drug to the colon. On reaching the colon, they undergo assimilation by microorganisms or enzymatic degradation.

A single hydrolase is usually inadequate for the degradation of a carrier, but most hydrolases have unspecific activities, i.e., they split the chains of polymers that are not their typical substrates. For example, chitosan is susceptible to lipases, pectinases, amylases among others [257–260].

Chitosan would not seem suitable as a pH-sensitive polymer: because it is soluble at acidic pH values, and becomes insoluble approximately at pH 6.5. Nevertheless, an enteric coating can protect chitosan from the acidity of the stomach. When the preparation reaches the intestine, the enteric layer dissolves due to high pH and the drug-bearing chitosan core remains exposed to the bacterial enzymes, thus releasing the drug. Chitosan, in fact, is digested largely by secreted rather than cell-associated bacterial enzymes.

Chitooligomers set free by hydrolases become carbon sources for the growth of intestinal bacteria. *Lactobacillus lactis* utilizes the oligomers $(GlcNAc)_{1-6}$, the monomer and dimer being bifidogenic substances [261–263].

7
Administration Routes

7.1
Oral Route

Chitosan for oral administration to humans is generally recognized as safe. In vitro, chitosan has been reported to bind bile acids The role of the accompanying anion is important: for instance chitosan orotate salt has enhanced capacity for bile acids [11, 264–267].

Within the small intestine, bile-acid binding interferes with micelle formation. Nauss et al. [268] reported that, in vitro, chitosan binds bile acid micelles in toto, with consequent reduced assimilation of all micelle components, i.e., bile acids, cholesterol, monoglycerides and fatty acids. Moreover, in vitro, chitosan inhibits pancreatic lipase activity [269]. Dissolved chitosan may further depress the activity of lipases by acting as an alternative substrate [270].

Because chitosan forms insoluble salts upon reaction with bile acids (cholate, taurocholate etc.), one might expect that the chitosan bile salts further collect lipids by hydrophobic interaction [191, 271, 272]. Bile acids, once sequestered by chitosan, are no longer available as emulsifiers for the correct formation of the emulsion necessary for the digestion of lipids. It is known that the pancreatic lipases require a certain dimension of the oil droplets in the emulsion in order to hydrolyze triglycerides, thus, when the bile acids become scarce, their capacity as emulsifiers leads to inadequate emulsions and then to scarce hydrolysis of triglycerides. Ample information on digestive lipases supports these views [273–275]. The presence of bile salts in one case activate the bile-salt-dependent lipases, and in the other case provide the emulsion necessary to the pancreatic lipases for enzymatic activity. When the bile salt availability decreases due to chitosan ingestion, lipases become unable to work adequately and assimilation of lipids by the organism decreases sharply.

Chitosan is also effective in lowering serum cholesterol concentration and hypertension, in subjects with a restricted diet. Of course, the quality and the chemical form of chitosan should be adequate to the scope.

7.2
Nasal Route

The nasal tissue is highly vascularized and provides efficient systemic absorption. Compared with oral or subcutaneous administration, nasal administration enhances bioavailability and improves safety and efficacy. Chitosan enhances the absorption of proteins and peptide drugs across nasal and intestinal epithelia. Gogev et al. demonstrated that the soluble formulation of glycol chitosan has potential usefulness as an intranasal adjuvant for recombinant viral vector vaccines in cattle [276].

The interaction of chitosan with the cell membrane results in general in the structural reorganization of tight junction-associated proteins leading to enhanced drug permeability [277, 278]. A nasal formulation with improved absorption of macromolecules and water-soluble drugs is still a challenge because of the short retention time in the nasal cavity due to the efficient physiological clearance mechanisms [279]. When the control was cleared rapidly with a half-life of 21 min, half-lives of 41 and 84 min were recorded for chitosan solution and microspheres respectively. Illum et al. [280] discussed influenza, pertussis, diphtheria and the vaccination based on chitosan powder and solution.

Hamman et al. [281, 282] tested five trimethyl chitosans with different degrees of quaternization as nasal delivery systems: the degree of quaternization had a major role in the absorption enhancement of this polymer across the nasal epithelia in a neutral environment.

The immunogenicity of conjugate vaccine against group C meningococci was enhanced when delivered intranasally with chitosan derivatives. The use of trimethyl chitosan as a delivery system allowed the reduction of each of the components for the induction of antibody and bactericidal responses to group C meningococci conjugate vaccine delivered intranasally at titers similar to or higher than those induced by parenteral immunization. Data of this kind are useful for the design of efficacious mucosal vaccines and their safety [283].

Aspden et al. [284] investigated the effect on mucociliary transport velocity of 0.25% solutions of five different types of chitosan, with varying molecular weights and degrees of deacetylation. The five types of chitosan tested were shown to have no toxic effect on the frog palate clearance mechanism. The cilia beat frequency in guinea pigs after nasal administration of chitosan solution was also studied for 28 days and none of the chitosans used showed any effect on the cilia beat frequency suggesting no harm in using various types of chitosan for nasal delivery applications [279]. A review describes the pharmaceutical perspectives [285]; another one discusses the intranasal vaccination against plague, tetanus, and diphtheria [286].

7.3
Ophthalmic Preparations

The chitosan-based systems for improving the retention and biodistribution of drugs applied topically onto the eye have been reviewed. Besides its low toxicity and good ocular tolerance, chitosan exhibits favorable biological behavior, such as bioadhesion- and permeability-enhancing properties, and also interesting physicochemical characteristics, which makes it a unique material for the design of ocular drug delivery vehicles. The review summarizes the techniques for the production of chitosan gels, chitosan-coated colloidal systems and chitosan nanoparticles, and describes their mechanism of action upon contact with the ocular mucosa. The results reported until now have provided evidence of the potential of chitosan gels for enhancing and prolonging the retention of drugs on the eye surface. On the other hand, chitosan-based colloidal systems were found to work as transmucosal drug carriers, either facilitating the transport of drugs to the inner eye (chitosan-coated colloidal systems containing indometacin) or their accumulation into the corneal/conjunctival epithelia (chitosan nanoparticles containing cyclosporin) [287].

The ability of chitosan as an ocular delivery device has been evaluated by comparing chitosan formulations with the commercial collyrium Tobrex®. Chitosan gel has been tested for the increased precorneal drug residence times, expecting it to slow down the drug elimination by lachrymal flow by increasing solution viscosity and by interacting with the negative charges of the mucus. At least a three-fold increase of the corneal residence time was achieved in the presence of chitosan when compared to control [288].

The ability of chitosan hydrochloride to enhance the transcorneal permeability of the drug has been demonstrated [289]. Polyethylene oxide (PEO) was used as a base material to which ofloxacin-containing chitosan microspheres prepared by spray-drying were added and powder compressed resulting in circular inserts (6 mm).

Co-administration of ofloxacin and chitosan in eyedrops increased the bioavailability of the antibiotic [290]. Trimethyl chitosan was more effective because of its solubility (plain chitosan precipitates at the pH of the tear fluid). On the other hand, N-carboxymethyl chitosan did not enhance the corneal permeability; nevertheless it mediated zero-order ofloxacin absorption, leading to a time-constant effective antibiotic concentration [291]. Also N,O-carboxymethyl chitosan is suitable as an excipient in ophthalmic formulations to improve the retention and the bioavailability of drugs such as pilocarpine, timolol maleate, neomycin sulfate, and ephedrine. Most of the drugs are sensitive to pH, and the composition should have an acidic pH, to enhance stability of the drug. The delivery should be made through an anion exchange resin that adjusts the pH at around 7 [292]. Chitosan solutions do not lend themselves to thermal sterilization. A chitosan suspension, however, can be sterilized at 121 °C for 15 min and then treated with sterile HCl solution. This preparation was found to be particularly useful for pharmaceutical and ophthalmic compositions [293].

The semi-interpenetrating polymeric networks obtained by the radical-induced polymerization of N-isopropylacrylamide in the presence of chitosan using tetraethyleneglycoldiacrylate as the cross-linker were used as controlled release vehicles for pilocarpine hydrochloride [294].

7.4
Wound-Dressing Materials

Peculiar characteristics of chitins and chitosans are hemostatic action, anti-inflammatory effect, biodegradability, biocompatibility, besides antimicrobial activity, retention of growth factors, release of glucosamine and N-acetylglucosamine monomers and oligomers, and stimulation of cellular activities [11, 12, 295–297].

The hemostatic bandage Syvek Patch has been introduced in the recent past for the control of bleeding at vascular access sites in interventional cardiology and radiology procedures. Syvek fibers contain fully acetylated, high-molecular-weight chitin in a crystalline, three-dimensional beta structure array, and is isolated from the large-scale culture and processing of a marine diatom. Two new products, the Clo-Sur and ChitoSeal, both based on chitosan, have recently become available also as patch hemostats. Structural, chemical, and biological comparisons of Syvek and chitosan reveal a number of important differences. Superior performances of Syvek in promoting hemostasis were claimed [298].

Partially deacetylated chitin is degraded by enzymes such as lysozyme, N-acetyl-D-glucosaminidase and lipases [297, 299]. It is not excluded that NO also plays a role in a chemoenzymatic degradation process. Histological findings on wounded skin dressed with this chitin, indicated that collagen fibers 7 days post-operation were fine in the wounds and more mature than in the control: their arrangement was similar to that in normal skin. The tensile strength was clearly superior compared to controls. At 7 days, the wounds were completely re-epithelialized, granulation tissues were almost replaced with fibrosis and hair follicles were almost healed [69].

Evidence has been collected that significant portions of chitin-based dressings are depolymerized, and that oligomers are further hydrolyzed to N-acetylglucosamine, a common aminosugar in the body, which enters the innate metabolic pathway to be incorporated into glycoproteins [300]. The biodegradability is independent of the degree of acetylation of the chitins. This implies that lysozyme is not the only enzyme involved in the degradation, or that the substrate is modified in situ [301].

Chitooligomers act as templates for hyaluronan synthesis. Hyaluronan has been shown to promote cell motility, adhesion and proliferation, and to have important roles in morphogenesis, inflammation, and wound repair [302–304].

The in vitro biocompatibility of wound dressings in regards to fibroblasts has been assessed and compared with three commercial wound dressings made of collagen, alginate and gelatin. Methylpyrrolidinone chitosan and collagen were found to be the most compatible materials [305, 306].

The use of wheat germ agglutinin, a lectin, for modifying chitosan and enhancing the cell–biomaterial interaction was examined by Wang et al. [307]. The percentage of living fibroblast cells on the surfaces of tissue culture polystyrene control, wheat germ agglutinin-modified chitosan, and plain chitosan films increased to 99%, 99%, and 85%, respectively, after seeding for 48 h. DNA staining revealed that a portion of fibroblasts cultivated on chitosan films were undergoing apoptosis. In contrast, fibroblasts growing on wheat germ agglutinin-modified chitosan film surfaces did not show any indication of apoptosis. The number of fibroblast cells was the highest on the wheat germ agglutinin-modified chitosan surfaces, followed by the polystyrene and unmodified chitosan surfaces. Wheat germ agglutinin and other lectin molecules enhance the cell–biomaterial interaction via oligosaccharide-mediated cell adhesion and improve cell adhesion and proliferation, the two key issues in tissue engineering.

Chitosan has been associated with other biopolymers and with synthetic polymer dispersions to produce wound dressings. Biosynthetic wound dressings composed of a spongy sheet of chitosan and collagen, laminated with a gentamicyn sulphate-impregnated polyurethane membrane, have been produced and clinically tested with good results.

Chitin- and chitosan-based biomaterials are endowed with biochemical significance not encountered in cellulose, starch and other polysaccharides. They can be considered as a primer on which the normal tissue architecture is organized. Key factors in the rebuilding of physiologically effective tissues exerted by chitosans are an enhanced vascularization and a continuous supply of chitooligomers to the wound that stimulate correct deposition, assembly and orientation of collagen fibrils, and are incorporated into the extracellular matrix components. The chitooligomers are released by lysozyme and N-acetyl-β-D-glucosaminidase [308–310].

Macrophages from laboratory animals are activated by chemically modified chitin [311, 312]. Chitosan activates macrophages for tumoricidal activity and for the production of interleukin-1. Oligochitosan had an in vitro stimulatory effect on the release of tumor necrosis factor-alpha and interleukin-1 beta in macrophages. Moreover, oligochitosan could be taken up by macrophages: Scatchard analysis of 2-aminoacridone-oligochitosan in macrophages indicated that its internalization was mediated by a specific receptor on macrophage membrane with $K_d = 2.1 \times 10^{-5}$ M. Oligochitosan internalization is mediated by a macrophage lectin receptor like with mannose specificity [313].

Chitosan also shows immunopotentiating activity: the mechanism involves, at least in part, the production of interferon-gamma and the stimulatory effect on nitric oxide production. Chitosan-based dressings also modulate peroxide production [312, 314–318].

Chitin/chitosan and their oligomers and monomers exert effects on the migration of mouse peritoneal macrophages and the rat macrophage cell line. In direct migratory assay using the blind well chamber method, the migratory activity of the mouse peritoneal macrophages was enhanced significantly by chitin oligomers and chitosan oligomers, but reduced by chitin, chitosan and glucosamine. The migratory activity of rat macrophage was increased significantly by chitin, chitosan, chitin oligomers and glucosamine [61, 319].

When examining this kind of data, one realizes how difficult is to make general statements on the effects of chitosans on the cells; another example is the antimicrobial effect of chitosans that varies with molecular size, degree of substitution and other parameters depending on the particular microorganism considered.

Overproduction of prostaglandin E-2 and proinflammatory cytokines contributes to immunosuppression and cytotoxicity during wound healing. Chitosans (M_w = 50, 150, or 300 g/mol) significantly inhibit the overproduction of prostaglandin-E2 as well as cyclooxygenase-2 protein expression and activity accompanied by attenuation of pro-inflammatory cytokines production such as tumor necrosis factor-alpha and interleukin-1beta, but increase of the anti-inflammatory cytokine, IL-10. The beneficial effect of chitosan on wound healing may be associated, at least partly, with the inhibition of PGE(2) production by suppressing cyclooxygenase-2 induction and activity as well as attenuation of the proinflammatory/anti-inflammatory cytokines ratio in activated macrophages [59, 320].

Chitin/chitosan items administered intravenously to mice become bound to macrophage plasma membrane mannose/glucose receptors that mediate their interiorization. Fibroblast growth factor-2 stimulates the proliferation of fibroblasts and capillary endothelial cells, thus promoting correct wound repair via angiogenesis; however, this factor has short half-life in vivo and high diffusibility. A way to keep it on the desired position is to use a photo-cross-linkable chitosan in the form of a viscous solution that becomes an insoluble gel upon UV irradiation, and has better adhesion than fibrin glue. High M_w chitosan was reacted with *p*-azidobenzoic acid and lactobionic acid and, upon exposure to UV light for 10 sec, was cross-linked by reaction of azido and amino groups. The factors retained in the hydrogel remained biologically active and were released upon in vivo degradation of the chitosan. Wound contraction was induced and wound closure was accelerated in healing-impaired diabetic mice compared to controls. Histological examination demonstrated advanced granulation tissue formation, capillary formation and epithelialization in wounds treated with fibroblast growth factor-2 (FGF-2)-bearing chitosan hydrogels in mice. The application of a chitosan hydrogel as an occlusive dressing was recommended because chitosan would be more effective in this form to protect and contract the wound in a suitably moist healing environment [321, 322].

Because the consistent delivery of basic fibroblast growth factor (bFGF) is problematic, the stability of bFGF incorporated into a chitosan film for providing sustained release of bFGF has been studied on wound healing in genetically diabetic mice. Hydroxypropyl chitosan facilitated wound repair, and the bFGF incorporated into chitosan film was stable. The granulation tissue was more abundant in the bFGF-chitosan group, but proliferation of fibroblasts and increase of capillaries were mostly due to the hydroxypropyl chitosan itself [323].

7.4.1
Wound Materials and Specific Uses

Several chitin-based wound dressings are available commercially since a few years, mainly in oriental countries: they are mentioned in Table 2, see also Fig. 7. However, they are not available in Europe, which could be due to marketing choices.

Table 2 Chitin-based dressings

Tegasorb®. 3M.
The dressing contains chitosan particles that swell absorbing exudate and producing a soft gel. A layer of waterproof Tegaderm®film dressing covers the hydrocolloid.
For leg ulcers, sacral wounds, chronic wounds.
Reportedly superior to Comfeel and Granuflex.

Beschitin®. Unitika.
Non-woven material manufactured from chitin.
Available in Japan since 1982.

Chitipack S®. Eisai Co.
Sponge-like chitin from squid.
Favours early granulation, no scar formation.
For traumatic wounds, surgical tissue defects.

Chitipack P®. Eisai Co.
Dispersed and swollen chitin supported on polyethylenetherephthalate.
For the treatment of large skin defects. Favours early granulation.
Suitable for defects difficult to suture.

Chitopack C®. Eisai Co.
Cotton-like chitosan obtained by spinning chitosan acetate salt into a coagulating bath of ethylene glycol, ice and NaOH; fibers washed with water and methanol.
Complete reconstruction of body tissue, rebuilding of normal subcutaneous tissue, and regular regeneration of skin.

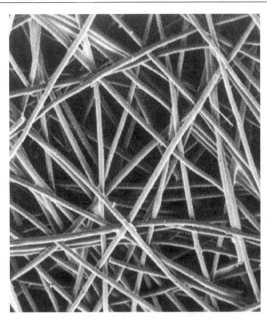

Fig. 7 Beschitin©, a non-woven wound-dressing material manufactured from chitin by Unitika, Tokyo, Japan

Most of the experimental results on chitosans have been obtained with freeze-dried modified chitosans, such as methylpyrrolidinone chitosan, and reacetylated chitosan, as mentioned below.

Chitin can be associated chemically to other polymeric substances. For instance, the 1 : 4 chitin-poly(acrylic acid) combination can be used as a wound dressing. Acrylic acid was linked to chitin, acting as the grafting site on the chain that was further polymerized to form a network. The degree of swelling of chitin–poly(acrylic acid) films was 30–60 times of their original weights depending upon the monomer feed content. The chitin–poly(acrylic acid) film with 1 : 4 weight ratio of chitin:acrylic acid, possessed optimal physical properties. The cytocompatibility of the film was investigated with a cell line of L929 mouse fibroblasts. The L929 cells proliferated and attached well onto the film [324].

Associations can be of physical nature too. Chitosan blends with hydrophilic polymers including polyvinylalcohol, polyethyleneoxide and polyvinylpyrrolidone, were investigated as candidates for oral gingival delivery systems. Chitosan blends were superior to chitosan alone in terms of comfort, ease of processing, film quality, and flexibility [325].

The efficacy of chitosan in the treatment of leg ulcers stems from its anti-inflammatory action and stimulation of epithelialization. Chitosans stimulate the granulation process and the epidermis formation, thus accelerating healing, even in the elderly suffering from chronic venous insufficiency, the prolonged treatment with 5-methylpyrrolidinone chitosan led to complete healing in a number of difficult cases [326].

Reacetylated chitin gels were used to treat leg and decubitus ulcers in paraplegic subjects [309]. Selected preparative conditions permitted to obtain a self-sustaining gel useful for this use. The treatment periods were 63–182 days and complete healing was obtained.

7.4.2
Nerve Regeneration

Chitosan was used as nerve conduit having the ability to facilitate nerve cell attachment, differentiation and growth. Advances have been made by Suzuki et al. [327] with crab tendon chitin tubes harvested from *Macrocheira kaempferi*, demineralized, partially deacetylated and molded in triangular cross-section. Laminin, a glycoprotein that promotes neural cell attachment, differentiation and neurite outgrowth was chemically bound to the inner tubular surface. The thus prepared tendon tube is hydrophilic and water permeable, but its pore size does not permit cells to pass through. With the polysaccharide aligned structure preserved, it has remarkable mechanical resistance and its three flat surfaces favor cell attachment. An acute inflammatory reaction occurred for 2 weeks but the overall tissue response was mild. Encapsulation of the tube was not observed. Although cationic groups on a material surface evoke a severe inflammation reaction upon implantation, the surface of the chitosan tube is not recognized as a foreign material. Laminin disappeared when the tube disintegrated 12 weeks post-operation, leaving the newly formed neural tissue that bridged the surgically amputated peripheral nerve in rats.

Poly-L-lysine-blended chitosan, collagen-blended chitosan and albumin-blended chitosan were also considered, with collagen control material. Culture of PC12 cells and fetal mouse cerebral cortex cells on these biomaterials was used to evaluate their nerve cell affinity. The composite materials, had significantly improved nerve cell affinity compared to chitosan. Poly-L-lysine-blended chitosan exhibited the best nerve cell affinity and was a better material in promoting neurite outgrowth than collagen [328].

Bridge grafting into the sciatic nerve of rats was carried out using chitosan tubes having either a circular or triangular cross-section, as well as triangular tubes combined with laminin. The mechanical strength of triangular tubes was found to be higher than circular tubes. In fact, the results of histological findings, as well as mechanical properties, suggested that a triangular tube shape with a hydroxyapatite coating benefits nerve regeneration. The effect of laminin peptides to enhance the growth of regenerating axons has been found comparable with intact laminin-1 [329, 330].

7.4.3
Cartilage and Bone Regeneration

The reconstruction of the periodontal tissue with chitosan was a prelude to the discovery of the osteoinductive properties of chitosan [331]. Surgical wounds from wisdom tooth avulsions were treated with freeze-dried methylpyrrolidinone chitosan that promoted bone regeneration. Methylpyrrolidinone chitosan was useful in apicoectomy as well. None of the patients reported adverse effects over three years of observation [332].

The chitosan osteoconductive properties make it attractive as a bioactive coating to improve osteointegration of orthopaedic and craniofacial implant devices. Coatings made of 91% de-acetylated chitosan were chemically bonded to titanium coupons via silane-glutaraldehyde. The chitosan coatings were chemically bonded to the titanium substrate and the bond strengths (1.5–1.8 MPa) were not affected by gas sterilization. The gas-sterilized coatings exhibited little dissolution over 8 weeks in cell-culture solution, and the attachment and growth of osteoblast cells was greater on the chitosan-coated samples than on the uncoated titanium [333].

Bone defects surgically produced in sheep and rabbit models, have been treated with freeze dried methylpyrrolidinone chitosan [334–336]. In view of improving bone tissue reconstitution with chitosan associated with calcium phosphate. Microscopic and histological analyses showed the presence of an osteogenic reaction moving from the rim of the surgical lesion toward the center. In control lesions, dense fibrous tissue, without the characteristic histoarchitecture of bone was observed.

Dicarboxymethyl chitosan and 6-oxychitin sodium salt, applied to femoral surgical defects for 3 weeks produced a good histoarchitectural order in the newly formed bone tissue. The spongious trabecular architecture was restored in the defect site. The association of the chitin derivatives with the osteoblasts seemed to be the best biomaterial in terms of bone tissue recovery [128].

7.4.4
Bone Substitutes

The so-called bioactive ceramics have been attractive because they spontaneously bond to living bone, however, they are much more brittle and much less flexible than natural bone. Previous studies reported that the essential condition for ceramics to show bioactivity is formation of a biologically active carbonate-containing apatite on their surfaces after exposure to the body fluid [337]. Calcium sulfate was also used [338].

The chitosan-bonded hydroxyapatite bone-filling paste was made as follows. Chitosan (0.5 g) was dissolved in malic acid (0.5 g) solution made with saline, and a chitosan film was formed by mixing this solution with hydroxyapatite powder (2 g), followed by neutralization with 5% sodium polyphosphate. To help cells and blood vessels to penetrate this material, the tensile strength and elongation were optimized [166, 167, 339].

Dicarboxymethyl chitosan mixed with calcium acetate and disodium hydrogen phosphate in suitable ratios yielded clear solutions from which an amorphous material (ca. 50% inorganic) was isolated [176] and was used for the treatment of bone lesions in experimental surgery and in dentistry. Bone tissue regeneration was promoted in sheep, leading to complete healing of otherwise non-healing surgical defects. Radiographic evidence of bone regeneration was observed in human patients undergoing apicoectomies and avulsions. The dicarboxymethyl chitosan-calcium phosphate chelate favored osteogenesis while promoting bone mineralization.

The in situ precipitation route towards obtaining composites of polymer and calcium phosphate is similar to the strategy employed in naturally occurring biocomposites and may prove a viable method for the synthesis of bone substitutes [340].

Chitosan membranes can also be superficially modified, for instance with 3-isocyanatopropyl triethoxysilane. Silanol groups and calcium salt acted as nucleation sites and accelerator, respectively, for the formation of apatite crystals: therefore, this chitosan membrane is a bioactive guided bone-regeneration material thanks to its apatite-forming ability [341].

Chitosan microsphere-encapsulated human growth hormone seems to be quite effective in early bone consolidation in distraction osteogenesis [342].

The restorative effects of orally administered N-acetyl-D-glucosamine and glucuronic acid on experimentally produced cartilaginous injuries have been studied in rabbits. The massive proliferation of matured cartilaginous tissue was observed, surrounded by proliferating chondroblast cells. In the regenerated tissue, matured cartilage substrate was also observed, whilst in the control and glucose groups the injured parts were covered by fibrous connective tissues [343].

Chitosan freeze-dried fleeces support chondrocyte attachment and synthesis of extracellular matrix [344]. Chitosan was used to assist the spontaneous tissue repair of the meniscus [345]. The repair of the cartilage and the prevention of its degradation in osteoarthritis is, however, possible with the association of glucosamine sulfate salt and chondroitine sulfate, the latter being particularly effective [346].

Exogenous glucosamine can be transported into the cell and acted on by a kinase which phosphorylates it at the 6 position, enabling it to enter the pathway of glycosaminoglycan synthesis. Exogenous glucosamine is readily incorporated into hyaluronan by cultured fibroblasts. It seems that the glucosamine 6-phosphate synthesis is the rate-controlling step [347].

Enhanced availability of exogenous glucosamine is beneficial for the synthesis of hyaluronan because endogenous glucosamine is insufficient to achieve the high concentration of UDP-N-acetyl glucosamine necessary to optimize synthase activity. Thus, the administration of glucosamine (as well as chitosan) in general is beneficial for wound healing.

The biological significance of chitosan biomaterials in the human body depends largely on the actions that certain hydrolases exert on them. The resulting chitooligomers stimulate various cells, while the released monomers are phosphorylated and incorporated into hyaluronan, keratan sulphate and chondroitin sulphate, components of the intracellular matrix and connective tissue [348].

8
Conclusions

A number of important papers could not be cited in this chapter, due to the length limitations and the specific target of the chapter. For example, the antimicrobial activity of chitosans [349], the chitinolytic enzymes, the preparation of cosmetics, and the occurrence of chitin in fungi [350] are some of the subjects not dealt with specifically here, notwithstanding their importance.

Many years ago chitin was seen as a scarcely appealing natural polymer due to the variety of origins, isolation treatments and impurities, but the works of several analytical chemists and the endeavor of an increasing number of companies have qualified chitins and chitosans for sophisticated applications in the biosciences. Chemistry today offers a range of finely characterized modified chitosans for use in the biomedical sciences. Moreover, surprising roles of these polysaccharides and related enzymes are being "unexpectedly" discovered [351].

References

1. Richards AG (1951) The Integument of Arthropods. Univ Minnesota Press, Minneapolis
2. Rudall KM (1963) Adv Insect Physiol 1:257
3. Jeuniaux C (1963) Chitine et chitinolyse. Masson, Paris
4. Hepburn HR (1976) The Insect Integument. Elsevier, Amsterdam
5. Neville AC (1975) Biology of the Arthropod Cuticle. Springer-Verlag, Berlin Heidelberg New York
6. Muzzarelli RAA (1977) Chitin. Pergamon: Oxford
7. Muzzarelli RAA, Pariser ER (eds) (1978) Proc 1st Int Conf Chitin Chitosan. MIT, Cambridge USA
8. Neville AC (1993) Biology of Fibrous Composites: development beyond the cell membrane. Cambridge Univ Press, New York
9. Gooday GW (1990) In: Kuhn PJ, Trinci APJ, Jung MJ, Goosey MW, Copping LG (eds) Biochemistry of Cell Walls and Membranes in Fungi. Springer-Verlag, Germany p 61
10. ASTM Standard Guide F 2103–01, 2001

11. Jollès P, Muzzarelli RAA (eds) (1999) Chitin and Chitinases. Birkhauser, Basel
12. Muzzarelli RAA, Jeuniaux C, Gooday GW (eds) (1986) Chitin in Nature and Technology. Plenum, New York NY
13. Muzzarelli RAA (1985) In: Aspinall GO (ed) The Polysaccharides. Academic, New York p 3
14. Horst MN, Walker AN (1993) In: Muzzarelli RAA (ed) Chitin Enzymology. Atec, Italy 109
15. Vincent JFV, Wegst UGK (2004) Arthropod Struct Devel 33:187
16. Tharanathan RN, Kittur FS (2003) Chitin, the undisputed biomolecule of great potential. Crit Reviews Food Science Nutr 43:61
17. Dutta PK, Dutta J, Tripathi VS (2004) J Sci Ind Res 63:20
18. Mitsumata T, Suemitsu Y, Fujii K, Fujii T, Taniguchi T, Koyama K (2003) Polymer 44:7103
19. Shinn AH, Smith TJ (2003) Biopharm Int Appl Technol Biopharm Devel 16:34
20. Wang SB, Chen AZ, Weng LJ, Chen MY, Xie XL (2004) Macromol Biosci 4:27
21. Lyu SY, Kwon YJ, Joo HJ, Park WB (2004) Arch Pharm Res 27:118
22. Taqieddin E, Amiji M (2004) Biomaterials 25:1937
23. Xiao CB, Zhang JH, Zhang ZJ, Zhang LN (2003) J Appl Polym Sci 90:1991
24. Surini S, Akiyama H, Morishita M, Takayama K, Nagai T (2003) STP Pharma Sci 13:265
25. Zanchetta P, Lagarde N, Guezennec (2003) J Calcified Tissue Int 73:232
26. Lee SB, Lee YM, Song KW, Park MH (2003) J Appl Polym Sci 90:925
27. Drury JL, Mooney DJ (2003) Biomaterials 24:4337
28. Surini S, Akiyama H, Morishita M, Nagai T, Takayama K (2003) J Control Rel 90:291
29. Kim SJ, Shin SR, Lee KB, Park YD, Kim SI (2004) J Appl Polym Sci 91:2908
30. Pang FJ, He CJ, Wang QR (2003) J Appl Polym Sci 90:3430
31. Twu YK, Huang HI, Chang SY, Wang SL (2003) Carbohydr Polym 54:425
32. Li HB, Du YM, Xu YM (2004) J Appl Polym Sci 91:2642
33. Ponce-Jimenez MD, Toral FALD, Fornue ED (2002) J Am Inst Conserv 41:243
34. Shimizu Y, Dohmyou M, Yoshikawa M, Takagishi T (2004) Textile Res J 74:34
35. Yamamoto H, Ohkawa K, Nakamura E, Miyamoto K, Komai T (2003) Bull Chem Soc Jap 76:2053
36. Ohkawa K, Kitagawa T, Yamamoto H (2004) Macromol Mater Eng 289:33
37. Murata Y, Hirai D, Kofuji K, Miyamoto E, Kawashima S (2004) Biol Pharm Bull 27:440
38. Ehrenfreund-Kleinman T, Domb AJ, Golenser J (2003) J Bioact Compat Polym 18:323
39. Magnin D, Dumitriu S, Chornet E (2003) J Bioact Compat Polym 18:355
40. Magnin D, Lefebvre J, Chornet E, Dumitriu S (2004) Carbohydr Polym 55:437
41. Masci G, Husu I, Murtas S, Piozzi A, Crescenzi V (2003) Macromol Biosci 3:455
42. Kurita K, Akao H, Yang J, Shimojoh M (2003) Biomacromolecules 4:1264
43. Lai SN, Locci E, Saba G, Husu I, Masci G, Crescenzi V, Lai A (2003) J Polym Sci Part A – Polym Chem 41:3123
44. Guo XL, Yang KS, Hyun JY, Kim WS, Lee DH, Min KE, Park LS, Seo KH, Kim YI, Cho CS, Kang IK (2003) J Biomater Sci – Polym Ed 14:551
45. Zhang C, Ping QN, Ding Y, Cheng Y, Shen J (2004) J Appl Polym Sci 91:659
46. Kato Y, Onishi H, Machida Y (2004) Biomaterials 25:907
47. Yao FL, Chen W, Wang H, Liu HF, Yao KD, Sun PC, Lin H (2003) Polymer 44:6435
48. Mi FL, Shyu SS, Lin YM, Wu YB, Peng CK, Tsai YH (2003) Biomaterials 24:5023
49. Suyatma NE, Copinet A, Tighzert L, Coma V (2004) J Polym Environ 12:1
50. Kim YH, Choi JW, Lee EY (2003) Polymer Korea 27:405
51. Qin CQ, Du YM, Zhang ZQ, Liu Y, Xiao L, Shi XW (2003) J Appl Polym Sci 90:505

52. Xu YM, Du YM, Huamg RH, Gao LP (2003) Biomaterials 24:5015
53. Curti E, De Britto D, Campana SP (2003) Macromol Biosci 3:571
54. Snyman D, Govender T, Kotze AF (2003) Pharmazie 58:705
55. Ruan YH, Dong YM, Wu MS, Zeng MQ, Wang SJ (2003) Chem Res Chin Univ 19:512
56. Lee KY, Mooney DJ (2003) Adv Control Drug Del: Sci Technol Prod 846:73
57. De Campos AM, Sanchez A, Gref R, Calvo P, Alonso MJ (2003) Eur J Pharm Sci 20:73
58. Alonso MJ, Sanchez A (2003) J Pharm Pharmacol 55:1451
59. Chou TC, Fu E, Shen EC (2003) Biochem Biophys Comm 308:403
60. Aerts JM, Hollak C, Boot R, Groener A (2003) Macrophage Ther Target 158:193
61. Okamoto Y, Inoue A, Miyatake K, Ogihara K, Shigemasa Y, Minami S (2003) Macromol Biosci 3:587
62. Fujita M, Ishihara M, Simizi M, Obara K, Ishizuka T, Saito Y, Yura H, Morimoto Y, Takase B, Matsui T, Kikucki M, Maehara T (2004) Biomaterials 25:699
63. Harding SE (2003) Biochem Soc Trans 31:1036
64. Kurita K (2001) Prog Polym Sci 26:1921
65. Revel JF, Marchessault RH (1993) Int J Biol Macromol 15:329
66. Noishiki Y, Takami H, Nishiyama Y, Wada M, Okada S, Kuga S (2003) Biomacromolecules 4:896
67. Terbojevich M, Carraro C, Cosani A, Marsano E (1988) Carbohydr Res 180:73
68. Bianchi E, Ciferri A, Conio G, Marsano E (1990) Mol Cryst Liq Cryst Lett 7:111
69. Cho YW, Cho YN, Chung SH, Yoo G, Ko SK (1999) Biomaterials 20:2139
70. Lu SJ, Song XF, Cao DY, Chen YP, Yao KD (2004) J Appl Polym Sci 91:3497
71. Muzzarelli RAA, Barontini G, Rocchetti R (1978) Biotechnol Bioeng 20:87
72. Muzzarelli RAA, Tanfani F, Scarpini GF (1980) Biotechnol Bioeng 22:885
73. Roy I, Sardar M, Gupta MN (2003) Biochem Eng J 16:329
74. Sashiwa H, Fujishima S, Yamano N, Kawasaki N, Nakayama A, Muraki E, Sukwattanasinitt M, Pichyangkura R, Aiba S (2003) Carbohydr Polym 51:391
75. Sashiwa H, Fujishima S, Yamano N, Kawasaki N, Nakayama A, Muraki E, Hiraga K, Oda K, Aiba S (2002) Carbohydr Res 337:761
76. Fu JY, Wu SM, Chang CT, Sung HY (2003) J Agric Food Chem 51:1042
77. Roy I, Gupta MN (2003) Biocatal Biotransform 21:297
78. Rogovina SZ, Akopova TA (1994) Polym Sci Ser A 36:487
79. Park JH, Cho YW, Chung H, Know IC, Jeong SY (2003) Biomacromolecules 4:1087
80. Holappa J, Nevalainen T, Savolainen J, Soininen P, Elomaa M, Safin R, Suvanto S, Pakkanen T, Masson M, Loftsson T, Jarvinen T (2004) Macromolecules 37:2784
81. Yamaguchi I, Itoh S, Suzuki M, Sakane M, Osaka A, Tanaka J (2003) Biomaterials 24:2031
82. Yui T, Kobayashi H, Kitamura S, Imada K (1994) Biopolymers 34:203
83. Lavertu M, Xia Z, Serreqi AN, Berrada M, Rodrigues A, Wang D, Buhmann MD, Gupta A (2003) J Pharm Biomed Anal 32:1149
84. Muzzarelli RAA, Rocchetti R, Stanic V, Weckx M (1997) In: Muzzarelli RAA, Peter MG (eds.) Chitin Handbook. Atec, Italy, p. 109
85. Reed A, Northcote A (1981) Anal Biochem 116:53
86. Muzzarelli RAA (1998) Anal Biochem 260:255
87. Valentin R, Molvinger K, Quignard F, Brunel D (2003) New J Chem 27:1690
88. Steckel H, Mindermann-Nogly F (2004) Eur J Pharm Biopharm 57:107
89. Kurita K (2001) Prog Polym Sci 26:1921
90. Tsuchida E, Abe K (1982) Adv Polym Sci 45:83
91. Kubota N, Kikuchi Y (1999) Dumitriu S (ed) Polysaccharides. Dekker: New York 595
92. Kim SJ, Yoon SY, Lee KB, Park YD, Kim SI (2003) Solid State Ionics 164:199

93. Mansouri S, Lavigne P, Corsi K, Benderdour M, Beaumont E, Fernandes JC (2004) Eur J Pharm Biopharm 57:1
94. Anwer K, Rhee BG, Mendiratta SK (2003) Critical Rev Ther Drug Carrier Sys 20:249
95. Park IK, Kim TH, Kim SI, Akaike T, Cho CS (2003) J Dispersion Sci Technol 24:489
96. Kim TH, Ihm JE, Choi YJ, Nah JW, Cho CS (2003) J Control Rel 93:389
97. Ishii T, Okahata Y, Sato T (2001) Biochim Biophys Acta – Biomembranes 1514:51
98. Ko JA, Park HJ, Hwang SJ, Park JB, Lee JS (2002) Int J Pharm 249:165
99. Pan Y, Li Yj, Zhao H-Y, Zheng J-M, Xu H, Wei G, Hao J-S, Cui F-D (2002) Int J Pharm 249:139
100. Calvo P, Remunan-Lopez C, Vila-Jato JL, Alonso MJ (1997) J Appl Polym Sci 63:125
101. Calvo P, Remunan-Lopez C, Vila-Jato JL, Alonso MJ (1997) Pharm Res 14:1431
102. Fernández-Urrusuno R, Romani D, Calvo P, Vila-Jato JL, Alonso MJ (1999) STP Pharma Sci 9:429
103. Janes KA, Fresneau MP, Marazuela A, Fabra A, Alonso MJ (2001) J Control Rel 73:255
104. De Campos AM, Sánchez A, Alonso MJ (2001) Int J Pharm 224:159
105. Janes KA, Calvo P, Alonso M (2001) J Adv Drug Del Rev 47:83
106. Sakiyama T, Takata H, Kikuchi M, Nakanishi K (1999) J Appl Polym Sci 73:2227
107. Jiang H, Su W, Brant M, De Rosa ME, Bunning TJ (1999) J Polym Sci Part B – Polym Phys 37:769
108. Andersson M, Lofroth JE (2003) Int J Pharm 257:305
109. Lin WC, Liu TY, Yang MC (2004) Biomaterials 25:1947
110. Torrado S, Prada P, de la Torre PM, Torrado S (2004) Biomaterials 25:917
111. Fingerman N, Nagabhushanan R, Thompson MF, Inoue K (1998) Recent Advances in Marine Biotechnology, vol. 2. Oxford & IBH: New Delhi
112. Varma AJ, Deshpande SV, Kennedy JF (2004) Carbohydr Polym 55:77
113. Muzzarelli RAA, Tanfani F, Emanuelli M, Bolognini L (1985) Biotechnol Bioeng 27:1115
114. Guan HM, Cheng XS (2004) Polym Adv Technol 15:89
115. Kim HS (2004) J Ind Eng Chem 10:273
116. Muzzarelli RAA, Tubertini O (1969) Talanta 16:1571
117. Gotoh T, Matsushima K, Kikuchi KI (2004) Chemosph 55:135
118. Hirano S, Yoshida S, Takabuchi N (1993) Carbohydr Polym 22:137
119. Komai T, Kaifu K, Matsushita M, Koshino I, Kon T (1982) Polymer J 14:803
120. Tokura S, Nishi N, Nishimura S, Ikeuchi Y (1984) In: Zikakis JP (ed.) Chitin Chitosan and related Enzymes. Academic, New York, p. 303
121. Kakizaki M, Shoji T, Tsutsumi A, Hideshima T (1986) In: Muzzarelli RAA, Jeuniaux C, Gooday GW (eds) Chitin in Nature and Technology. Plenum, New York, p. 398
122. Komai T, Kaifu K, Matsushita M, Koshino I, Kon T (1986) In: Muzzarelli RAA, Jeuniaux C, Gooday GW (eds) Chitin in Nature and Technology. Plenum, New York, p. 497
123. Urbanczyk GW (1997) In: Goosen MFA (ed) Applications of Chitin and Chitosan. Technomic, Basel, p. 281
124. Szosland L (1997) In: Muzzarelli RAA (ed) Chitin Handbook. Atec, Italy, p. 53
125. Muzzarelli C, Francescangeli O, Tosi G, Muzzarelli RAA (2004) Carbohydr Polym 56:137
126. Muzzarelli C, Stanic V, Gobbi L, Tosi G, Muzzarelli RAA (2004) Carbohydr Polym 57:73
127. Muzzarelli RAA, Muzzarelli C, Cosani A, Terbojevich M (1999) Carbohydr Polym 39:361

128. Mattioli-Belmonte M, Nicoli-Aldini N, DeBenedittis A, Sgarbi G, Amati S, Fini M, Biagini G, Muzzarelli RAA (1999) Carbohydr Polym 40:23
129. Bragdt PL, van Bekkum H, Besemer AC (2004) Topics Catalysis 27:49
130. Toffey A, Samaranayake G, Frazier CE, Glasser WG (1996) J Appl Polym Sci 60:75
131. Bernkop-Schnurch A, Hornof M, Guggi D (2004) Eur J Pharm Biopharm 57:9
132. Muzzarelli RAA, Delben F, Tomasetti M (1994) Agro-Food Ind High Tech 5:35
133. Rinaudo M, Le Dung P, Gey C Milas M (1992) Int J Biol Macromol 14:122
134. Lapasin R, Stefancic S, Delben F (1998) Agro-Food Ind High Tech 7:12
135. Zhou J, Elson C, Lee TDG (2004) Surgery 135:307
136. Polnok A, Borchard G, Verhoef JC, Sarisuta N, Junginger HE (2004) Eur J Pharm Biopharm 57:77
137. van der Merwe SM, Verhoef JC, Kotze AF, Junginger HE (2004) Eur J Pharm Biopharm 57:85
138. Polnok A, Verhoef JC, Borchard G, Sarisuta N, Junginger HE (2004) Int J Pharm 269:303
139. Lim SH, Hudson SM (2004) Carbohydr Res 339:313
140. Yoshida H, Nishihara H, Kataoka T (1994) Biotechnol Bioeng 43:1087
141. Grobouillot AR, Champagne CP, Darling GD, Poncelet D, Neufeld RJ (1993) Biotechnol Bioeng 42:1157
142. Hall LD, Yalpani M (1980) J Chem Soc Chem Comm 1980:1153
143. Yalpani M, Hall LD (1984) Macromolecules 17:272
144. Morimoto M, Saimoto H, Shigemasa Y (2002) Trends Glycosci Glycotechn 14:205
145. Morimoto M, Saimoto H, Usui H, Okamoto Y, Minami S, Shigemasa Y (2001) Biomacromolecules 2:1133
146. Li X, Tsushima Y, Morimoto M, Saimoto H, Okamoto Y, Minami S, Shigemasa Y (2000) Polym Adv Technol 11:176
147. Li X, Morimoto M, Sashiwa H, Saimoto H, Okamoto Y, Minami S, Shigemasa Y (1999) Polym Adv Technol 10:455
148. Kato Y, Onishi H, Machida Y (2001) J Control Release 70:295
149. Kato Y, Onishi H, Machida Y (2001) Int J Pharm 226:93
150. Kato Y, Onishi H, Machida Y (2004) Biomaterials 25:907
151. Kato Y, Onishi H, Machida Y (2000) Biomaterials 21:1579
152. Onishi H, Machida Y (1999) Biomaterials 20:175
153. Kato Y, Onishi H, Machida Y (2003) Macromol Res 11:382
154. Son YJ, Jang JS, Cho YW, Chung H, Park RW, Kwon IC, Kim IS, Park JY, Seo SB, Park CR, Jeong SY (2003) J Control Release 91:135
155. Park IK, Yang J, Jeong HJ, Bom HS, Harada I, Akaike T, Kim S, Cho CS (2003) Biomaterials 24:2331
156. Park IK, Kim TH, Park YH, Shin BA, Choi ES, Chowdhury EH, Akaike T, Cho CS (2001) J Control Release 76:349
157. Park IK, Ihm JE, Park YH, Choi YJ, Kim SI, Kim WJ, Akaike T, Cho CS (2003) J Control Release 86:349
158. Roy R, Tropper DF, Romanowska A, Letellier M, Cousineau L, Meunier SJ, Boratynski J (1991) Glycoconjugate J 8:75
159. Sashiwa H, Makimura Y, Shigemasa Y, Roy R (2000) Chem Comm 909
160. Sashiwa H, Thompson JM, Das SK, Shigemasa Y, Tripathy S, Roy R (2000) Biomacromolecules 1:303
161. Sashiwa H, Shigemasa Y, Roy R (2001) Bull Chem Soc Jpn 74:937
162. Falini G, Fermani S (2004) Tissue Eng 10:1
163. Xu HHK, Takagi S, Quinn JB, Chow LC (2004) J Biomed Mater Res Part A 68A:725

164. Baran ET, Tuzlakoglu K, Salgado AJ, Reis RL (2004) J Mater Sci Mater Med 15:161
165. Zhang Y, Zhang MQ (2004) J Mater Sci Mater Med 15:255
166. Maruyama M, Ito M (1996) J Biomed Mater Res 32:527
167. Ito M, Hidaka Y (1997) In: Muzzarelli RAA, Peter MG (eds.) Chitin Handbook. Atec, Italy, p. 373
168. Granja PL, Silva AIN, Borges JP, Barrias CC, Amaral IF (2004) Bioceramics 16:573
169. Yin YJ, Zhao F, Song XF, Yao KD, Lu WW, Leong JC (2000) J Appl Polym Sci 77:2929
170. Yamaguchi I, Tokuchi K, Fukuzaki H, Koyama Y, Takakuda K, Monma H, Tanaka T (2001) J Biomed Mater Res 55:20
171. Yokogawa Y, Paz Reyes J, Mucalo MR, Toriyama M, Kawamoto Y, Suzuki T, Nishizawa K, Nagata F, Kamayama T (1997) J Mater Sci Mater Med 8:407
172. Leroux L, Hatim Z, Freche M, Lacout JL (1999) Bone 5:31
173. Viala S, Freche M, Lacout JL (1996) Carbohydr Polym 29:197
174. Viala S, Freche M, Lacout JL (1998) Ann Chim Sci Mat 23:69
175. Wan ACA, Khor E, Hastings GW (1998) J Biomed Mater Res 41:541
176. Muzzarelli RAA, Ramos V, Stanic V, Dubini B, Mattioli-Belmonte M, Tosi G, Giardino R (1998) Carbohydr Polym 36:267
177. Zhang Y, Zhang MQ (2001) J Biomed Mater Res 55:304
178. Falini C, Fermani S, Ripamonti A (2001) J Inorg Biochem 84:255
179. Muzzarelli RAA, Ilari P, Xia W, Pinotti M, Tomasetti M (1994) Carbohydr Polym 24:294
180. Kumar G, Smith PJ, Payne GF (1999) Biotechnol Bioeng 63:154
181. Govar CJ, Chen TH, Liu NC, Harris MT, Payne GF (2003) Biocatal Polym Sci 840:231
182. Payne GF, Chaubal MV, Barbari TA (1996) Polymer 37:4643
183. Muzzarelli RAA, Tanfani F (1985) Carbohydr Polym 5:297
184. Chen TH, Small DA, Wu LQ, Rubloff GW, Ghodssi R, Vazquez-Duhalt R, Bentley WE, Payne GF (2003) Langmuir 19:9382
185. Wu LQ, Chen TH, Wallace KK, Vazquez-Duhalt R, Payne GF (2001) Biotechnol Bioeng 76:325
186. Vachoud L, Chen TH, Payne GF, Vazquez-Duhalt R (2001) Enzyme Microbial Technol 29:380
187. Chen TH, Embree HD, Brown EM, Taylor MM, Payne GF (2003) Biomaterials 24:2831
188. Aberg CM, Chen TH, Payne GF (2002) J Polym Environ 10:77
189. Muzzarelli RAA, Cucchiara M, Muzzarelli C (2002) J Appl Cosmetol 20:201
190. Ohya T, Cai R, Nishizawa H, Hara K, Ouchi T (2000) STP Pharma Sci 10:77
191. Lee KY, Kim JH, Kwon LC, Jeong SY (2000) Colloid Polym Sci 278:1216
192. Kim YH, Gihm SH, Park CR (2001) Bioconjugate Chem 12:932
193. Chang J, Zhongguo SY (1996) J Biomed Eng 15:102
194. Sinha VR, Singla AK, Wadhawan S, Kaushik R, Kumria R, Bansal K, Dhawan S (2004) Int J Pharm 274:1
195. Rege PR, Shukla DJ, Block LH (1999) Int J Pharm 181:49
196. He P, Davis SS, Illum L (1999) Int J Pharm 187:53
197. Davis SS (1999) J Microencapsul 16:343
198. Rege PR, Garmise RJ, Block LH (2003) Int J Pharm 252:41
199. Rege PR, Garmise RJ, Block LH (2003) Int J Pharm 252:53
200. De la Torre PM, Enobakhare Y, Torrado G, Torrado S (2003) Biomaterials 24:1499
201. Huang YC, Yeh MK, Chiang CH (2002) Int J Pharm 242:239
202. Huang YC, Yeh MK, Cheng SN, Chiang CH (2003) J Microencapsul 20:459
203. Shi XY, Tan TW (2002) Biomaterials 23:4469
204. Sabnis SS, Rege PR, Block LH (1997) Pharm Dev Technol 2:243

205. Giunchedi P, Juliano C, Gavini E, Cossu M, Sorrenti M (2002) Eur J Pharm Biopharm 53:233
206. Muzzarelli C, Tosi G, Francescangeli O, Muzzarelli RAA (2003) Carbohydr Res 338:2247
207. Jollès P (1996) (ed.) Lysozymes Model Enzymes. Birkhauser, Basel
208. Kumar MNVR, Bakowsky U, Lehr CM (2004) Biomaterials 25:1771
209. Genta I, Giunchedi P, Pavanetto F, Conti B, Perugini P, Conte U (1997) In Muzzarelli RAA, Peter MG (eds.) Chitin Handbook. Atec, Italy, p. 391
210. Muzzarelli RAA, Ilari P, Xia W, Pinotti M, Tomasetti M (1994) Carbohydr Polym 24:294
211. Kumbar SG, Kulkarni AR, Aminabhavi TM (2002) J Microencapsul 19:173
212. Kriznar B, Mateovic T, Bogataj M, Mrhar (2003) A Chem Pharm Bull 51:359
213. Zhang L, Guo J, Peng XH, Jin Y (2004) J Appl Polym Sci 92:878
214. Seol YJ, Lee YJ, Park YJ, Lee YM, Rhyu IC, Lee SJ, Han SB, Chung CP (2004) Biotechnol Lett 26:1037
215. Qu X, Wirsen A, Albertsson AC (1999) J Appl Polym Sci 74:3193
216. Qu X, Wirsen A, Albertsson AC (1999) J Appl Polym Sci 74:3186
217. Muzzarelli RAA, Barontini G, Rocchetti R (1976) Biotechnol Bioeng 18:1445
218. Crescenzi V, Imbriaco D, Velasquez C, Dentini M, Ciferri A (1995) Macromol Chem Phys 196:2873
219. Xu HHK, Takagi S, Quinn JB, Chow LC (2004) J Biomed Mater Res 68A:725
220. Chenite A, Chaput C, Wang D, Combes C, Buschmann MD, Hoemann CD, Leroux JC, Atkinson BL, Binette F (2000) Biomaterials 21:2155
221. Ruel-Garièpy E, Chenite A, Chaput C, Guirguis S, Leroux JC (2000) Int J Pharm 203:89
222. Chenite A, Buschmann M, Wang D, Chaput C, Kandani N (2001) Carbohydr Polym 46:39
223. Jarry C, Chaput C, Chenite A, Renaud MA, Buschmann M, Leroux JC (2001) J Biomed Mater Res 58:127
224. Peesan M, Rujiravanit R, Supaphol P (2003) Polymer Testing 22:381
225. Srinivasa PC, Ramesh MN, Kumar KR, Tharanathan RN (2004) J Food Eng 63:79
226. Zotkin MA, Vikhoreva GA, Kechek'yan As (2004) Polymers Science Series B 46:39
227. Yao FL, Chen W, Wang H, Liu HF, Yao KD, Sun PC, Lin H (2003) Polymer 44:6435
228. Yang CQ, Andrews BAK (1991) J Appl Poly Sci 48:1609
229. Tangpasuthadol V, Pongchaisirikul N, Hoven VP (2003) Carbohydr Res 338:937
230. Tang R, Du YM, Fan LH (2003) J Polym Sci Part B Polym Phys 41:993
231. Matienzo LJ, Winnacker SK (2002) Macromol Mater Eng 287:871
232. Wan Y, Creber KAM, Peppley B, Bui VT (2003) Polymer 44:1057
233. Wan Y, Creber KAM, Peppley B, Bui VT (2004) J Polym Sci Part B Polym Phys 42:1379
234. Wan Y, Creber KAM, Peppley B, Bui VT (2003) Macromol Chem Phys 204:850
235. Kolhe P, Kannan RM (2003) Biomacromolecules 4:173
236. Pen LT, Jiang YM (2003) Food Sci Technol 36:359
237. Srinivasa PC, Baskaran R, Ramesh MN, Prashanth KVH, Tharanathan RN (2002) Eur Food Res Technol 215:504
238. Romanazzi G, Nigro F, Ippolito A (2003) Postharvest Biol Technol 29:73
239. Galed G, Fernandez-Valle ME, Martinez A, Heras A (2004) Magn Reson Imaging 22:127
240. Tomihata K, Ikada Y (1997) Biomaterials 16:567
241. Lopez CR, Portero A, Vila-Jato JL, Alonso MJ (1998) J Control Rel 55:143

242. Macleod GS, Collett JH, Fell JT (1999) J Control Rel 58:303
243. Ofori-Kwakye K, Fell JT, Sharma HL, Smith AM (2004) Int J Pharm 270:307
244. Khan TA, Peh KK, Ch'ng HS (2000) J Pharm Pharm Sci 3:303
245. Dureja H, Tiwary AK, Gupta S (2001) Int J Pharm 213:193
246. Kumar MNVR (1999) Bull Mater Sci 22:905
247. Rathke TD, Hudson SM (1994) JMS Rev Macromol Chem Phys C34:375
248. Le Y, Anand SC, Horrocks R Indian (1997) J Fibre Textile Res 22:337
249. Hudson SM, Smith G (1998) In: Kaplan DL (ed.) Biopolymers from Renewable Resources. Springer, New York, p. 96
250. Lim SH, Hudson SM (2003) J Macromol Sci Part C Polym Rev C43:223
251. Agboh OC, Qin Y (1997) Polym Adv Technol 8:355
252. Rutherford FA, Austin PR (1977) In: Muzzarelli RAA, Pariser ER (eds) Proc First Int Conf Chitin Chitosan. MIT, Cambridge, USA, p. 182
253. Austin PR (1977) US Patent 4,059,457
254. Unitika Co Ltd (1984) Japanese Patent 59,068,347
255. Knaul JZ, Hudson SM, Creber KAM (1999) J Appl Polym Sci 72:1721
256. Knill CJ, Kennedy JF, Mistry J, Miraftab M, Smart G, Groocock MR, Williams HJ (2004) Carbohydr Polym 55:76
257. Orienti I, Cerchiara T, Luppi B, Bigucci F, Zuccari G, Zecchi V (2002) Int J Pharm 238:51
258. Ofori-Kwakye K, Fell JT (2001) Int J Pharm 226:139
259. Muzzarelli RAA, Xia W, Tomasetti M, Ilari P (1995) Enzyme Microb Technol 17:541
260. Roy I, Sardar M, Gupta MN (2003) Enzyme Microb Technol 32:582
261. Chen H-C, Chang C-C, Mau W-J, Yen L-S (2002) FEMS Microbiol Lett 209:53
262. Modler HW, Mekellar RC, Yaguchi M (1990) Can Inst Food Sci Technol 23:29
263. Lee AW, Park YS, Jung JS, Shin WS (2002) Anaerobe 8:319
264. Harrison TA (2002) Agro-Food Ind Hi-Tech 13:8
265. Lee JK, Kim SU, Kim JH (1999) Biosci Biotechnol Biochem 63:833
266. Sugano M, Fujikawa T, Hiratsuji Y, Nakashima K, Fukuda N, Hasegawa (1980) Am J Clin Nutr 33:787
267. Murata Y, Kojima N, Kawashima S (2003) Biol Pharm Bull 26:687
268. Nauss JL, Thompson JL, Nagyvary JJ (1983) Lipids 18:714
269. Han LK, Kimura Y, Okuda H (1999) Int J Obesity 23:174
270. Muzzarelli RAA (2000) (ed.) Chitosan per Os. Atec, Italy
271. Faldt P, Bergenstahl B, Claesson PM (1993) Colloid Surf A: Physicochem Eng Asp 71:187
272. Kim CH, Chun HJ (1999) Polym Bull 42:25
273. Weng W, Ling L, van Bennekum AM, Potter SH, Harrison EH, Blaner WS, Breslow JL Fisher EA (1999) Biochem 38:4143
274. Lombardo D (2001) Biochim Biophys Acta 1533:1
275. Miled N, Canaan S, Dupuis L, Roussel A, Riviere M, Carriere F, de Caro A, Cambillau C, Verger R (2000) Biochimie 82:973
276. Gogev S, de Fays K, Versali MF, Gautier S, Thiry E (2004) Vaccine 22:1946
277. Schipper NG, Olsson S, Hoogstraate JA, Deboer AG, Varum KM, Artursson P (1997) Pharm Res 14:923
278. Artursson P, Lindmark T, Davis SS, Illum L (1994) Pharm Res 11:1358
279. Soane RJ, Frier M, Perkins AC, Jones NS, Davis SS, Illum L (1999) Int J Pharm 178:55
280. Illum L, Jabbal-Gill I, Hinchcliffe M, Fisher AN, Davis SS (2001) Adv Drug Del Rev 51:81
281. Hamman JH, Stander Ma, Kotzé AF (2002) Int J Pharm 232:235

282. Hamman JH, Stander M, Junginger HE, Kotzé AF (2000) STP Pharma Sci 10:35
283. Baudner BC, Morandi M, Giuliani MM, Verhoef JC, Junginger HE, Costantino P, Rappuoli R, Del Giudice G (2004) J Infect Dis 189:828
284. Aspden TJ, Adler J, Davis SS, Skaugrud O, Illum L (1995) Int J Pharm 122:69
285. Kumar MNVR, Muzzarelli RAA, Muzzarelli C, Sashiwa H, Domb AJ (2004) Chem Rev 104:6017
286. Alpar HO, Eyles JE, Williamson ED, Somavarapu S (2001) Adv Drug Del Rev 51:173
287. Alonso MJ, Sanchez A (2003) J Pharm Pharmacol 55:1451
288. Felt O, Furrer P, Mayer JM, Plazonnet B, Buri P, Gurny R (1999) Int J Pharm 180:185
289. Di Colo GD, Zambito Y, Burgalassi S, Serafini A, Saettone MF (2002) Int J Pharm 248:115
290. Felt O, Gurny R, Buri P, Baeyens V (2001) AAPS Pharm Sci 3:34
291. Di Colo G, Zambito Y, Burgalassi S, Serafini A, Saettone MF (2003) Proc 30th Control Rel Soc 32
292. Reed KW, Yen SF (1995) WO Patent 97/06782
293. Yen SF, Sou M (1997) WO Patent 97/07139
294. Verestiuc L, Ivanov C, Barbu E, Tsibouklis J (2004) Int J Pharm 269:185
295. Muzzarelli RAA (1998) In: Dumitriu S (ed.) Polysaccharides. Marcel Dekker, New York, p. 569
296. Muzzarelli RAA (1996) In: Salamone JC (ed.) The Polymeric Materials Encyclopedia. CRC, Boca Raton
297. Tokura S, Azuma I (1992) (eds.) Chitin Derivatives in Life Sciences. Japan Soc Chitin, Sapporo
298. Fischer TH, Connolly R, Thatte HS, Schwaitzberg SS (2004) Microscopy Res Technol 63:168
299. Sashiwa H, Saito K, Saimoto H, Minami S, Okamoto Y, Matsuhashi A, Shigemasa Y (1993) In: Muzzarelli RAA (ed.) Chitin Enzymology. Atec, Italy, p. 177
300. Berthod F, Saintigny G, Chretien F, Hayek D, Collombel C, Damour O (1994) Clin Mater 15:259
301. Muzzarelli RAA (1993) Carbohydr Polym 20:7
302. Varki A (1996) Proc Natl Acad Sci USA 93:4523
303. Bakkers J, Semino CE, Stroband H, Kune JW, Robbins PW (1997) Proc Natl Acad Sci USA 94:7982
304. Laurent TC (1998) (ed.) The chemistry biology and medical applications of hyaluronan and its derivatives. Portland, London
305. Muzzarelli RAA (1992) Carbohydr Polym 19:29
306. Berscht PC, Nies B, Liebendorfer A, Kreuter J (1995) J Mater Sci Mat Med 6:201
307. Wang YC, Kao SH, Hsieh HJ (2003) Biomacromolecules 4:224
308. Shigemasa Y, Minami S (1996) Biotechnol Genetic Eng Rev 13:383
309. Muzzarelli RAA, Biagini G (1993) In: Muzzarelli RAA (ed) Chitin Enzymology. Atec, Italy, p. 187
310. Muzzarelli RAA, Mattioli-Belmonte M, Pugnaloni A, Biagini G (1999) In: Jolles P, Muzzarelli RAA (eds.) Chitin and Chitinases. Birkhauser, Basel, p. 251
311. Nishimura K, Nishimura S, Nishi N, Saiki I, Tokura S, Azuma I (1984) Vaccine 2:93
312. Nishimura S, Nishi N, Tokura S (1986) Carbohydr Res 156:286
313. Feng J, Zhao LH, Yu QQ (2004) Biochem Biophys Res Comm 317:414
314. Shibata Y, Foster L Metzger WY, Myrvik QN (1997) Infect Immun 65:1734
315. Peluso G, Petillo O, Ranieri M, Santin M, Ambrosio L, Calabro D, Avallone B, Balsamo G (1994) Biomaterials 15:1215
316. Marletta MA (1993) J Biol Chem 268:12231

317. Wu KK (1995) Adv Pharmacol 33:179
318. Chung LY, Schmidt RJ, Hamlyn PF, Sagar BF (1998) J Biomed Mat Res 39:300
319. Howling GI, Dettmar PW, Goddard PA, Hampson FC, Dornish M, Wood EJ (2001) Biomaterials 22:2959
320. Ueno H, Nakamura F, Murakami M, Okumura M, Kadosawa T, Fujinaga T (2001) Biomaterials 22:2125
321. Ono K, Saito Y, Yura H, Ishikawa K, Kurita A, Akaike T, Ishihara M (2000) J Biomed Mater Res 49:289
322. Obara K, Ishihara M, Ishizuka T, Fujita M, Ozeki Y, Maehara T, Saito Y, Yura H, Matsui T, Hattori H, Kikuchi M, Kurita A (2003) Biomaterials 24:3437
323. Mizuno K, Yamamura K, Yano K, Osada T, Saeki S, Takimoto N, Sakurai T, Nimura Y (2003) J Biomed Mater Res Part A 64A:177
324. Tanodekaew S, Prasitsilp M, Swasdison S, Thavornyutikarn B, Pothsree T, Pateepasen R (2004) Biomaterials 25:1453
325. Khoo CGL, Frantzich S, Rosinski A, Sjostrom M, Hoogstraate J (2003) Eur J Pharm Biopharm 55:47
326. Mancini S, Muzzarelli RAA (1998) In: Mancini S (ed.) Trattato di Flebologia. Utet, Torino
327. Suzuki M, Ito S, Yamaguchi I, Takakuda K, Kobayashi H, Shinomiya K, Tanaka J (2003) J Neurosci Res 72:646
328. Cheng MY, Cao WL, Cao Y, Gong YD, Zhao NM, Zhang XF (2003) J Biomater Sci Polym Ed 14:1155
329. Itoh S, Suzuki M, Yamaguchi I, Takakuda K, Kobayashi H, Shinomiya K, Tanaka J (2003) Artif Organs 27:1079
330. Itoh S, Yamaguchi I, Suzuki M, Ichinose S, Takakuda K, Kobayashi H, Shinomiya K, Tanaka J (2003) Brain Res 993:111
331. Muzzarelli RAA, Biagini G, Pugnaloni A, Filippini O, Baldassarre V, Castaldini C, Rizzoli C (1989) Biomaterials 10:598
332. Muzzarelli RAA, Biagini G, Bellardini M, Simonelli C, Castaldini C, Fratto G (1993) Biomaterials 14:39
333. Bumgardner JD, Wiser R, Gerard PD, Bergin P, Chestnutt B, Marini M, Ramsey V, Elder SH, Gilbert JA (2003) J Biomater Sci Polym Ed 14:423
334. Mattioli-Belmonte M, Biagini G, Muzzarelli RAA, Castaldini C, Gandolfi MG, Krajewski A, Ravaglioli A, Fini M, Giardino R (1995) J Bioact Compat Polym 10:249
335. Muzzarelli RAA, Mattioli-Belmonte M, Tietz C, Biagini R, Ferioli G, Brunelli MA, Fini M, Giardino R, Ilari P, Biagini G (1994) Biomaterials 15:1075
336. Muzzarelli RAA, Zucchini C, Ilari P, Pugnaloni A, Mattioli-Belmonte M, Biagini G, Castaldini C (1993) Biomaterials 14:925
337. Cho BC, Kim TG, Yang JD (2005) J Craniofacial Surg 16:213
338. Miyazaki T, Ohtsuki C, Tanihara M, Ashizuka M (2004) Bioceramics 16:545
339. Ito M (1991) Biomaterials 12:41
340. Muzzarelli RAA, Muzzarelli C (2002) J Inorg Biochem 92:89
341. Rhee SF (2004) Bioceramics 16:501
342. Cho BC, Kim JY, Lee JH, Chung HY, Park JW, Roh KH, Kim GU, Kwon IC, Jang KH, Lee DS Park NW, Kim IS (2004) J Craniofacial Surg 15:299
343. Tamai Y, Miyatake K, Okamoto Y, Takamori Y, Sakamoto K, Minami S (2003) Carbohydr Polym 54:251
344. Nettles DI, Elder SH, Gilbert JA (2002) Tissue Eng 8:1009
345. Muzzarelli RAA, Bicchiega V, Biagini G, Pugnaloni A, Rizzoli R (1992) J Bioact Compat Polym 7:130

346. Matheson AJ, Perry CM (2003) Drugs Aging 20:1041
347. Karzel K, Domenjoz R (1971) Pharmacology 5:337
348. Muzzarelli RAA (1997) Cell Biol Life Sci 53:131
349. Rabea EI, Badawy MET, Stevens CV, Smagghe G, Steurbaut W (2003) Biomacro-
     molecules 4:1457
350. Rast DM, Baumgartner D, Mayer C, Hollenstein GO (2003) Phytochemistry 64:339
351. Couzin J (2004) Science 304:1577

Adv Polym Sci (2005) 186: 211–254
DOI 10.1007/b136821
© Springer-Verlag Berlin Heidelberg 2005
Published online: 1 September 2005

Analysis of Polysaccharides by Ultracentrifugation. Size, Conformation and Interactions in Solution

Stephen E. Harding

NCMH Physical Biochemistry Laboratory, University of Nottingham,
School of Biosciences, Sutton Bonington LE12 5RD, UK
Steve.Harding@nottingham.ac.uk

Abstract The launch of the XL-I analytical ultracentrifuge by Beckman Instruments (Palo Alto, USA) in 1996 and subsequent development of penetrating software for the analysis of the optical records digitally recorded in this new generation instrument has made some exciting possibilities for the analysis of polysaccharides in a solution environment. We review these developments and investigate the application of the technique to the study of polysaccharide polydispersity, molecular weight analysis, conformation and flexibility analysis, the study of associative interactions, including large complex formation phenomena, and to the measurement of charge and charge-shielding phenomena.

Keywords Complex · Flexibility · Molecular weight distribution · Mucoadhesion · Sedimentation coefficient distribution

1
Introduction

The enhancement of our knowledge of the biophysical properties of polysaccharides in dilute and concentrated solution form is important for both our understanding of their function on a "macroscopic" scale in nature, and how they can be used and manipulated by Industry and Medicine [1]. In the *Food Industry* for example such biophysical information—namely molecular size, shape, polydispersity and interaction properties—alongside chemical composition data, is important for the proper understanding of polysaccharide behaviour as gelling agents, thickeners and as phase separation media. This information used alongside economic considerations can help us choose the right polysaccharide with the right properties for a particular product. The same is true for the *Pharmaceutical and Healthcare Industries*: this may be for example in the choice of the right polysaccharide as an encapsulation agent, as a mucoadhesive or slow release formulation in drug delivery, as a viscosity enhancer in toothpaste, eye and nose drops or as a film former in hair products. In the *Printing Industry*, knowledge of biophysical properties can help us choose the right polysaccharide for producing good clear print quality, and in the *Oil Industry* it facilitates the choice of the most suitable polysaccharide as a lubricant and water immobilisation agent. In *Medicine* the choice of the right dextran as a blood plasma substitute is strongly dictated by molecular weight considerations.

There are several techniques now at our disposal for obtaining this fundamental biophysical information about solutions of polysaccharides (Table 1 [2–7]), but as is well known these substances are by no means easy to characterise. These difficulties arise from their highly expanded nature in solution, their polydispersity, (not only with respect to their molecular weight but also for many with respect to composition), the large variety of conformation and in many cases their high charge and in some their ability to stick together [1, 8]. All of these features can complicate considerably the interpretation of solution data.

Table 1 Hydrodynamic techniques for characterising the biophysical properties of polysaccharides in solution

Technique	Biophysical information	Comment	Ref.
Analytical ultracentrifugation (AUC)	Molecular weight M, molecular weight distribution, $g(M)$ vs. M, polydispersity, sedimentation coefficient, s, and distribution, $g(s)$ vs. s: solution conformation and flexibility. Interaction & complex formation phenomena. Molecular charge	No columns or membranes required	[2]
Static (multi-angle laser) light scattering (SLS or MALLs)	Molecular weight, radius of gyration, R_g: solution conformation and flexibility.	Solutions need to be clear of supramolecular aggregates. R_g more sensitive to conformation than s.	[3]
Size exclusion chromatography coupled to static light scattering (SEC-MALLs)	Molecular weight, molecular weight distribution. Polydispersity. Radius of gyration and distribution: solution conformation and flexibility.	Method of choice for molecular weight work.	[4]
Dynamic light scattering (DLS)	Translational diffusion coefficient, hydrodynamic or Stokes radius R_h: branching information (when R_h used with R_g)	Fixed 90° angle instruments not suitable for polysaccharides, Multi-angle instrument necessary.	[3]
Small angle X-ray scattering	R_g, chain contour length, L: solution conformation and flexibility	If M is also known, can provide mass per unit length M_L	[5]
Viscometry	Intrinsic viscosity $[\eta]$: conformation and flexibility.	$[\eta]$, like R_g, much more sensitive than s. Accurate concentration estimates required.	[6]
Surface plasmon resonance (SPR)	Interaction strength (molar dissociation constant, K_d).	One member of the reacting pair needs to be immobilised onto an "inert" surface. Not suitable for self-association analysis.	[7]

In this article we choose to focus on analytical ultracentrifugation as a primary method for obtaining fundamental physical information about polysaccharides in solution mainly because of its diversity, its absolute nature and inherent fractionation ability without the need for separation columns or membranes [2]. We will consider the salient features behind the modern delivery of the various types of sedimentation analysis, the instrumentation involved and analysis software, and we will indicate where combination with the other solution techniques listed in Table 1 is appropriate, as well as with imaging techniques like electron microscopy and atomic force microscopy. We refrain from unnecessary mathematical detail which would otherwise render the description opaque to the non-specialist reader but instead provide the source references where such detail can be found. The main polysaccharides we will use as examples are two neutral groups of polysaccharide, namely starch (amylose and amylopectin), galactomannans (guar gum, tara gum and locust bean gum) and some polyelectrolytes, namely the polyanionic heparin and the polycationic chitosan group (of various degrees of acetylation).

2
Types of Sedimentation Analysis

What sort of information can we obtain about polysaccharides in solution from sedimentation analysis in analytical ultracentrifuge? The information obtained depends on the type of sedimentation technique we apply—all possible with the same instrumentation [9–12]. *Sedimentation velocity* can provide us with information on the physical homogeneity of a sample, conformation and flexibility information—in some cases to surprising detail. It can also provide us with interaction information if, for example we assay for what is called "co-sedimentation" phenomena (i.e. species sedimenting at the same rate as another). At lower rotor speeds, *sedimentation equilibrium* can provide absolute molecular weight and molecular weight distribution information, supplementing the more popular technique (for polysaccharides) of size exclusion chromatography coupled to multi-angle laser light scattering: SEC/MALLs [13]. Another recent postulated use for the analytical ultracentrifuge is the *measurement of the molecular charge* on polysaccharides in solution, and this will also be considered [14].

There are two other adaptations of the analytical ultracentrifuge of relevance to polysaccharides: one is *gel sedimentation analysis* for the measurement of swelling pressures and related rheological parameters. Another is *diffusion analysis through matrices and interfaces*: although dynamic light scattering is now popularly used for the measurement of translational diffusion coefficients, the optical system on the analytical ultracentrifuge is proving very useful for the investigation of the diffusion of molecules through matrices, and towards and through interfaces between incompatible two phase systems (including aqueous two-phase polysaccharide systems). In this article—which focuses on the biophysical properties of polysaccharides in solution—we will not consider gel sedimentation or matrix/interface diffusion: both have been covered by two other recent and extensive reviews, both appearing in *Biotechnology and Genetic Engineering Reviews* [15, 16]. The actual formation of membranes at interfaces can also be followed by synthetic boundary methods in the ultracentrifuge, and this has been dealt with in a recent article by C. Wandrey and colleagues [17].

3
Instrumentation

An analytical ultracentrifuge (Fig. 1) is simply an ultracentrifuge with an appropriate optical system and data-capture facility for observing and recording solute distributions both during and at the end of the sedimentation process. The analytical ultracentrifuge is by no means a new concept—Svedberg's first oil-turbine driven instrument dates from the early 1920s. It is worth noting that Svedberg and his students subsequently published several papers on polysaccharides: these were mainly in the 1940s and were mostly on cellulose and cellulose derivatives (see [18] and references sited therein). The boom period of the technique was 1940–1970 with several commercially available analytical ultracentrifuges available, most notably the Beckman (Palo Alto, USA) Model E (see [19] and references therein). From the late 1970s analytical ultracentrifuges no longer became commercially available on the western side of the Iron Curtain (apart from a crude Schlieren adaptor for a Beckman L8

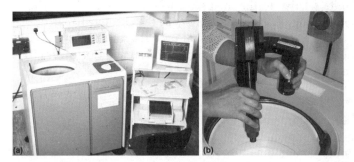

Fig. 1 A modern analytical ultracentrifuge. **a** Beckman Optima XLA/I, with full on-line data capture and analysis facility. **b** Its UV/visible monochromator and, for interference optics, the laser light source are contained in the rotor chamber and have to be installed and removed at the start/end of each run

preparative ultracentrifuge). In Eastern Europe the Hungarian Optical Works in Budapest continued to manufacture their high quality MOM analytical ultracentrifuge [20]. Following the collapse of the Berlin Wall in 1990 a new generation instrument appeared in the form of the Beckman Optima XLA ultracentrifuge, with on-line ultra-violet(UV)/visible absorption optics, giving a direct record of solute concentration (in absorption units), $c(r)$ versus radial displacement r [21]. This, however, had limited relevance for polysaccharides because of the lack of chromophore these substances possess in the near-UV (250–300 nm) and visible region. Chemical addition of a chromophore label permitted its use. Studies on two labelled dextrans: "blue" dextran (Cibacron Blue F3GA-dextran) [22] and di-iodotyrosine dextran [23] demonstrated that under these circumstances the UV absorption optical system on an analytical ultracentrifuge could successfully be used for measurements of sedimentation coefficient and molecular weight: with the latter substance some alteration in dextran conformation induced by the chromophore was evident. Those studies, however, were performed on the older MSE Centrican analytical ultracentrifuge. The first use of an XLA ultracentrifuge for the characterisation of a polysaccharide was by H. Cölfen and colleagues in 1996 who used an XLA to characterise the sedimentation coefficient and molecular weight of two chitosans labelled with the fluorophore 9-anthraldehyde [24]. This was subsequently followed by the launch by Beckman of the XLI ultracentrifuge with integrated refractive index (Rayleigh Interference) and UV/visible optics [25]: application to polysaccharides became possible on a routine basis.

The laser (wavelength 670 nm) on the XL-I instrument provides high-intensity, highly collimated light and the resulting interference patterns (between light passing through the solution sector and reference solvent sector of an ultracentrifuge cell) are captured by a CCD camera. A Fourier transformation converts the interference fringes into a record of concentration $c(r) - c(a)$ relative to the meniscus ($r = a$) as a function of radial displacement r from the axis of rotation. The measurement is in terms of Rayleigh fringe units relative to the meniscus, $j(r)$, with $J(r) = j(r) + J(a)$, $J(r)$ being the absolute fringe displacement and $J(a)$ the absolute fringe displacement at the meniscus. For a standard optical path length cell ($l = 1.2$ cm) with laser wavelength $\lambda = 6.70 \times 10^{-5}$ cm, a simple conversion exists from $J(r)$ in "fringe shift" units to $c(r)$ in g/ml:

$$
\begin{aligned}
c(r) &= J(r)\lambda/\{(\mathrm{d}n/\mathrm{d}c)l\} \\
&= \{5.58 \times 10^{-5}/(\mathrm{d}n/\mathrm{d}c)\}J(r)
\end{aligned}
\tag{1}
$$

with similar conversions for $j(r)$ and $J(a)$. $\mathrm{d}n/\mathrm{d}c$ is the (specific) refractive index increment, which depends on the polysaccharide, solvent and wavelength. A comprehensive list of values for a range of macromolecules has recently been published [26]. Some of the values taken from [26] for polysaccharides are listed in Table 2: it can be seen that in aqueous systems most

Table 2 Refractive index increments of some polysaccharides (data taken from [26] and references cited therein). DS: degree of substitution. λ: wavelength

Polysaccharide	Solvent	λ (nm)	Temp (°C)	dn/dc (ml/g)
Alginate (magnesium)	MgCl$_2$ (aq.)	546	20–25	0.158
Alginate (potassium)	Aq. solution	546	20–25	0.165
Alginate (sodium)	Aq. solution	546	20–25	0.165
Amylopectin	DMSO/ water (90/10 w/w)	632.8	25	0.074
Amylose	DMSO/water (50/50 vol)	546	25	0.112
Amylose	DMSO/water (90/10 vol)	546	25	0.062
Amylose	4.2M GuHCl	546	25	0.118
Amylose acetate	Nitromethane	436	20	0.0835
Amylose tributyrate	Ethyl acetate	546	25	0.098
Amylose tricarbaanilate	Acetone	546	27	0.228
Amylose tripropionate	Ethyl acetate	546	25	0.092
Beta-glucan	Water	488	25	0.151
Carboxymethyl cellulose (several DS)	0.1 M NaNO$_3$	632.8	25	0.163
Cellulose	Acetone	546	25	0.111
Cellulose diacetate	Acetone	546	25	0.112
Cellulose nitrate (12.9% N)	Acetone	546	25	0.115
Cellulose nitrate (14.0% N)	Acetone	436	25	0.09
Cellulose propionate	Acetone	632.8	25	0.111
Cellulose propionate	MEK	632.8	25	0.096
Cellulose tributyrate	DMF	546	41	0.0442
Cellulose tricaproate	DMF	546	41	0.0442
Cellulose tricarbanilate	Acetone	546	27	0.218
Cellulose tricarbanilate	Dioxane	546	25	0.165
Cellulose trinitrate	Acetone	546	25	0.112
Cellulose trinitrate	Ethyl acetate	546	30	0.102
Chitosan	Acetate buffer	633	25	0.181*
Dextran	Acetate buffer	633	25	0.150
Dextran	Water	436	25	0.152
Dextran	NaCl (aq.)	436	25	0.136
Dextran	0.5 M NaCl	546	25	0.147
Dextran	1.0 M NaCl	546	25	0.144
Ethyl cellulose	Methanol	546	25	0.13
Gum arabic	HCl (aq.)	436	25	0.152
Heparin (various commercial)	Acetate buffer	690	25	0.129–0.134
Hyaluronic acid	Water	546	25	0.166–0.170
Hydroxypropyl cellulose	Water	546	25	0.146
Hydroxypropyl cellulose	Water	578	25	0.143
Hydroxypropyl starch (DS 0.5)	Water	546	20	0.152

Table 2 (continued)

Polysaccharide	Solvent	λ (nm)	Temp (°C)	dn/dc (ml/g)
Hydrolysed waxy maize starch (amylopectin)	Water	546	20	0.152
Kappa-carrageenan	0.1 M LiCl	633	25	0.111
Kappa-carrageenan	0.1 M LiCl	633	60	0.115
Pectin (citrus)	Water	633	25	0.146
Pullulan	Phosphate buffer	633	30	0.137
Schizophyllan	Water	546	25	0.145
Sodium carboxymethyl amylose	0.35 M NaCl	546	35	0.133

* depends on the degree of acetylation

Table 3 Refractive index increment and partial specific volume of hyaluronan ($M \sim 10^5$ g/mol) determined at different salt concentrations (from [28])

Solvent NaCl (mol l^{-1})	Refractive index increment (ml/g)			Partial specific volume (ml/g)
	436 nm	546 nm	633 nm	
0.25	0.167	0.165	0.164	0.57
0.20	0.167	0.165	0.164	0.57
0.05	0.168	0.166	0.165	0.56
0.01	0.171	0.168	0.166	0.53
0.005	0.172	0.169	0.167	0.51
0.001	0.178	0.175	0.173	0.47

values lie between 0.14–0.16 ml/g, although for non-aqueous systems the values can range enormously from 0.044–0.218 ml/g. The data for example for κ-carrageenan suggest little temperature dependence although that for dextrans suggests a significant dependence on wavelength. A study on the polycationic chitosan [27] suggested that the degree of substitution of some polysaccharides can strongly influence dn/dc, particularly if ionic groups are involved. Preston and Wik [28] have explored in detail the effect of ionic strength and wavelength on dn/dc for the polyanion hyaluronate (Table 3). These results show that if a user needs, for whatever reason, an accurate value for dn/dc for a polysaccharide he should measure it directly in the particular buffer used for the ultracentrifuge experiments. It is worth emphasising that converting fringe shift concentrations $\{j(r)$ or $J(r)\}$ to weight concentrations is not normally necessary for most applications. In addition, for sedimentation velocity work it is possible to work with $j(r)$ or $c(r) - c(a)$, i.e. concentrations relative to the meniscus without having to worry about measuring the offset or meniscus concentration $J(a)$ or $c(a)$ to convert to absolute $J(r)$ or $c(r)$.

4
Polysaccharide Polydispersity and Simple Shape Analysis by Sedimentation Velocity

With sedimentation velocity we measure the change in solute distribution across a solution in an ultracentrifuge cell as a function of time. An example of such a change is given in Fig. 2a for potato amylose [29].

Traditional analysis methods on these measurements have been based around recording the movement of the radial position of the boundary r_b with time t, from which a sedimentation coefficient, s (sec or Svedbergs, S, where $1\,S = 10^{-13}$ sec) can be obtained (see, e.g. [30]):

$$s = (\mathrm{d}r_b/\mathrm{d}t)/\omega^2 r_b \tag{2}$$

ω is the angular velocity $(\mathrm{rad}/s) = (2\pi/60) \times \mathrm{rpm}$. Since the measured s will be affected by the temperature, density and viscosity of the solvent in which it is dissolved it is usual to normalise to standard conditions—namely the density and viscosity of water at 20.0 °C to yield $s_{20,w}$ [30].

$$s_{20,w} = \{(1 - \bar{v}\rho_{20,w})/(1 - \bar{v}\rho_0)\}.\{\eta_0/\eta_{20,w}\}s_{T,b} \tag{3}$$

ρ_0 and η_0 are the densities and viscosities of the solvent, $\rho_{20,w}$ and $\eta_{20,w}$ the corresponding values at 20.0 °C in water. $\bar{v}$ is the partial specific volume, which for neutral polysaccharides can often be reasonably estimated from the carbohydrate content [31] and takes on values between 0.5–0.7 ml/g for

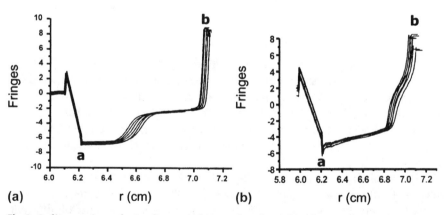

(a) r (cm) (b) r (cm)

Fig. 2 Sedimentation velocity diagrams for starch polysaccharides. **a** Sedimenting boundary for potato amylose, scanned at different times. Sample concentration was 8 mg/ml in 90% in dimethyl sulphoxide. Rotor speed was 50 000 rpm at a temperature of 20 °C. **b** Sedimenting boundary for wheat starch (containing amylose and the faster moving amylopectin), scanned at different times. Total sample concentration was 8 mg/ml in 90% in dimethyl sulphoxide. Rotor speed was 35 000 rpm at a temperature of 20 °C. The direction of sedimentation in both **a** and **b** is from left to right. From [29]

neutral polysaccharides in aqueous solvent. In cases where estimates based on composition cannot be reasonably made—such as for polyanionic and polycationic materials—$\bar{v}$ can be measured by densimetry [32] coupled with an accurate concentration measurement. For example Preston and Wik [28] have shown that $\bar{v}$ varies from 0.47 to 0.57 from ionic strengths 0.001 to 0.25 mol L^{-1} (see Table 3).

A computer algorithm SEDNTERP [33, 34] has been developed for facilitating the correction in Eq. 3. There is no longer any need (except for unusual solvents) to look up solvent densities and viscosities in the Chemical Rubber Handbook or other data books—a user of the algorithm just has to specify the buffer composition and the temperature of the measurement and the correction is done automatically.

The s and $s_{20,w}$ obtained from Eqs. 1–3 will be *apparent* values because of the effects of solution non-ideality, deriving from co-exclusion and—for charged polysaccharides—polyelectrolyte effects [30]. To eliminate the effects of non-ideality it is necessary to measure either s or $s_{20,w}$ for a range of different cell loading concentrations c, and perform an extrapolation to zero concentration. For polysaccharides this has been conventionally achieved from a plot of $1/s$ (or $1/s_{20,w}$) versus c [30]:

$$\{1/s\} = \{1/s^0\} \cdot \{1 + k_s c\} \tag{4}$$

a relation valid over a limited range of concentration. k_s is the Gralén coefficient named after Svedberg's doctoral student who introduced this term in his 1944 thesis on the analysis of cellulose and its derivatives [35].

For a wider span of concentrations, a more comprehensive description of concentration dependence has been given by Rowe [36, 37]:

$$s = s^0 \left\{ 1 - \left[k_s c - \left(((cv_s)^2 (2\phi_p - 1)) / \phi_p^2 \right) \right] / [k_s c - 2cv_s + 1] \right\} \tag{5}$$

v_s (ml/g) is the "swollen" specific volume of the solute [volume (ml) of a polysaccharide swollen though solvent association per gram of the anhydrous molecule] and ϕ_p is the maximum packing fraction of the solute (~ 0.4 for biological solutes [37]). A least-squares proFit (Quantum Soft, Zurich, Switzerland) algorithm has been developed for fitting s vs. c data to Eq. 5 and examples of fitting this relation to wheat starch amylose and amylopectin are given in Fig. 3 [29].

For a polydisperse solution—the hallmark of solutions of polysaccharides—s (and s^0) will be a weight average [30, 38, 39]. If the solution contains more than one discrete (macromolecular) species—e.g. a mixture of different polysaccharides, the polydispersity will be manifested by asymmetry in the sedimenting boundary or, if the species have significantly different values for $s_{20,w}$, discrete boundaries are resolved (Fig. 2b [29]).

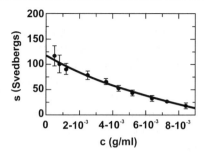

Fig. 3 Concentration dependence of the sedimentation coefficient for wheat amylopectin. The data have been fitted to Eq. 5 (see text) yielding $s^0 = (120 \pm 10)S$, $k_s = (170 \pm 60)$ml/g and $v_s = (40 \pm 4)$ml/g. From [29]

4.1
Sedimentation Coefficient Distributions: DCDT and SEDFIT

Since the appearance of the new generation analytical ultracentrifuges in 1990 (XL-A) and 1996 (XL-I), the acquisition of multiple on-line data acquisition has resulted in some important advances in the software for recording and analysing not only the change in boundary position with time but the change in the whole radial concentration profile, $c(r, t)$ with time t. This advance has in particular facilitated the measurement of *distributions* of sedimentation coefficient [40–43]. The (differential) distribution of sedimentation coefficients $g(s)$ can be defined as the population (weight fraction) of species with a sedimentation coefficient between s and $s + ds$ [40]. A plot of $g(s)$ versus s then defines the distribution. Integration of a peak or resolved peaks of a $g(s)$ vs. s plot can then be used to calculate the weight average s of the sedimenting species and their partial loading concentrations.

4.2
Time Derivative Analysis: DCDT

The simplest way computationally of obtaining a sedimentation coefficient distribution is from time derivative analysis of the evolving concentration distribution profile across the cell [40, 41]. The time derivative at each radial position r is $(\partial\{c(r, t)/c_0\}/\partial t)_r$ where c_0 is the initial loading concentration. Assuming that a sufficiently small time integral of scans are chosen so that $\Delta c(r, t)/\Delta t = \partial c(r, t)/\partial t$ the apparent weight fraction distribution function $g^*(s)$ {n.b. sometimes written as $g(s^*)$} can be calculated

$$g^*(s) = \{\partial\{c(r, t)/c_0\}/\partial t\} \cdot \{(\omega^2 t^2)/\ln(a/r)\} \cdot \{(r/a)^2\} \qquad (6)$$

and the abscissa

$$s = \{\ln(a/r)\} \cdot 1/\{\omega^2 t\} \qquad (7)$$

where a is the radial position of the meniscus. In this way the evolving $c(r, t)$ versus r profiles are transformed into an apparent sedimentation coefficient distribution plot. The asterisk is used to indicate that the distributions calculated using this equation are "apparent" in the sense they are not corrected for diffusion effects (nor, as is often forgotten, non-ideality). Depending on how fast the diffusion process is (for large polysaccharides this can be small compared to the sedimentation process) this will cause an overestimate of the width of a sedimentation coefficient distribution, but not on the weight average sedimentation coefficient. This procedure forms the basis of the *DCDT* algorithm developed by W. Stafford [40] later refined by J. Philo as *DCDT*+ [41, 44]. The success of the method depends on data sampling at high frequency and a large spectrum of radial positions so as to reduce random noise. It has the great facility that time independent noise is eliminated in the differential.

Figure 4 [29] shows the $g^*(s)$ versus profiles for potato amylose and the amylose/amylopectin mixture from wheat starch corresponding to the concentration versus radial displacement data of Fig. 3. The s data used in the concentration dependence plot of Fig. 3 for wheat amylopectin comes from $g^*(s)$ vs. s analysis data of Fig. 2b and similar. The concentrations shown in the abscissa in Fig. 4 have been obtained from the total starch loading concentration normalised by the weight fraction of the amylopectin component estimated from the $g^*(s)$ vs. s profiles.

Various improvements of the DCDT [40, 41] approach have been considered to correct for the effects of diffusion, based largely around extrapolation to $(1/\sqrt{t}) \to 0$.

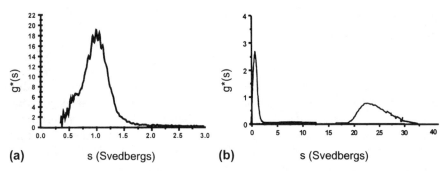

(a)   s (Svedbergs)   **(b)**   s (Svedbergs)

Fig. 4 Sedimentation velocity $g^*(s)$ profiles for starch polysaccharides using DCDT+. The profiles correspond to the radial displacement plots of Fig. 2. **a** Potato amylose, sample concentration 8 mg/ml in 90% in dimethyl sulphoxide. Rotor speed was 50 000 rpm at a temperature of 20 °C. **b** Wheat starch (containing amylose, left peak and the faster moving amylopectin, right peak), (total) sample concentration 8 mg/ml in 90% dimethyl sulphoxide. Rotor speed was 35 000 rpm at a temperature of 20 °C. From [29]

4.3
Lamm Equation Approach: SEDFIT

More recently, attention has turned to direct modelling of the evolution of the concentration distribution with time for obtaining the sedimentation coefficient distribution. The notation used in this approach is a little different [42, 43], $c(s)$ being chosen to represent the (differential) distribution of sedimentation coefficients (not as $g(s)$ previous [30, 40, 41]) although it is still defined as the population of species with a sedimentation coefficient between s and $s + ds$. Despite the perhaps confusing use of the symbol $c(s)$ it does not have the units of weight concentration, but weight concentration per sedimentation coefficient (e.g. $g.ml^{-1}.S^{-1}$). The distribution has been related to the experimentally measured evolution of the concentration profiles throughout the cell by a Fredholm integral equation

$$a(r, t) = \int_{s_{min}}^{s_{max}} c(s) \cdot \chi(s, D, r, t)\, ds + a_{TI}(r) + a_{RI}(t) + \varepsilon \qquad (8)$$

In this relation $a(r, t)$ is the experimentally observed signal, ε represents random noise, $a_{TI}(r)$ represents the time invariant systematic noise and $a_{RI}(t)$ the radial invariant systematic noise: Schuck [42] and Dam and Schuck [43] describe how this systematic noise is eliminated. χ is the normalised concentration at r and t for a given sedimenting species of sedimentation coefficient s and translational diffusion coefficient D: it is normalised to the initial loading concentration so it is dimensionless.

The evolution with time of the concentration profile $\chi(s, D, r, t)$ in a sector shaped ultracentrifuge cell is given by the Lamm [45] equation:

$$\frac{\partial \chi}{\partial t} = (1/r) \cdot \frac{\partial}{\partial r}\left[rD\frac{\partial \chi}{\partial r} - s\omega^2 r^2 \chi \right] \qquad (9)$$

Although only approximate analytical solutions to this partial differential equation have been available for $\chi(s, D, r, t)$, accurate numerical solutions are now possible using finite element methods first introduced by Claverie and coworkers [46] and recently generalized to permit greater efficiency and stability [42, 43]: the algorithm SEDFIT [47] employs this procedure for obtaining the sedimentation coefficient distribution.

To solve Eq. 8 to obtain $c(s)$ as a function of s requires s_{min} and s_{max} to be carefully chosen and adjusted accordingly: inappropriate choice can be diagnosed by an increase of $c(s)$ towards the limits of s_{min} or s_{max}. However, a stable solution can only be obtained if the contribution from diffusion broadening is dealt with.

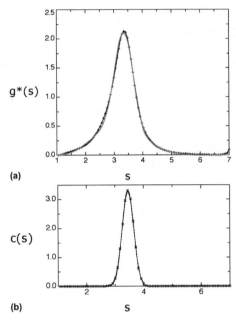

Fig. 5 Sedimentation concentration distribution plots for guar gum using SEDFIT.
a $g^*(s)$ vs. s **b** $c(s)$ vs. s. A Gaussian fit to the data (*lighter line*) is also shown in (**a**). Rotor
speed was 40 000 rpm at 20.0 °C, concentration was 0.75 mg/ml in 0.02% NaN$_3$. The guar
had been heated at 160 °C for 10 min at a pressure of 3 bar. From [49]

SEDFIT tries to do this by using a dependence of D on s, and it does so by
making use of a link involving the translational frictional ratio f/f_0:

$$D(s) = \left\{ \left(\sqrt{2} \right) /(18\pi) \right\} \cdot k_B t \cdot s^{-1/2} \left(\eta_0 (f/f_0)_w \right)^{-3/2} \left((1 - \bar{v}\rho_0)/\bar{v} \right)^{-1/2} \qquad (10)$$

f is the frictional coefficient of a species and f_0 the corresponding value for
a spherical particle of the same mass and (anhydrous) volume (see e.g. [48]).
k_B is the Boltzmann constant. Although of course a distribution of s implies also a distribution in D and f/f_0, for protein work advantage is taken
of the fact that the frictional ratio is a relatively insensitive function of
concentration: a single or weight average f/f_0 is taken to be representative
of the distribution. It is open to question for large linear macromolecules
like many polysaccharides. If this assumption is valid Eq. 8 can be numer-
ically inverted—i.e. solved—to give the sedimentation coefficient distribution, with the position and shape of the $c(s)$ peak(s) more representative
of a true distribution of sedimentation coefficient. $(f/f_0)_w$, where the sub-
script w denotes a weight average, is determined iteratively by non-linear
regression, optimizing the quality of the fit of the $c(s)$ as a function of
$(f/f_0)_w$. It has been shown by extensive simulation that non-optimal values of
$(f/f_0)_w$ have little effect on the position of the $c(s)$ peaks, although effect the

width and resolution, i.e. the correct s value is reported. Regularization [47] can be used which provides a measure of the quality of fit from the data analysis.

SEDFIT also offers the option of evaluating the distribution corresponding to non-diffusing particles, viz $D \sim 0$, i.e. the diffusive contribution to Eq. 8 is small compared to the sedimentation contribution. In this case Eq. 8 can be inverted without any assumptions concerning f/f_0. If diffusive affects are significant it will be an apparent sedimentation coefficient distribution, i.e. $g^*(s)$ vs. s, analogous to that produced by the DCDT procedure, and the correct s value for a peak is still reported. This procedure, called the "least squares" $g^*(s)$ procedure to distinguish it from DCDT has the advantage over DCDT in that errors caused by the approximation $\Delta t \sim dt$ are absent. Figure 5 gives a comparison of the least squares $g^*(s)$ vs. s and $c(s)$ vs. s distribution for guar gum [49]. Least squares $g^*(s)$ vs. s is currently our method of choice for polysaccharides.

4.4
Molecular Weight and Conformation

The sedimentation coefficient s^0, or its normalized form $s^0_{20,w}$ is a function of the conformation and flexibility of a macromolecule (via its translational frictional property) and its mass. So if we are going to obtain conformation and flexibility information we need to know the molecular weight (molar mass) and *vice verse*.

It is possible to get molecular weight from the sedimentation coefficient if we assume a conformation or if we combine with other measurements, namely the translational diffusion coefficient via the Svedberg equation [50]

$$M = RT \cdot \{s^0/D^0\}/(1 - \bar{v}\rho_0) \tag{11}$$

where ρ_0 is the solvent density (if s and D are their normalized values $s^0_{20,w}$, $D^0_{20,w}$, ρ_0 will be the density of water at 20.0 °C, 0.9981 g/ml).

Equation 11 has been popularly used for example to investigate the molecular weights of carboxymethylchitins [51–53], glycodendrimers [54, 55], galactomannans [56], beta-glucans [57, 58] and alginates [59].

The translational diffusion coefficient in Eq. 11 can in principle be measured from boundary spreading as manifested for example in the width of the $g^*(s)$ profiles: although for monodisperse proteins this works well, for polysaccharides interpretation is seriously complicated by broadening through polydispersity. Instead special cells can be used which allow for the formation of an artificial boundary whose diffusion can be recorded with time at low speed (~ 3000 rev/min). This procedure has been successfully employed for example in a recent study on heparin fractions [5]. Dynamic light scattering has been used as a popular alternative, and a good demonstration of how this can be performed to give reliable D data has been given by Burchard [3].

Whereas the s^0 is a weight average, the value returned from dynamic light scattering for D^0 is a z-average. As shown by Pusey [60] combination of the two via the Svedberg Eq. 11 yields the weight average molecular weight, M_w, although it is not clear what type of average for M is returned if an estimate for D^0 is made from ultracentrifuge measurements.

Another useful combination that has been suggested is $s^0_{20,w}$ with k_s [36, 37]

$$M_w = N_A \left[6\pi\eta s^0_{20,w}/(1 - \bar{v}\rho_0) \right]^{3/2} \left[(3\bar{v}/4\pi)(k_s/2\bar{v}) - (v_s/\bar{v}) \right]^{1/2} \tag{12}$$

s, k_s and v_s can be obtained from fitting s vs. c data to Eq. 5. The method was originally developed for single solutes and where charge effects can be neglected (either because the macromolecular solute is uncharged, or because the double layer or polyelectrolyte behaviour has been "compressed" by addition of neutral salt). For quasi-continuous distributions, such as polysaccharides one can apply Eqs. 5 and 12 to the data, provided that for every concentration one has a "boundary" to which a weight-averaged s value can be assigned. If the plot of $1/s$ vs. c is essentially linear over the data range, then specific interaction can be excluded, the solute system treated as a simple mixture and Eqs. 5 and 12 can be applied.

For wheat starch amylopectin for example (cf. Fig. 3) a value for M_w of $\sim 30 \times 10^6$ g/mol is estimated [29]. This equation is only approximate—any contributions from molecular charge to the concentration dependence parameter k_s are assumed to be negligible or suppressed—but is nonetheless very useful when other methods—especially for very large polysaccharides like amylopectin—are inapplicable. The method also yields an estimate for the swollen specific volume v_s: for example Majzoobi has obtained a value of (40 ± 4) ml/g for wheat starch amylopectin [29]. For polydisperse materials such as polysaccharides the question is what sort of average M value is yielded by this method? In the absence of any obvious analytical solution, computer simulation has been used to determine the form of the average. In work to be published, A.J. Rowe and coworkers have shown that even for "unfavourable" simulated mixtures (e.g. multi-modal, no central tendency), the average M value yielded is very close to an M_w (i.e. weight-averaged M). To put this in quantitative terms, the departure from M_w is generally $< 1\%$ of the way towards M_z. This is trivial, in terms of the errors present in the raw data. Thus, there is an exact procedure which can be defined for the evaluation of M(average) in a polydisperse solute system under the defined conditions, and simulation demonstrates that for all practical purposes the outcome is an M_w.

A sedimentation coefficient *distribution*—either $c(s)$ versus s or $g^*(s)$ vs. s—for a polysaccharide can also be converted into an apparent molecular weight distribution if the conformation of the polysaccharide is known or can be assumed, via a power law or "scaling" relation:

$$s^0 = K''M^b \tag{13}$$

where the power law exponent is dependent on the conformation of the macromolecule, with the limits, in the case of non-draining molecules $b \sim$ 0.15 for a rigid rod and ~ 0.67 for a compact sphere. A flexible coil shape molecule has a $b \sim 0.4 - 0.5$ [61, 62]. An early example of this transformation has been given for a heavily glycosylated mucin glycoprotein with polysaccharide like properties [63] based on a $g^*(s)$ vs. s distribution given by Pain [64]). The assumption was made that the contribution from diffusion broadening of these large molecules was negligible in comparison to sedimentation. Incorporation of Eq. 13 instead of Eq. 10 into the $c(s)$ vs. s evaluation process is now being considered for polysaccharides.

Conversely, if we know the molecular weight we can make inferences about the conformation of polysaccharides in solution using Eq. 13 and other power-law relations. We will consider this in more detail after we have considered further molecular weight measurement by absolute (i.e. without assumptions concerning conformation) procedures.

5
Polysaccharide Molecular Weight Analysis by Sedimentation Equilibrium

Whereas in a sedimentation velocity experiment at relatively high rotor speed—for a polysaccharide say 40 000–50 000 rev/min—the sedimentation rate and hence sedimentation coefficient are a measure of the size and shape of the molecule, at much lower speeds, say 10 000 rev/min or less in a sedimentation equilibrium experiment, the forces of sedimentation and diffusion on the macromolecule become comparable and instead of producing a sedimenting boundary a steady state equilibrium distribution of macromolecule is attained with a low concentration at the air/solution meniscus building up to a high concentration at the cell base. This final steady state pattern [65] is a function only of molecular weight and related parameters (non-ideal virial coefficients and association constants were appropriate) and not on molecular shape since at equilibrium there is no net transport or frictional effects: sedimentation equilibrium in the analytical ultracentrifuge is an absolute way of estimating molecular weight.

5.1
Molecular Weight Information Obtained

Since polysaccharides are by their very nature polydisperse, the value obtained will be an average of some sort. With Rayleigh interference and, where appropriate, UV-absorption optics, the principal average obtained is the weight average, M_w [30]. Although relations are available for obtaining also number average M_n and z-average M_z data these latter averages are difficult to obtain with any reliable precision. Direct recording of the concentration gradient dc/dr versus radial displacement r using refractive index gradient or "Schlieren" optics however facilitates the measurement of M_z (see [66]). Such an optical system is unfortunately not present on the present generation XL-A or XL-I ultracentrifuges except for in-house adapted preparative XL ultracentrifuges [67]. Schlieren optics are also present on the older generation Model E analytical ultracentrifuge: this is one of the main reasons why we at the NCMH keep and maintain one of these instruments [68].

An important consideration with polysaccharides is that at sedimentation equilibrium there will be a redistribution not only of total concentration of polysaccharide throughout the cell (low concentration at the meniscus building up to a higher concentration at the cell base) but also a redistribution of species of different molecular weight, with a greater proportion of the higher molecular weight part of the distribution appearing near the cell base. In obtaining a true weight (or number, z averages) it is therefore important to consider the *complete* concentration distribution profile throughout the ultracentrifuge cell. As with our description of sedimentation velocity, for clarity we will confine our consideration only to the extraction of the two most directly related parameters: the weight average molecular weight and the molecular weight distribution. The extraction of other parameters—such as point average data—are avoided here but can be found in other articles (see [9, 11, 69, 70]).

5.2
Obtaining the Weight Average Molecular Weight: MSTAR

As stated above UV-absorption optics – when they can be applied – have the advantage that the recorded absorbances $A(r)$ as a function of radial position are (within the Lambert–Beer law limit of $A(r) \sim 1.4$) directly proportional to the weight concentration $c(r)$ in g/ml. Although the multiple fringes in interference optics give a much more precise record of concentration, we stress again, these are *concentrations relative to the meniscus*. That is we obtain directly from the optical records a profile of $c(r) - c(a)$ versus radial displacement r, with the meniscus at $r = a$. In fringe displacement units this is $J(r) - J(a)$, which we write as $j(r)$ for short. To obtain molecular weight information we need $J(r)$ and hence some way of obtaining $J(a)$ is required: this is not such a requirement for sedimentation velocity where relative concentrations are sufficient. At this stage it is worth pointing out there are three types of sedimentation equilibrium experiment which depend on the conditions used (namely speed and solution column length in the centrifuge cell):

(i) the low speed method [71] in which the solute (polysaccharide) concentration at the meniscus {$c(a)$ or, in fringe displacement units, $J(a)$} remains finite,

(ii) the high speed or "meniscus depletion" method [72] in which the concentration at the meniscus is effectively zero,

(iii) the intermediate speed method [73] in which the solute concentration at the meniscus remains small but finite.

Extraction of molecular weight data from the optical records is considerably easier for the meniscus depletion method since no estimate for $J(a)$ is required. This procedure has in the past been popular for protein work because it not only avoids the problem of $J(a)$ measurement but also facilitates the evaluation of number average molecular weight data. Sadly, this is not generally applicable to solutions of polysaccharides because of their polydispersity: at sedimentation equilibrium a polydisperse solution will tend to redistribute its species so there is a greater proportion of higher molecular weight species near the cell base. Any attempt to deplete the meniscus (rich in the lower molecular weight part of the distribution) of polysaccharide is almost guaranteed to result in loss of optical registration of the interference fringes near the bottom of the cell, leading to underestimates for M. A further difficulty in attempting to achieve meniscus depletion conditions is that the effective thermodynamic non-ideality of the system—which can also lead to an underestimate for M—is increased at higher rotor speed: this is discussed later in this article.

So methods (i) or (iii) are required. Further, since the greater the difference between the concentration at the meniscus $J(a)$ and the cell base $J(b)$ the smaller will be the effect of errors in $J(a)$ on the final molecular weight evaluated, for polysaccharides the intermediate speed method, (iii) should be the method of choice. This means that a procedure for evaluating $J(a)$ is required. It was recently shown by Hall and coworkers [74] that simply floating it as another variable in the procedure for extracting M is not valid, particularly for polydisperse or interacting systems. A convenient procedure for extracting $J(a)$ and then M was given by Creeth and Harding in 1982 [73], and is briefly summarised now.

The fundamental equation of sedimentation equilibrium can be manipulated to define a new function with dimensions of molar mass (g/mol) called $M^*(r)$. $M^*(r)$ at a radial position r is defined by

$$M^*(r) = \frac{j(r)}{\left\{ kJ(a) \left(r^2 - a^2 \right) + 2k \int_a^r r \cdot j(r) \, dr \right\}} \tag{14}$$

where $k = (1 - \bar{v}\rho_0)\omega^2/2RT$, with ρ_0 the solvent density. Equation 14 has the limiting form

$$\lim_{r \to a} \left\{ j(r)/\left(r^2 - a^2\right) \right\} = k \cdot M^*(a) \cdot J(a). \tag{15}$$

A plot of $j(r)/(r^2 - a^2)$ vs. $\{1/(r^2 - a^2)\} \cdot \int_a^r r \cdot j(r) dr$ therefore has a limiting slope of $2kM^*(a)$ and an intercept $kM^*(a).J(a)$. Hence $J(a)$ is determinable from $2 \times$ (intercept/limiting slope).

It is worth pointing out that Teller and coworkers [75] have considered other manipulations involving representations similar to Eqs. 14 and 15 for obtaining $J(a)$. Other methods of obtaining $J(a)$ have been considered in detail by Creeth and Pain [76]. More recently Minton [77] has given an almost identical procedure, although he appears to have missed the original Creeth and Harding article published 12 years previous [73]. Once $J(a)$ has been found M^* as a function of radial position r can be defined.

A particularly useful property of the M^* function is that at the cell base $(r = b)$,

$$M^*(b) = M_{\text{w,app}} \tag{16}$$

the apparent weight average molecular weight of the polysaccharide [73]. It will be an apparent value because it will be affected by thermodynamic non-ideality (molecular co-exclusion and, for charged polysaccharides, polyelectrolyte behaviour), which needs to be corrected for (see below). The M^* procedure for obtaining $M_{\text{w,app}}$ can be thought of in terms of the "PACMAN" computer game: $M^*(r)$ becomes closer and closer to $M_{\text{w,app}}$ as the integral consumes more and more data as the cell base is approached $(r \to b)$. Optical distortion effects at the cell base means that a short extrapolation of $M^*(r)$ to $M^*(= b)$ is required, but this normally poses no difficulty. An example is given in Fig. 6. Practical details behind the MSTAR algorithm [78] upon which this procedure is based can be found in [69].

It is worth pointing out here that another popular algorithm for analysing molecular weight from sedimentation equilibrium is *NONLIN* [79]. Whereas this is useful for the analysis of protein systems (monodisperse or associating), for a polydisperse system like polysaccharides it is unsuitable: the estimate for $M_{\text{w,app}}$ obtained refers to only to a selected region of the ultracentrifuge cell, and provides no rigorous procedure for dealing with the meniscus concentration problem.

One can see the M^* procedure has a parallel to either $g^*(s)$ vs. s or $c(s)$ vs. s in sedimentation velocity where the data are transformed from radial displacement space [concentration, $c(r)$ versus r] to sedimentation coefficient space [$g^*(s)$ or $c(s)$ versus s]. Here we are transforming the data from concentration space [concentration relative to the meniscus $j(r)$ versus r] to molecular weight space [$M^*(r)$ versus r].

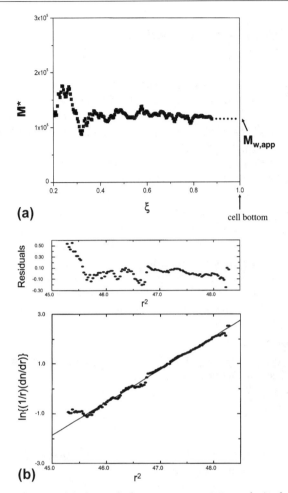

Fig. 6 Sedimentation equilibrium of chitosans **a** MSTAR analysis for $M_{w,app}$, from optical registration of the concentration distribution using Rayleigh interference optics on the XL-I ultracentrifuge and Eqs. 14 and 15 (see text) for chitosan G213. $M_{w,app} = 110\,000$ g/mol. ξ, is a normalised radial displacement squared parameter: $\xi = (r^2 - a^2)/(b^2 - a^2)$ where r is the radial displacement from the axis of rotation at a given point in the ultracentrifuge cell, and a and b the corresponding radial positions at the solution meniscus and cell base respectively. $\xi = 0$ at the meniscus and $= 1$ at the cell base. Loading concentration 1.0 mg/ml in 0.2 M acetate buffer. **b** Analysis for $M_{z,app}$ using optical registration of the concentration gradient distribution using the Schlieren optics on the Model E ultracentrifuge for chitosan G214 at 0.3 mg/ml in 0.2 M acetate buffer. The line fitted is to $\ln\{(1/r)(dn/dr)\} = \{M_{z,app}(1 - \bar{v}\rho_0)\omega^2/2RT\}r^2$. $M_{z,app} = 215\,000$ g/mol. Rotor speed = 10 000 rpm. From [144]

5.3
Correcting for Thermodynamic Non-Ideality:
Obtaining M_w from $M_{w,app}$

For polysaccharides, non-ideality arising from co-exclusion volume and poly-electrolyte effects, can be a serious problem (Table 4), and, if not corrected for, can lead to significant underestimates for M_w. It was possible with the older generation Model E ultracentrifuges—which could accommodate long (30 mm) optical path length cells—to work at very low solute loading concentrations (0.2 mg/ml). At these concentrations for some polysaccharides—particularly neutral ones or those of molecular weight < 100 000 g/mol—the non-ideality effect could be neglected: the estimate for $M_{w,app}$ was within a few percent of the true or "ideal" M_w. However, the new generation XL-I can only accommodate a maximum 12 mm optical path length cell with a minimum concentration requirement of 0.5 mg/ml: lower concentrations produce insufficient fringe displacement for meaningful analysis. This is another reason why we have kept running a Model E ultracentrifuge in the NCMH: the ability to use the longer path length cells makes a large difference to the severity of the non-ideality problem as Table 4 [80–88] shows. The term $(1 + 2BM_wc)$, where B is the thermodynamic second virial coefficient, represents the factor by which the apparent molecular weight measured at a finite concentration $c = 0.2$ mg/ml and 0.5 mg/ml underestimates the true or ideal M_w. One can see that whereas the earlier lower limit (0.2 mg/ml) for many

Table 4 Comparative non-ideality of polysaccharides

	$10^{-6} \times M$ (g mol^{-1})	$10^4 \times B$ (ml mol g^{-2})	BM (ml/g)	$1 + 2BMc$[a]	$1 + 2BMc$[b]	Ref.
Pullulan P5	0.0053	10.3	5.5	1.002	1.006	[80]
Pullulan P50	0.047	5.5	25.9	1.010	1.026	[80]
Xanthan (fraction)	0.36	2.4	86	1.035	1.086	[81]
β-glucan	0.17	6.1	104	1.042	1.104	[82]
Dextran T-500	0.42	3.4	143	1.057	1.143	[83]
Pullulan P800	0.76	2.3	175	1.070	1.175	[80]
Chitosan (Protan 203)	0.44	5.1	224	1.090	1.224	[84]
Pullulan P1200	1.24	2.2	273	1.109	1.273	[80]
Pectin (citrus fraction)	0.045	50.0	450	1.180	1.450	[85]
Scleroglucan	5.7	0.50	570	1.228	1.570	[86, 87]
Alginate	0.35	29.0	1015	1.406	2.015	[88, 91]

[a] Based on a loading concentration of 0.2 mg/ml.
[b] Based on a loading concentration of 0.5 mg/ml.

cases led to only small errors, with the new instruments (limit 0.5 mg/ml) leading to severe underestimates in virtually all cases.

For polyelectrolytes the second virial coefficient is very sensitive to ionic strength. Preston and Wik [28] have shown a tenfold increase in B—from ~ 50 ml mol g^{-2} to ~ 500 ml mol g^{-2}—upon decreasing the ionic strength from 0.2 down to 0.01 mol l^{-1}.

However, it is usually easy, if a little laborious, to correct for non-ideality. Measurement of $M_{w,app}$ over a range of loading concentrations is necessary followed by an extrapolation to zero concentration using an equation of the form [76]

$$\{1/M_{w,app}\} = \{1/M_w\} + 2Bc$$
$$= \{1/M_w\}(1 + 2BM_wc) \qquad (17)$$

correct to first order in concentration. The availability of four and eight-hole rotors in the XL-A and XL-I means that several concentrations can be run simultaneously. Further multiplexing is possible with the use of phantis-style 6-channel ultracentrifuge cells [72], which permit the simultaneous measurement of three solution/reference solvent pairs, although these tend to return $M_{w,app}$ values of lower accuracy. Figure 7 shows an example for xanthan [90].

The second virial coefficient B in Eq. 17 refers to the static case. In the ultracentrifuge the measured value can show a speed dependence [39], an effect which can be minimized by using low speeds and short solution columns. If present it will not affect the value of M_w after extrapolation to zero concentration.

In some extreme cases, third or even higher virial coefficient(s) may be necessary to adequately represent the data: for example κ-carrageenan [90] and alginate [91]. In a further study on alginates, Straatman and Borchard [92] demonstrated excellent agreement between M_w and B values ob-

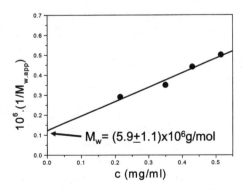

Fig. 7 Concentration extrapolation of $1/M_{w,app}$ to obtain M_w for xanthan. Rotor speed = 3000 rpm. Phosphate chloride buffer (pH = 6.5, I = 0.3). From [89]

tained from sedimentation equilibrium and light scattering methods. Preston and Wik [28] have shown for a different polyelectrolyte—hyaluronan—how increasing the ionic strength can keep B to a minimum.

5.4
Distributions of Molecular Weight

Direct inversions of the concentration distribution profiles to obtain molecular weight distribution information are generally impossible because of complications involving non-ideality. Successful attempts have been given but only for simple discrete forms of polydispersity (two to three macromolecular species [93]).

The simplest procedure for avoiding these complications [94] is to use sedimentation equilibrium in conjunction with gel permeation chromatography (GPC). Fractions of relatively narrow (elution volume) band width are isolated from the eluate and their M_w values evaluated by low speed sedimentation equilibrium in the usual way: the GPC columns can thereby be "self-calibrated" and elution volume values converted into corresponding molecular weights—a distribution can therefore be defined in a way which avoids the problem of using inappropriate standards for GPC: the value of multiplexing is clearly indicated. This procedure has been successfully applied to for example dextrans, alginates and pectins: for pectins excellent agreement with analogous procedures involving classical light scattering coupled to GPC has been obtained [95].

5.5
Number and z-Averages: Polydispersity Index

Molecular weights of polysaccharides in solution can also be measured by osmotic pressure and light scattering. Osmotic pressure yields the number average molecular weight, which can be usefully used with M_w from sedimentation equilibrium as a measure of polydispersity: Preston and Wik [28] have done this for example with hyaluronic acid. The ratio M_w/M_n the "polydispersity index" is often given as a measure of polydispersity, and can be related to the width of a molecular weight distribution via the well-known Herdan [96] relation:

$$\{\sigma_n/M_n\} = [(M_w/M_n) - 1]^{1/2} \tag{18}$$

where σ_n is the number-average standard deviation of the distribution, whatever form it may take. The problem with osmotic pressure is that since the sensitivity depends on the number as opposed to weight concentrations of solute, for polysaccharides of $M_n > 100\,000$ it becomes difficult to apply [97]. An alternative strategy is to measure the z-average molecular weight from sedimentation equilibrium using the Schlieren optical system, which records the concentration gradient (in the form of the refractive index gradient dn/dr) as a function of radial position r, rather than Rayleigh Interference [66]: once again it is a pity that the modern XL-I analytical ultracentrifuge currently does not have this facility. Figure 6b shows a determination of the $M_{z,app}$ from the slope of a plot of $\ln\{(1/r)(dn/dr)\}$ versus r^2 for a chitosan under the same solvent conditions as Fig. 6a. Like $M_{w,app}$, $M_{z,app}$ has to be corrected for non-ideality. The form of the extrapolation (correct to first order in c) is [77]

$$\{1/M_{z,app}\} = \{1/M_z\} + 4Bc$$
$$= \{1/M_z\}(1 + 4BM_z c). \tag{19}$$

Once M_z/M_w has been obtained, the corresponding Herdan relation [96] for M_z/M_w is:

$$\{\sigma_w/M_w\} = [(M_z/M_w) - 1]^{1/2} \tag{20}$$

where σ_w is the weight-average standard deviation of the distribution, whatever form it may take. It is not to be confused with the same symbol unfortunately introduced some 20 years later to represent a reduced weight average molecular weight [98]: use of the original Rinde [99] parameter A_w for the latter is recommended.

Fujita [38] showed that for a log-normal distribution of molecular weights (the usual case for polysaccharides) $M_z/M_w \equiv M_w/M_n$.

5.6
SEC/MALLs and the New Role of Sedimentation Equilibrium

The measurement of the angular dependence of the total intensity of light scattered by solutions of polysaccharides provides, like sedimentation equilibrium, a direct and absolute way of measuring the weight average molecular weight, again if allowance for thermodynamic non-ideality is made (nb. some researchers tend to prefer "A_2" as the notation for the 2nd virial coefficient rather than B). Although opinions varied, prior to 1990 (see e.g., [100]) there was a good case for suggesting sedimentation equilibrium as the preferred method of choice for the measurement of molecular weights, simply because of the less stringent requirements on sample clarity: with light scattering it is essential that solutions are free of supra-molecular aggregates.

The inclusion of a flow cell into a light scattering photometer facilitated the coupling on-line to a gel permeation chromatography column and SEC-MALLs ("size exclusion chromatography coupled to multi angle laser light scattering") has now revolutionised the measurement of molecular weight and molecular weight distribution [4, 101]. The combined effect of the SEC columns and a pre- or "guard column" can provide clear fractionated samples to the light scattering cell, facilitating not only measurement of M_w for the whole distribution, but also the distribution itself. Prior ultracentrifugation of the polysaccharide solution ($\sim$ 40 000 rpm for 30 mins) is still advisable. The first polysaccharide studies were published in 1991 [102–104] and it is now regarded by many as the method of choice for polysaccharide molecular weight determination. Furthermore, the angular dependence of the scattered light facilitates measurement of R_g as a function of elution volume and hence molecular weight: this provides conformation information about the polysaccharide [105].

Nonetheless, uncertainties can sometimes remain particularly if materials have been incompletely clarified or there are problems with the columns (the form of the angular dependence data can usually tell us if things are not well). Sedimentation equilibrium offers a powerful and valuable independent check on the results generated from SEC-MALLs: although it takes a longer time to generate a result, and molecular weight distributions are considerably more difficult to obtain, agreement of M_w from sedimentation equilibrium with M_w from SEC-MALLs gives the researcher increased confidence in some of the other information (molecular weight distribution and R_g–M dependence) coming from the latter.

6
Polysaccharide Conformation Analysis

The sedimentation coefficient s^0 provides a useful indicator of polysaccharide conformation and flexibility in solution, particularly if the dependence of s^0 on M_w is known [62]. There are two levels of approach: (i) a "general" level in which we are delineating between overall conformation types (coil, rod, sphere); (ii) a more detailed representation where we are trying to specify particle aspect ratios in the case of rigid structures or persistence lengths for linear, flexible structures.

6.1
The Wales–van Holde Ratio

The simplest indicator of conformation comes not from s^0 but the sedimentation concentration dependence coefficient, k_s. Wales and Van Holde [106] were the first to show that the ratio of k_s to the intrinsic viscosity, $[\eta]$ was a measure of particle conformation. It was shown *empirically* by Creeth and Knight [107] that this has a value of ~ 1.6 for compact spheres and non-draining coils, and adopted lower values for more extended structures. Rowe [36, 37] subsequently provided a derivation for rigid particles, a derivation later supported by Lavrenko and coworkers [10]. The Rowe theory assumed there were no free-draining effects and also that the solvent had sufficient ionic strength to suppress any polyelectrolyte effects. A value of 1.6 was evaluated for spheres, reducing to ~ 0.2 for long rod shape molecules.

Lavrenko and coworkers [10] also examined in detail the effects of free draining of solvent during macromolecular motion, demonstrating that this also had the effect of lowering $k_s/[\eta]$. A hydrodynamic intra-chain interaction or "draining" parameter has been defined [108] with limits $X = \infty$ for the non-free draining case and $X = 0$ for the free-draining case. A relation was given between $k_s/[\eta]$ and X [10, 108]:

$$\{k_s/[\eta]\} = 8X/(3 + 8X) \tag{21}$$

This relation evidently leads to theoretical limits for $k_s/[\eta] = 0$ for free draining and 1 for non-free draining. The consequences of this are that unless the draining characteristics of the chain are properly known one has to be cautious in making conclusions about particle asymmetry, since it has been claimed that draining affects can mimic increase in asymmetry in lowering the $k_s/[\eta]$. Notwithstanding, many non-spherical molecules have empirical values for $k_s/[\eta]$ greater than 1.0: pullulans for example, considered as a random coil have been shown to have $k_s/[\eta] \sim 1.4$ (see [80]). Berth and coworkers [109] have argued that the very low $k_s/[\eta]$ values for chitosans are due to draining effects rather than a high degree of extension. Lavrenko and coworkers [10] have compiled an extensive list of $k_s/[\eta]$ values for a large number of other polysaccharides, complementing a list given by Creeth and Knight [107]: values are seen to range from 0.1 (potato amylose in 0.33 M KCl) to 1.8 (a cellulose phenylcarbamate in 1,4 dioxane), with some polysaccharides showing a clear dependence on molecular weight.

6.2
Power Law or Scaling Relations

The relation linking the sedimentation coefficient with the molecular weight for a homologous polymer series given above is (see [61, 111]):

$$s^0 = K''M^b \tag{13}$$

(nb Lavrenko and coworkers [10] call the exponent $1 - b$). This relation is similar to the well-known Mark–Houwink–Kuhn–Sakurada relation linking the intrinsic viscosity with molecular weight:

$$[\eta] = K'M^a \tag{22}$$

and also a relation linking the radius of gyration R_g with molecular weight.

$$R_g = K'''M^c \tag{23}$$

The power law or "MHKS" exponents a, b, c have been related to conformation [61, 62] (Table 5).

Table 5 Power Law exponents (from [61])

	a	b	c
Sphere	0	0.67	0.33
Coil	0.5–0.8	0.4–0.5	0.5–0.6
Rod	1.8	0.15	1.0

The coefficients in Table 5 correspond to the non-draining case. If draining effects are present then these will change the values for a and b – see, for example [110]. For example it has been shown that a varies from 0.5 (non-draining case) to 1 (draining), again mimicking the effects of chain elongation. For homologous, linear types of polymer the power law indices are intercorrelated by the relations given by Tsvetkov, Eskin and Frenkel [62]:

$$b \sim (1 - c), \quad a \sim (2 - 3b) \quad \text{and} \quad c \sim (a + 1)/3 \tag{24}$$

Another scaling relation exists between the sedimentation coefficient and k_s [see [10]]

$$k_s = K''''(s^0)^{\text{æ}} \tag{24a}$$

Lavrenko et al give æ and K''' for a range of polysaccharides [10].

Various relations have been proposed linking the various power law exponents for a homologous series under specified conditions [10, 62] such as:

$$\text{æ} = (2 - 3b)/b \tag{25}$$

6.2.1
General Conformation: Haug Triangle and Conformation Zoning

Delineation of the three general conformation extremes (random coil, compact sphere, rigid rod) as indicated by the simple power or scaling laws and Wales–van Holde ratio, have been conveniently represented in the well-known Haug triangle (Fig. 8, see [61]). An extension of this idea was given by Pavlov et al. [112, 113]) who suggested five general conformation types or "zones", all of which could be distinguished using sedimentation measurements. The zones were: A (extra rigid rod), B (almost rigid rod), C (semi-flexible coil), D (random coil), E (globular/branched). A and B are distinguished by B having a very limited amount of flexibility. The zones were constructed empirically using a large amount of data (s, k_s) accumulated for polysaccharides of "known" conformation type, and plotted a scaling relation normalised with mass per unit length (M_L) measurements. (Fig. 9) The latter parameter can be obtained from knowledge of molecular weight from sedimentation equilibrium or light scattering and the chain length L from small

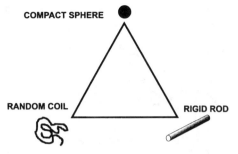

Fig. 8 The Haug triangle. The three extremes of conformation (*compact sphere, random coil* and *rigid rod*) are placed at the apices of a triangle. The conformation of a given macromolecule is represented by a locus along the sides of the triangle between these extremes. Knowledge of the power law exponents (see text) can help to give us an idea of the conformation type. From [61]

angle X-ray scattering, X-ray fibre diffraction or high resolution NMR: Pavlov and colleagues give a comprehensive comparison of methods for heparin [5]. If the molecular weight is known, M_L can also be estimated from electron microscopy [114]. Measurement of a data set (s, k_s, M_L) of any target polysaccharide would then establish its conformation type. The limiting slopes of ~ 4 (extra rigid rod) and ~ 0 (globular/sphere) were shown to be theoretically reasonable. It should be stressed that this procedure is only a guide to conformation type. Along with other procedures using k_s for conformation studies it assumes that charge contributions to the concentration have been suppressed by the supporting electrolyte, and also that draining effects are not significant or similar for the target polysaccharide and those polysaccharides whose data were used to set up the zones. Other normalised scaling relations have been suggested based on viscometry methods [113].

6.3
Rigid Cylindrical Structures

Once a general conformation type or "preliminary classification" has been established it is possible to use sedimentation data to obtain more detailed information about polysaccharide conformation. For example, the low value of $k_s/[\eta] \sim 0.25$ found for the bacterial polysaccharide xylinan has been considered to be due to asymmetry [115]. If we then assume a rigid structure the approximate theory of Rowe [36, 37] can be applied in terms of a prolate ellipsoid of revolution to estimate the aspect ratio p ($\sim L/d$ for a rod, where L is the rod length and d is its diameter) ~ 80.

For a cylindrical rod an expression also exists for the sedimentation coefficient [116]:

$$s^0 = \{M(1 - \bar{v}\rho_0)/(3\pi\eta_0 N_A L)\} \cdot \{\ln(L/d) + \gamma\} \tag{26}$$

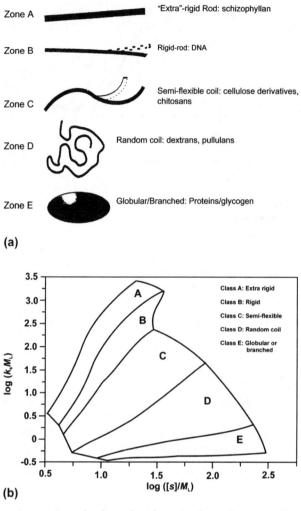

Fig. 9 Conformation zoning of polysaccharide. **a** Conformation zones **b** Empirical plots for various polysaccharides of known conformation type used to form the zone plot. Measurement of s, k_s and M_L for a target polysaccharide will then help define its conformation zone or type. Redrawn and based on [112, 113]

where γ is a function of p and has a limiting value of ~ 0.386 for very long rods ($p \rightarrow \infty$). Replacing L by the (molar) mass per unit length $M_L = M/L(\text{g mol}^{-1}\,\text{cm}^{-1})$ this becomes

$$s^0 = \{M_L(1 - \bar{v}\rho_0)/(3\pi\eta_0 N_A)\} \cdot \{\ln M - \ln M_L - \ln d + \gamma\} \tag{27}$$

For the cases of finite p (in the range 2–20) the currently accepted expression for $\gamma(p)$ is that of Tirado and Garcia de la Torre [117]

$$\gamma(p) = 0.312 + (0.561/p) + (0.100/p^2) \tag{28}$$

Above $p > 10$ the limiting value ($\gamma = 0.386$) can be used.

From Eqs. 27 and 28 we can obtain an estimate for the rod length L if we know M or M_L (see above discussion) and have an estimate for the diameter d. As pointed out by Garcia de la Torre [118] the choice for d is not so critical since it comes into the equations as the logarithm. It applies only to polysaccharides which are known to be rods.

6.4
Semi-Flexible Chains: Worm-Like Coils

Most linear polysaccharides are not rigid rods at all but are semi-flexible structures. The conformation and hydrodynamics of semi-flexible chains are most usefully represented by worm-like chains (see [119–122]), in which the bending flexibility is represented by the persistence length L_p. This is an intrinsic property of a linear macromolecule: the greater the L_p the greater the rigidity and vice verse. More precisely, the conformation and flexibility of a macromolecular chain depends directly on L/L_p, the ratio of the contour length to the persistence length. For $L/L_p \ll 1$ the conformation is rod-like and Eqs. 26–28 can be applied. For $L/L_p \gg 0$ the conformation approaches that of a random coil [119–122]. This can be best seen from the dependence of the radius of gyration on chain length, as clearly described by Freire and Garcia de la Torre [122]:

$$R_g^2 = \{L \cdot L_p/3\} \cdot \left\{1 - (3L/L_p) + \left(6L_p^2/L^2\right) + 6\left(L_p^3/L^3\right)\left(1 - e^{-L/L_p}\right)\right\} \tag{29}$$

In the limit $L_p/L \sim 0$, R_g is proportional to $L^{1/2}$ (n.b. this is misprinted in [122])—the classical dependence for a random coil—whereas when $L_p/L \gg 1$, the classical relation for a rod is obtained: $R_g = L/\sqrt{12}$.

The sedimentation coefficient for wormlike chains was first worked out by Hearst and Stockmayer [123], later improved by Yamakawa and Fujii [124] to give this expression for s^0:

$$s^0 = \{M(1 - \bar{v}\rho_0)/(3\pi\eta_0 N_A L)\}$$
$$\cdot \left\{1.843 \ln(L/2L_p)^{1/2} + \alpha_2 + \alpha_3(L/2L_p)^{-1/2} + ...\right\} \tag{30}$$

If the persistence length L_p is much larger than the mean chain diameter, d, Yamakawa and Fujii gave limiting values for $\alpha_2 = -\ln(d/2L_p)$ and $\alpha_3 = 0.1382$. Freire and Garcia de la Torre [122] have considered further these coefficients. The factor $2L_p$ appears rather than L_p simply because $2L_p$ is equivalent to the statistical Kuhn segment length λ^{-1}.

A fundamental problem with the sedimentation coefficient is that it is the least sensitive parameter to conformation when compared with the intrinsic viscosity $[\eta]$ and the radius of gyration R_g. This lower sensitivity is offset by the ease of measurement and the ability to obtain s^0 to a higher accuracy (to better than 1%) compared with the other parameters. Nonetheless it is advisable not to use s in isolation but in conjunction with R_g and $[\eta]$ vs. M: two recent examples are a comparative study using ultracentrifugation, viscometry and light scattering on the relative conformations and flexibilities of galactomannans (guar, tara gum and locust bean gum), after pressure assisted solubilisation procedures [49] and a study using ultracentrifugation, viscometry and small angle X-ray scattering to investigate the conformation and flexibility of heparin [5].

7
Associative Interactions

There are many instances where associative interactions involving polysaccharides, whether they be self-association, complex formation or with small ligands are important (see [1]). Examples of self-association are dimerisation or trimerisation of helical types of polysaccharides, such as schizophyllan, scleroglucan (we could also mention xanthan and κ-carrageenan although that has been the subject of some disagreement). Examples of complex formation include the use of cellulose derivatives as dental adhesives, and an example of small ligand interactions is the intercalation of iodine by amylose or amylopectin. There has been considerable attention focussed on the use of polysaccharide systems as encapsulation agents for flavours and drugs, and this invokes both macromolecular and small ligand interactions involving polysaccharides [1]. The following Chapter by Muzzarelli and Muzzarelli [125] refers to a large number of these and other types of interactions involving the polycationic chitosan family of polysaccharides.

The analytical ultracentrifuge would appear to offer considerable potential for the analysis of these and other types of interaction. Indeed one of the main reasons behind the renaissance of analytical ultracentrifugation in the 1990s [126, 127] was the simmering need by molecular biologists and protein chemists for non-invasive solution based methods for studying biomolecular interactions, particularly the weaker ones involved in molecular recognition phenomena (see, e.g., [128, 129]). The analytical ultracentrifuge, its clean, medium free (no columns or membranes) and absolute nature has indeed proven a highly attractive tool for characterising the stoichiometry, reversibility and strength (as represented by the molar dissociation constant, K_d) of an interaction between well-defined systems: protein-protein, protein-DNA, protein-small ligand [2]. With polysaccharides we are generally dealing with a different situation. Firstly, a polysaccharide does not have a single, clearly defined molecular weight: it is polydisperse with a distribution of molecular weights. Secondly, weak interactions ($K_d > 50\,\mu M$)—at least as far as we know—do not play a crucial functional role with polysaccharides as they do with proteins. Interactions, particularly involving polyelectrolytes of opposite charge (chitosan-alginate for encapsulation systems, chitosan-DNA for gene therapy) tend to be very strong or irreversible: the complexes tend to be much larger than for the simple associative protein-protein interactions. This means the main ultracentrifuge tool used for investigating protein-protein interactions, namely sedimentation equilibrium, has only limited applicability: sedimentation equilibrium has an upper limit of molecular weight of ~ 50 million g/mol. Examples of the use of the analytical ultracentrifuge to assay interactions involving polysaccharides include a study on mixtures of alginate with bovine serum albumin [130, 131], a study of galactomannan incubated with gliadin (as part of an ongoing investigation into the possible use of galactomannans to help intestinal problems) [132], chitosan with lysozyme [133] and synergistic interactions involving xanthan [134].

Polysaccharides can regulate weak interactions between protein molecules. A recent example is the effect of low molecular weight heparin molecules on the weak dimerisation of the plasminogen growth factor NK1, or at least a mutant thereof [135].

For large irreversible complexes involving polysaccharides a more valid assay procedure is to use sedimentation velocity (which can cope with complexes as large as 10^9 g/mol), with the change in sedimentation coefficient (s, normalised to standard conditions or not) as our marker for complex formation. If we so wish we can then convert this to a change in molecular weight if we assume a conformation and use the power law relation (Eq. 13). Alternatively, we can simply use s directly as our size criterion (this is not unusual: it is used for example in ribosome size representations, 30S, 50S etc., or in seed globulin, the 7S, 11S soya bean globulins etc [136, 137]).

7.1
Polysaccharide Mucoadhesive Interactions

A good example of where sedimentation velocity has played a valuable role in assaying large polysaccharide complexes is in the assessment of polysaccharides as mucoadhesives [138–143]: a drug administered orally or nasally tends to be washed away from the site of maximum absorption by the bodies natural clearance mechanisms before being absorbed. Incorporating the drug into a polysaccharide material which interacts with epithelial mucus in a controllable way has been proposed as a method of increasing the residence time and enhancing the absorption rate. The key macromolecule in mucus is mucin glycoprotein—a linear polypeptide backbone with linked saccharide chains to the extent that $> 80\%$ of the molecule is carbohydrate [63].

The carbohydrate has sites for ionic interaction (clusters of sialic acid or sulphate residues) and also hydrophobic interaction (clusters of hydrophobic methyl groups offered by fucose residues). Sedimentation velocity has been a valuable tool in the selection of appropriate mucoadhesives and in the characterisation of the complexes [138–143].

The approach is to first of all obtain mucin to a high degree of purity and to characterise the mucin and potential mucoadhesive. This is done by sedimentation velocity $[g^*(s)]$ analysis and sedimentation equilibrium (M^*) analysis—according to the procedures described above—together with SEC-MALLs [145, 146].

The reactants are then mixed in various proportions, and the sedimentation ratio ($s_{complex}/s_{mucin}$)—the ratio of the sedimentation coefficient of the complex to that of the pure mucin itself—is used as the measure for mucoadhesion. The ultra-violet absorption optics on the XL-A or XL-I ultracentrifuge have been used as the main optical detection system. Although the polysaccharide is generally invisible in the near UV (~ 280 nm), at the concentrations normally employed the mucin—in uncomplexed and complexed form—is detectable.

7.2
Mucoadhesion Involving Guar, Alginate, Carboxymethyl Cellulose, Xanthan and DEAE-Dextran

Experiments on a series of neutral and polyanionic polysaccharides revealed no significant change in the sedimentation coefficient (sedimentation ratio $s_{complex}/s_{mucin} \approx 1$) [138, 140, 141, 143] reinforcing macroscopic observations on whole mucus using tensiometry [147]. The polycationic derivative DEAE-dextran gave sedimentation ratios of 1.1–1.9 [138, 140, 141, 143] (Table 6) depending on the mixing ratio and temperature. This was rather modest considering the high charge density on the polymer with lots of potential sites for interaction with the ionized sialic acid groups on the mucin. This disappointment also reflects the disappointing result from the tensiometry analyses [147]. The $\alpha(1 \rightarrow 3)$ branches of the dextran appear to be responsible for considerable steric hindrance, preventing access to the charged mucin groups.

7.3
Mucoadhesion Experiments Involving Chitosans

A contrasting picture is seen for chitosans. Chitosans—as considered in detail in the following Chapter—are derivatives of chitin (after an alkali extraction procedure) and are available in large quantities from the shells of crabs, lobsters and other crustaceans. Pure chitin is poly-N-acetylglucoasmine. The N-acetyl groups are de-acetylated in chitosan to an extent represented by ei-

Table 6 Sedimentation coefficient ratio. ($s_{complex}/s_{mucin}$) as an index of adhesiveness (Based on data from [138, 140, 143, 145])

Polysaccharide mucoadhesive	$s_{complex}/s_{mucin}$	Conditions
Alginate	1	pH 6.8, 20 °C
Carboxmethyl cellulose	1	pH 6.8, 20 °C
Guar	1	pH 6.8, 20 °C
Xanthan	1	pH 6.8, 20 °C
DEAE-dextran	1.1–1.9*	pH 6.8, 20 °C
	1.2–1.4*	pH 6.8, 37 °C
Chitosan ($F_A \approx 0.11$)	48	pH 6.5, 20 °C
	34	pH 6.5, 37 °C
	15	pH 4.5, 20 °C
	38	pH 4.5, 37 °C
	22	pH 2.0, 20 °C
	12	pH 2.0, 37 °C
	26	pH 4.5, 20 °C + 3 mM bile salt
	35	pH 4.5, 37 °C + 3 mM bile salt
	18	pH 4.5, 20 °C + 6 mM bile salt
	14	pH 4.5, 37 °C + 6 mM bile salt
Chitosan ($F_A \approx 0.42$)	31	pH 4.5, 20 °C
	44	pH 4.5, 37 °C

*Depending on the mixing ratio

ther the degree of deacetylation or the degree of acetylation, F_A, with $F_A = 1$ being pure chitin and $F_A = 0$–0.6 representing the range of soluble chitosans. We stress here that chitosans are only readily soluble at pH values of 6.5 or less, and this factor has to be borne in mind in the formulation of any mucoadhesive product involving these substances (see [125]). Interestingly, whereas mucins present two types of residue for potential mucoadhesive interaction (the charged acidic groups on sialic acid and any sulphonated residues, and the hydrophobic methyl groups on fucose residues) chitosans present a similar opportunity (the charged NH_3^+ groups on deacetylated N-acetyl groups and also the hydrophobic acetyls on non-deacetylated residues). The results are quite spectacular [138–145] (Table 6). A highly charged chitosan ("sea cure" 210+) of $F_A \sim 0.11$ has impressive sedimentation ratios of 15–38 depending on the temperature. Interestingly for a lower-charged chitosan of $F_A \sim 0.42$, values of 31–44 were returned, reinforcing the view that both electrostatic and hydrophobic effects are important. The demonstration of large-size interaction products by the analytical ultracentrifuge used in this manner is reinforced by images from the powerful imaging techniques of electron microscopy and atomic force microscopy. Conventional transmission electron microscopy clearly demonstrates large complexes of the order of

~ 1 μm in size [148], and if we label the chitosan with gold we can see that the chitosan is distributed throughout the complex with "hot spots" in the interior [149]. Images from atomic force microscopy, visualized in topographic and phase modes, again shows complexes of this size. Control experiments revealed a loose coiled structure for pig gastric mucin and a shorter, stiffer conformation for the chitosan, consistent with solution measurements [150].

7.4
Effect of the Environment on the Extent of Interaction

The beauty of the analytical ultracentrifuge is that since it is a pure solution technique with no columns or membranes it allows us to easily alter the surrounding solvent conditions. For example [138, 140], if we vary the pH we see that the sedimentation ratio is ~ 34–48 at pH ~ 6.5 but is still significant at pH 2.0 ($s_{complex}/s_{mucin} \sim 12$–22: Table 6), below the pKa of the sialic acid groups on the mucin—which not only suggests the importance of the electrostatic contribution but indicates the existence of significant hydrophobic types of interaction. Attempts to investigate the effects of bile salts and differing ionic strengths down the alimentary tract also yield very much the same picture [138, 140]. At 0 mM bile salt $s_{complex}/s_{mucin}$ was found to be ~ 18–21, whereas at 6 mM the interaction is still significant, with $s_{complex}/s_{mucin} \sim 14$–18.

7.5
Sedimentation Fingerprinting

We would dearly love to perform these types of experiments on human small intestinal mucin if we could only get them in sufficient quantities in purified form. There has, however, been success in performing experiments on human mucin extracted from different parts of the stomach, namely the cardia, corpus and antrum regions. Although available in miniscule quantities we can assay mucoadhesiveness of chitosan on these by using a modification of the approach using the analytical ultracentrifuge described above, called Sedimentation Fingerprinting. In this method, introduced six years ago [151], the Schlieren optical system is used to record the concentration (refractive index) gradient dn/dr as a function of radial position r in the ultracentrifuge cell. The area under a "Schlieren peak" provides a measure of the sedimenting concentration. Alternatively, if interference optics on the XL-I ultracentrifuge are used, the area under a $g^*(s)$ versus s plot would provide similar concentration information. Although the mucins from a human stomach are at too low a concentration to be detected we can assay for interaction from the loss of area under the chitosan peak caused by interaction. In this way Deacon and coworkers [151] showed it was possible to demonstrate significant differences in mucoadhesive interactions for different regions of the stomach (Fig. 10). This type of information obtained with the ultracentrifuge reinforced with other data is helping us design effective mucoadhesive systems. An example is the use of tripolyphosphate to cross-link chitosan into a sphere [152]: if this is done in the presence of a drug, the drug can be encapsulated. Tripolyphosphate-linked chitosans have been shown to give good mucoadhesion [153], and although nothing has been published yet the principle of co-sedimentation could be used as a successful assay here as it has been in the assay of non-polysaccharide-based encapsulation systems [153, 155]. This and other aspects are considered in a recent review on mucoadhesion [138].

7.6
Measurement of Charge and Charge Screening

The final application we would like to highlight is a recent application of the analytical ultracentrifuge to our understanding of the behaviour of polysaccharides in solution is the measurement of charge on polyelectrolyte polysaccharides, and the extent of charge screening through interaction with low molecular weight electrolyte [14]. In recent years there has been a tendency to identify the charges on polysaccharides and polynucleotides with the values calculated from the chemical structure. However, the phenomenon of charge-screening (or counterion condensation) has long been an established feature of polyelectrolyte theory [156–158]. Relatively small magnitudes of the effective net charges of polysaccharides [159–161] and other carbohydrate-based polymers with a significant amount of charged groups—such as nucleic acids [162, 163]—afford examples with direct relevance to biology.

Winzor and coworkers have employed measurements of the Donnan distribution of small ions in dialysis equilibrium [14] to reinforce earlier evidence of charge-screening effects in polysaccharide anions [164, 165]. These researchers used the absorption optical system of a Beckman XL-I ultracentrifuge to monitor the distribution of ions in polysaccharide solutions

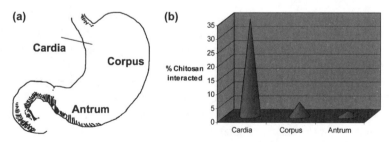

Fig. 10 Sedimentation analysis of the comparative mucoadhesiveness of a chitosan (sea cure 210+, $F_A \sim 0.11$) to mucins from different parts of the stomach **a** Mucin source **b** Histogram. Adapted from [151]

dialyzed against sodium phosphate buffer (pH 6.8, I 0.08) supplemented with 0.2 mM chromate as an indicator ion. Three polyanionic polysaccharides were chosen for analysis. After extensive dialysis against the same chromate-supplemented buffer to establish the Donnan equilibrium distribution of small ions the difference between chromate concentrations in the polysaccharide and diffusate solutions was monitored by means of the absorption optical system: the ultracentrifuge is merely being used as a double-beam spectrophotometer when a sufficiently low speed (3000 rpm) is used to ensure uniformity of solution composition throughout the cell. As in classical difference spectroscopy, diffusate was placed in the reference sector of the cell to allow direct measurement of the absorbance difference from a scan at 375 nm.

In equilibrium dialysis of a solution of a polyanion (valence Z_P negative) with molar concentration C_P against a solution of uni-univalent electrolyte CA (C = cation, A = anion) with molar concentration C_{CA} it was shown that the requirement for equal chemical potentials of the salt in the polyanion (α) and diffusate (β) phases results in the following relation

$$\left(\gamma_\pm^\alpha / \gamma_\pm^\beta\right)\left(C_i^\alpha / C_i^\beta\right)^{1/Z_i} = 1 - Z_P C_P^\alpha / (2I^\beta) + \dots \tag{31}$$

Comparison of the concentrations of either the cation or the anion in the two phases thus has potential for evaluating the polyanion valence provided that estimates of the mean ion activity coefficient ($\gamma_\pm$) are available. Furthermore, as realized by Svensson [165], expression of the Donnan distribution of small ions in this manner has two advantages in that (i) Eq. 31 applies to each type of small ion in situations where the supporting electrolyte is not restricted to single cationic and anionic species; and (ii) multivalence of a small ion is also accommodated.

In a typical equilibrium dialysis study of charged polysaccharides an indicator ion, L (chromate), is included in the supporting electrolyte medium (phosphate buffer, pH 6.8, I 0.08) to allow assessment of the effective net charge of the polyanions via a modified form of Eq. 31, namely,

$$\left(C_L^\alpha / C_L^\beta\right)^{1/Z_L} = 1 - f Z_P C_P / (2I^\beta) + \dots \tag{32}$$

where Z_L is the chromate valence (– 2). The factor f is included to allow expression of the effective net charge as a fraction of the nominal valence Z_P that is indicated by the chemical structure of the saccharide repeat unit: $f = Z_{eff}/Z_P$.

For dextran sulphate, heparin and polygalacturonate the effective net charges were shown by Winzor and coworkers to be only one-third ($f \sim 0.3$) of those deduced from the chemical structures [14]—a reflection of charge screening (counterion condensation) in aqueous polyelectrolyte solutions. Whereas the extent of charge screening for the first two polysaccharides agrees well with theoretical prediction, the disparity in the corresponding comparison for polygalacturonate was deemed to reflect partial esterification of carboxyl groups, whereupon the experimental parameter refers to the effective charge per hexose residue rather than the effective fractional charge of each carboxyl group. It is therefore wrong to calculate the charge on a polysaccharide in solution on the basis merely of the numbers or number density of charged residues like COO^- or SO_3^{2-}, even in cases where the pH is such that the groups are fully ionised.

8
Comment

The analytical ultracentrifuge has emerged from its Cinderella status as for polysaccharide characterisation, catalyzed by the new generation instrumentation with on-line data capture and some major advances in software for data capture and analysis. In a review article written a decade ago [12] it was predicted that the imminent launch of the XL-I ultracentrifuge would have a considerable impact with regards sedimentation velocity analysis of polysaccharide heterogeneity, polydispersity and conformation. This has indeed been the case. The role of the ultracentrifuge for sedimentation equilibrium analysis of polysaccharide molecular weight is still very important, but more as providing a valuable independent check on the reliability of results coming out of SEC/MALLs. It has been shown to have considerable potential for the analysis of interactions involving polysaccharides, even though the types of interaction involved are quite different from the better defined reversible types of interaction encountered in protein biochemistry where analytical ultracentrifugation has made a huge impact. One would therefore infer that an analytical ultracentrifuge would be a valuable part of the armoury of any researcher interested in understanding the size, conformation and interaction properties of polysaccharides in the environment where many occur naturally—in the solution state. And there is no need for columns or membranes or a requirement to immobilise material onto a surface: it is a free solution technique. Its versatility is even more impressive: this article has only focussed on its use for analysis of solutions: we haven't touched upon its uses for the analysis of the rheological and thermodynamic properties of polysaccharide gels [15], phase diffusion and interfacial transport phenomena involving polysaccharides [16], and the formation of films and membranes [17].

References

1. Tombs MP, Harding SE (1997) An Introduction to Polysaccharide Biotechnology. Taylor and Francis, London
2. Harding SE, Rowe AJ, Horton JC (eds) (1992) Analytical Ultracentrifugation in Biochemistry and Polymer Science. Royal Society of Chemistry, Cambridge, UK

3. Burchard W (1992) In: Harding SE, Sattelle DB, Bloomfield VA (eds) Laser Light Scattering in Biochemistry. Royal Society of Chemistry, Cambridge, UK, 3
4. Wyatt PJ (1992) In: Harding SE, Sattelle DB, Bloomfield VA (eds) Laser Light Scattering in Biochemistry. Royal Society of Chemistry, Cambridge, UK, 35
5. Pavlov G, Finet S, Tatarenko K, Korneeva E, Ebel C (2003) Eur Biophys J 32:437
6. Harding SE (1997) Prog Biophys Mol Biol 68:207
7. Van der Merwe A (2001) In: Harding SE, Chowdhry BZ (eds) Protein-Ligand Interactions: Hydrodynamics and Calorimetry. Oxford University Press, Oxford, UK, 137
8. Lapasin R, Pricl S (1995) Rheology of Industrial Polysaccharides. Theory and Applications. Blackie, London
9. Harding SE (1992) In: Harding SE, Rowe AJ, Horton JC (eds) Analytical Ultracentrifugation in Biochemistry and Polymer Science. Royal Society of Chemistry, Cambridge, UK, 495
10. Lavrenko PN, Linow KJ, Görnitz E (1992) In: Harding SE, Rowe AJ, Horton JC (eds) Analytical Ultracentrifugation in Biochemistry and Polymer Science. Royal Society of Chemistry, Cambridge, UK, 517
11. Harding SE (1993) Gums and Stabilisers for the Food Industry 8:55
12. Harding SE (1995) Carbohyd Polym 28:227
13. Jumel K, Fiebrig I, Harding SE (1996) Int J Biol Macromol 18:133
14. Winzor DJ, Carrington LE, Deszczynski M, Harding SE (2004) Biomacromolecules 5:2456
15. Cölfen H (1999) Biotech Genet Eng Rev 16:87
16. Harding SE, Tombs MP (2002) Biotech Genet Eng Rev 19:55
17. Wandrey C, Grigorescu G, Hunkeler D (2002) Prog Colloid Polym Sci 119:91
18. Jullander I (1987) In: Ranby B (ed) Physical Chemistry of Colloids and Macromolecules. The Svedberg Symposium. Blackwell Scientific, Oxford, UK, 29
19. Schachman (1992) In: Harding SE, Rowe AJ, Horton JC (eds) Analytical Ultracentrifugation in Biochemistry and Polymer Science. Royal Society of Chemistry, Cambridge, UK, 3
20. Görnitz E, Linow KJ (1992) In: Harding SE, Rowe AJ, Horton JC (eds) Analytical Ultracentrifugation in Biochemistry and Polymer Science. Royal Society of Chemistry, Cambridge, UK, 26
21. Giebeler R (1992) In: Harding SE, Rowe AJ, Horton JC (eds) Analytical Ultracentrifugation in Biochemistry and Polymer Science. Royal Society of Chemistry, Cambridge, UK, 16
22. Errington N, Harding SE, Rowe AJ (1992) Carbohyd Polym 17:151
23. Errington N, Harding SE, Illum L, Schacht E (1992) Carbohyd Polym 18:289
24. Cölfen H, Harding SE, Vårum KM (1996) Carbohyd Polym 30:5
25. Furst AJ (1997) Europ Biophys J 25:307
26. Theisen C, Johann C, Deacon MP, Harding SE (2000) Refractive Increment Data Book for Polymer and Biomolecular Scientists. Nottingham University Press, Nottingham, UK
27. Anthonsen MW, Vårum KM, Smidsrød O (1993) Carbohyd Polym 22:193
28. Preston BN, Wik KO (1992) In: Harding SE, Rowe AJ, Horton JC (eds) Analytical Ultracentrifugation in Biochemistry and Polymer Science. Royal Society of Chemistry, Cambridge, UK, 549
29. Majzoobi M (2004) PhD Thesis, University of Nottingham, Nottingham, UK
30. Schachman H (1959) Ultracentrifugation in Biochemistry. Academic Press, New York

31. Gibbons RA (1972) In: Gottschalk A (ed) Glycoproteins: Their Composition, Structure and Function. Elsevier, Amsterdam, 5A, 31
32. Kratky O, Leopold H, Stabinger H (1973) Meth Enzymol 27D:98
33. Laue TM, Shah BD, Ridgeway TM, Pelletier SL (1992) In: Harding SE, Rowe AJ, Horton JC (eds) Analytical Ultracentrifugation in Biochemistry and Polymer Science. Royal Society of Chemistry, Cambridge, UK, 90
34. http://www.jphilo.mailway.com/download.htm
    and http://www.rasmb.bbri.org/rasmb/windows/sednterp-philo/
35. Gralén N (1944) Sedmentation and Diffusion Measurements on Cellulose and Cellulose Derivatives. PhD Thesis, University of Uppsala, Uppsala, Sweden
36. Rowe AJ (1977) Biopolymers 16:295
37. Rowe AJ (1992) In: Harding SE, Rowe AJ, Horton JC (eds) Analytical Ultracentrifugation in Biochemistry and Polymer Science. Royal Society of Chemistry, Cambridge, UK, 394
38. Fujita H (1962) Mathematical Theory of Sedimentation Analysis. Academic Press, New York
39. Fujita H (1975) Foundations of Ultracentrifugal Analysis. Wiley, New York
40. Stafford W (1992) In: Harding SE, Rowe AJ, Horton JC (eds) Analytical Ultracentrifugation in Biochemistry and Polymer Science. Royal Society of Chemistry, Cambridge, UK, 359
41. Philo JS (2000) Anal Biochem 279:151
42. Schuck P (1998) Biophys J 75:1503
43. Dam J, Schuck P (2004) Meth Enzymol (in press)
44. http://www.jphilo.mailway.com/download.htm
45. Lamm O (1923) Ark Mat Astr Fys 21B(2):1
46. Claverie JM, Dreux H, Cohen R (1975) Biopolymers 14:1685
47. http://www.analyticalultracentrifugation.com/download.htm
48. Harding SE (1995) Biophys Chem 55:69
49. Patel T, Picout DR, Parlov G, Garcia de la Torre J, Ross-Murphy SB, Harding SE (2005) mss. submitted
50. Svedberg T, Pedersen KO (1940) The Ultracentrifuge. Oxford University Press, Oxford, UK
51. Korneeva EV, Vichoreva GA, Harding SE, Pavlov GM (1996) Abstr Am Chem Soc 212(1):75-cell
52. Pavlov GM, Korneeva EV, Harding SE, Vichoreva GA (1998) Polymer 39:6951
53. Pavlov GM, Korneeva EV, Vikhoreva GA, Harding SE (1998) Polym Sci Ser A 40; and (1998) Vysokomolekulyarnye Soed Ser A 40:2048
54. Pavlov GM, Korneeva EV, Nepogod'ev SA, Jumel K, Harding SE (1998) Polym Sci Ser A 40:1282; and (1998) Vysokomolekulyarnye Soed Ser A 40:2056
55. Pavlov GM, Korneeva EV, Jumel K, Harding SE, Meyer EW, Peerlings HWI, Stoddart JF, Nepogodiev SA (1999) Carbohyd Polym 38:195
56. Sharman WR, Richards EL, Malcolm GN (1978) Biopolymers 17:2817
57. Igarishi O, Sakurai Y (1965) Agr Biol Chem 29:678
58. Djurtoft R, Rasmussen KL (1955) Eur Brew Conv Congress, p 17
59. Wedlock DJ, Fasihuddin BA, Phillips GO (1987) Food Hydrocolloids 1:207
60. Pusey PN (1974) In: Cummings HZ, Pike ER (eds) Photon Correlation and Light Beating Spectroscopy. Plenum Press, New York, 387
61. Smidsrød O, Andresen IL (1979) Biopolymerkjemi. Tapir, Trondheim, Norway
62. Tsvetkov VN, Eskin V, Frenkel S (1970) Structure of Macromolecules in Solution. Butterworths, London

63. Harding SE (1989) Adv Carbohyd Chem 47:345
64. Pain RH (1980) Symp Soc Exp Biol 34:359
65. Svedberg T, Fåhraeus R (1926) J Am Chem Soc 48:430
66. Clewlow AC, Errington N, Rowe AJ (1997) Eur Biophys J 25:305
67. Mächtle W (1999) Prog Coll Polym Sci 113:1
68. http://www.nottingham.ac.uk/ncmh
69. Harding SE, Horton JC, Morgan PJ (1992) In: Harding SE, Rowe AJ, Horton JC (eds) Analytical Ultracentrifugation in Biochemistry and Polymer Science. Royal Society of Chemistry, Cambridge, UK, 275
70. Cölfen H, Harding SE (1997) Eur Biophys J 24:333
71. Van Holde KE, Baldwin RL (1958) J Phys Chem 62:734
72. Yphantis DA (1964) Biochemistry 3:297
73. Creeth JM, Harding SE (1982) J Biochem Biophys Meth 7:25
74. Hall DR, Harding SE, Winzor DJ (1999) Prog Coll Polym Sci 113:62
75. Teller DC, Horbett JA, Richards EG, Schachman HK (1969) Ann New York Acad Sci 164:66
76. Creeth JM, Pain RH (1967) Prog Biophys Mol Biol 17:217
77. Minton AP (1994) In: Schuster TM, Laue TM (eds) Modern Analytical Ultracentrifugation. Birkhäuser, Boston, 81
78. http://www.nottingham.ac.uk/ncmh/unit/method.html#Software
79. http://vm.uconn.edu/~wwwbiotc/uaf.html
80. Kawahara K, Ohta K, Miyamoto H, Nakamura S (1984) Carbohyd Polym 4:335
81. Sato T, Norisuye T, Fujita H (1984) Macromolecules 17:2696
82. Woodward JR, Phillips DR, Fincher GB (1983) Carbohyd Polym 3:143
83. Edmond E, Farquhar S, Dunstone JR, Ogston AG (1968) Biochem J 108:755
84. Muzzarelli RAA, Lough C, Emanuelli M (1987) Carbohyd Res 164:433
85. Berth G, Dautzenberg H, Lexow D, Rother G (1990) Carbohyd Polym 12:39
86. Lecacheux D, Mustiere Y, Panaras R, Brigand G (1986) Carbohyd Polym 6:477
87. Yanaki T, Kojima T, Norisuye T (1981) Polym J 13:1135
88. Wedlock DJ, Baruddin BA, Phillips GO (1986) Int J Biol Macromol 8:57
89. Dhami R, Harding SE, Jones T, Hughes T, Mitchell JR, To K-M (1995) Carbohyd Polym 27:93
90. Harding SE, Day K Dhami R, Lowe PM (1997) Carbohyd Polym 32:81
91. Horton JC, Harding SE, Mitchell JR, Morton-Holmes DF (1991) Food Hydrocolloids 5:125
92. Straatman A, Borchard W (2002) Prog Coll Polym Sci 119:64–69
93. Harding SE (1985) Biophys J 47:247
94. Ball A, Harding SE, Mitchell JR (1988) Int J Biol Macromol 10:259
95. Harding SE, Berth G, Ball A, Mitchell JR, Garcia de la Torre J (1991) Carbohyd Polym 16:1
96. Herdan G (1949) Nature 163:139
97. Tombs MP, Peacocke AR (1974) The Osmotic Pressure of Biological Macromolecules. Clarendon Press, Oxford
98. Roark D, Yphantis DA (1969) Ann New York Acad Sci 164:245
99. Rinde H (1928) The Distribution of the Sizes of Particles in Gold Sols Prepared According to the Nuclear Method. PhD Thesis, University of Uppsala, Uppsala, Sweden
100. Harding SE (1988) Gums and Stabilisers for the Food Industry 4:15
101. http://www.wyatt.com/
102. Horton JC, Harding SE, Mitchell JR (1991) Biochm Soc Trans 19:510
103. Rollings JE (1991) Biochem Soc Trans 19:493

104. Rollings JE (1992) In: Harding SE, Sattelle DB, Bloomfield VA (eds) Laser Light Scattering in Biochemistry. Royal Society of Chemistry, Cambridge, UK, 275
105. Wolff D, Czapla S, Heyer AG, Radosta S, Mischnick P, Springer J (2000) Polymer 41:8009
106. Wales M, Van Holde KE (1954) J Polym Sci 14:81
107. Creeth JM, Knight CG (1965) Biochim Biophys Acta 102:549
108. Freed KF (1976) J Chem Phys 65:4103
109. Berth G, Cölfen H, Dautzenberg H (2002) Prog Coll Polym Sci 119:50
110. Pavlov GM (2002) Prog Coll Polym Sci 119:84
111. Harding SE, Vårum KM, Stokke BT, Smidsrød O (1991) Adv Carbohyd Analysis 1:63
112. Pavlov GM, Rowe AJ, Harding SE (1997) Trends Analyt Chem 16:401
113. Pavlov GM, Harding SE, Rowe AJ (1999) Prog Coll Int Sci 113:76
114. Stokke BT, Elgsaeter A (1991) Adv Carbohyd Analysis 1:195
115. Harding SE, Berth G, Hartmann J, Jumel K, Cölfen H, Christensen BE (1996) Biopolymers 39:729
116. Broesma S (1960) J Chem Phys 32:1626
117. Tirado MM, Garcia de la Torre J (1979) J Chem Phys 71:2581
118. Garcia de la Torre J (1992) In: Harding SE, Rowe AJ, Horton JC(eds) Analytical Ultracentrifugation in Biochemistry and Polymer Science. Royal Society of Chemistry, Cambridge, UK, 333
119. Yamakawa H (1971) Modern Theory of Polymer Solutions. Harper and Row, New York
120. Bloomfield VA, Crothers DM, Tinoco L (1974) Physical Chemistry of Nucleic Acids. Harper and Row, New York
121. Cantor CR, Schimmel PR (1979) Biophysical Chemistry. Freeman, New York
122. Freire JJ, Garcia de la Torre J (1992) In: Harding SE, Rowe AJ, Horton JC (eds) Analytical Ultracentrifugation in Biochemistry and Polymer Science. Royal Society of Chemistry, Cambridge, UK, 346
123. Hearst JE, Stockmayer WH (1962) J Chem Phys 37:1425
124. Yamakawa H, Fujii M (1973) Macromolecules 6:407
125. Muzzarelli RAA, Muzzarelli C (2005) Adv Polym Sci Heinze T (ed) (this volume)
126. Schachman HK (1989) Nature 941:259
127. Schachman HK (1992) In: Harding SE, Rowe AJ, Horton JC (eds) Analytical Ultracentrifugation in Biochemistry and Polymer Science. Royal Society of Chemistry, Cambridge, UK, 3
128. Watson JD (1970) The Molecular Biology of the Gene (2nd Edn). Benjamin, New York
129. Silkowski H, Davis SJ, Barclay AN, Rowe AJ, Harding SE, Byron O (1997) Eur Biophys J 25:455
130. Harding SE, Jumel K, Kelly R, Gudo E, Horton JC, Mitchell JR (1993) In: Schwenke KD, Mothes R (eds) Food Proteins: Structure and Functionality. VCH, Weinheim, Germany, 216
131. Kelly R, Gudo ES, Mitchell JR, Harding SE (1994) Carbohyd Polym 23:115
132. Seifert A, Heinevetter L, Cölfen H, Harding SE (1995) Carbohyd Polym 28:239
133. Cölfen H, Harding SE, Vårum KM, Winzor DJ (1996) Carbohyd Polym 30:45
134. Mannion RO, Melia CD, Launay B, Cuvelier G, Hill SE, Harding SE, Mitchell JR (1992) Carbohyd Polym 19:91
135. Bandyopadhyay A (2004) Structural Biology of the Plasminogen-related Growth Factors and their Receptors. PhD Thesis, University of Cambridge, Cambridge, UK

136. Schwenke KD (1985) Eiweißquellen der Zukunft. Urania-Verlag, Leipzig, German Democratic Republic
137. Prakash V (1992) In: Harding SE, Rowe AJ, Horton JC (eds) Analytical Ultracentrifugation in Biochemistry and Polymer Science. Royal Society of Chemistry, Cambridge, UK, 445
138. Harding SE (2003) Biochem Soc Trans 31:1036
139. Deacon MP (1999) Polymer Bioadhesives for Drug Delivery. PhD Thesis, University of Nottingham, Nottingham, UK
140. Harding SE, Deacon MP, Fiebrig I, Harding SE (1999) Biotech Gen Eng Rev 16:41
141. Fiebrig I, Davis SS, Harding SE (1995) In: Harding SE, Hill SE, Mitchell JR (eds) Biopolymer Mixtures. Nottingham University Press, Nottingham, UK, 373
142. Fiebrig I (1995) Solution Studies on the Mucoadhesive Potential of Various Polymers for use in Gastrointestinal Drug Delivery Systems. PhD Thesis, University of Nottingham, Nottingham, UK
143. Anderson MT, Harding SE, Davis SS (1989) Biochem Soc Trans 17:1101
144. Fee M (2004) Evaluation of Chitosan Stability in Aqueous Systems. PhD Thesis, University of Nottingham, Nottingham, UK
145. Fiebrig I, Harding SE, Davis SS (1994) Prog Coll Polym Sci 94:66
146. Jumel K, Fogg FJJ, Hutton DA, Pearson JP, Allen A, Harding SE Eur Biophys J 25:477
147. Lehr CM, Bouwstra JA, Schacht EH, Junginger HE (1992) Int J Pharmaceut 78:43
148. Fiebrig I, Harding SE, Rowe AJ, Hyman SC, Davis SS (1995) Carbohyd Polym 28:239
149. Fiebrig I, Vårum KM, Harding SE, Davis SS, Stokke BT (1997) Carbohyd Polym 33:91
150. Deacon MP, McGurk S, Roberts CJ, Williams PM, Tendler SJB, Davies MC, Davis SS, Harding SE (2000) Biochem J 348:557
151. Deacon MP, Davis SS, White RJ, Nordman H, Carlstedt I, Errington N, Rowe AJ, Harding SE (1999) Carbohyd Polym 38:235
152. Zengshuan MA, Yeoh HH, Lim LY (2002) J Pharm Sci 91:1396
153. He P, Davis SS, Illum L (1998) Int J Pharm 166:75
154. Davison CJ, Smith KE, Hutchinson LEF, O'Mullane JE, Harding SE, Brookman L, Petrak K (1990) J Bioactive Compatible Polym 5:7
155. Morgan PJ, Harding SE, Petrak K (1990) Macromolecules 23:4461
156. Katchalsky A, Alexandrowicz Z, Kedem O (1996) In: Conway BE, Barradas RG (eds) Chemical Physics of Ionic Solutions. Wiley, New York, 295
157. Manning GS (1969) J Chem Phys 51:924
158. Manning GS (1978) Q Rev Biophys 11:179
159. Braswell E (1968) Biochim Biophys Acta 158:103
160. Preston BN, Snowden JM, Houghton KT (1972) Biopolymers 11:1645
161. Comper WD, Preston BN (1974) Biochem J 143:1
162. Creeth JM, Jordan DO (1949) J Chem Soc 1409
163. Mathieson AR, Matty SJ (1957) Polym Sci 23:747
164. Adair GS, Adair ME (1934) Biochem J 38:199
165. Svensson H (1946) Ark Kemi Mineral Geol 22A(10):1

Author Index Volumes 101–186

Subject Index

Printing: Krips bv, Meppel
Binding: Stürtz, Würzburg

RETURN TO: **CHEMISTRY LIBRARY**

100 Hildebrand Hall · 510-642-3753

LOAN PERIOD	1	2	3
4		5	

2 HOUR

ALL BOOKS MAY BE ~~RECALLED AFTER 7 DAYS.~~

~~Renewals may be reque~~sted ~~by phone, by using GLADIS,~~
~~type **inv** followed by your patron ID number.~~

DUE AS STAMPED BELOW.

FORM NO. DD 10
3M 5-04

UNIVERSITY OF CALIFORNIA, BERKELEY
Berkeley, California 94720–6000